Microbial Versatility in Varied Environments

Raghvendra Pratap Singh •
Geetanjali Manchanda •
Indresh Kumar Maurya •
Yunlin Wei
Editors

Microbial Versatility in Varied Environments

Microbes in Sensitive Environments

 Springer

Editors
Raghvendra Pratap Singh
Department of Research & Development
Uttaranchal University
Dehradun, Uttarakhand, India

Indresh Kumar Maurya
Department of Microbial Biotechnology
Panjab University
Chandigarh, Chandigarh, India

Geetanjali Manchanda
Department of Botany and Environmental
Studies
DAV University
Jalandhar, Punjab, India

Yunlin Wei
Faculty of Life Science and Technology
Kunming University of Science and
Technology
Xishan, Yunnan, China

ISBN 978-981-15-3030-2   ISBN 978-981-15-3028-9   (eBook)
https://doi.org/10.1007/978-981-15-3028-9

This Springer imprint is published by the registered company Springer Nature Singapore Pte Ltd.
The registered company address is: 152 Beach Road, #21-01/04 Gateway East, Singapore 189721, Singapore

Foreword

Gurukula Kangri Vishwavidyalaya
(NAAC "A" Grade Accredited Deemed to be University u/s 3 of UGC Act 1956)

Prof. R.C. Dubey

Email: *profrcdubey@gmail.com*
Contact: +919412157616

It gives me great pleasure to write the foreword of this book, *Microbial Versatility in Varied Environments*, edited by the renowned group of scientists in the fields of applied microbiology and molecular biology.

Over the past three decades, a plethora of microbes have been explored, and the resultant researches have revolutionized the field of microbiology and biotechnology. Studies on microbial life, their diversity, physiology, genetics, ecology, and biochemistry are vital, since these can reveal a lot in terms of the characteristics of biological processes, such as biochemical limits to macromolecular stability and genetic instructions for constructing macromolecules. These organisms have also provided a number of clues to the origin and evolution of life. Furthermore, a study of microbes in conflict environments has become vital in the field of research in astrobiology, since the microbes found in extreme environments may be analogous to potential life-forms in other planets. In the present times of climate change, the study of microbes living at the edge of life has become even more important. This book is an elegant compilation of chapters on these aspects. It highlights the life of microbes in varied environments ranging from high-temperature conditions to extreme low temperatures and high acidity to alkalinity as well as in conditions of oxygen stress. The chapters have been authored by scientists from different countries. The emphasis has been to give a wide perspective on the habitats of microbes that are thought to be unfavorable or least accessed for survival by humans. The theme of this book is highly significant since life in these environments can give vital clues about the origin and evolution of life on Earth, simulating the environment present billions of years ago. Additionally, the study of adaptation and survival of organisms in such environments can be important for finding life on other planets. This book is a great opportunity for molecular microbiologists for the bioprospecting of microbiota from extreme conditions and engineer the molecular information for agro-economic, industrial and clinical growth. This information shall be useful for students, researchers, and teachers interested in learning about evolution, microbial adaptations, and ecology in varied environments.

This proposed book primarily focuses on the microbes and their molecular adaptation in varied environmental conditions and is aimed at human resource development as well as on the socioeconomic implications of microbial advances. I congratulate the editors for orchestrating with global intellectual authorities on the subject to present their valuable ideas in an easy understandable form. I wish Dr. Raghvendra Pratap Singh every success with the launch of this book and thank him for his dedication to microbiology around the world.

Gurukula Kangri Vishwavidyalaya   R. C. Dubey
Haridwar, Uttarakhand, India
27/11/2019

Preface

In the present era, research in the fields of microbiology have achieved great heights. New discoveries, innovations, and advances in these fields have resulted in widespread applications of microorganisms in biotechnology, agriculture, medicine, food microbiology, and bioremediation. Since my first introduction to PCR, where I came to saw the miracles of *Taq polymerase* of *Thermus aquaticus* and *pfu* of *Pyrococcus furiosus*, I have been fascinated by the diverse lives of microbes, their extremely diverse ecology, their biochemistry, and their potential to survive in almost any environment. I have been passionate about to deciphering their interactions with other organisms and among themselves. These interactions provide vital clues about their roles in our environment and their importance for our well-being. It is vital to study the survival patterns and ecology of microbes in different environments as such studies reveal the strategies that these life-forms adopt for living in different conditions. The environments vary from high salt concentration lakes to extremely cold marine environment or roots of plants. Anywhere they live, the survival strategies of microbes vary. Given the importance and potential application of microbes and their products, the book explores the multidimensionality and versatility of microbes.

For this reason, I decided to compile the works of scientists from different countries in the fields of microbiology and biotechnology. This book is an extensive treaty on the subject and addresses questions such as: What makes the microbes survive in environments that are considered "not normal" by humans? What makes them survive the hostility in other life-forms? How have microbes thrived for millions of years and are still the most successful life-forms on Earth? This is despite the fact that they are unicellular simple living beings. In an attempt to answer these questions, the book reveals a lot about the fascinating world of microbes. In addition to the commercial application, the survival strategies highlight the evolutionary significance of microbial life. Furthermore, it throws light on the possible life on other planets.

Here, the editors deliver the updated comprehensive insight view of microbes from varied environments. The book is divided into two parts: Part I provides reliable and comprehensive information about microbial abundance and mechanism of adaptability in sensitive environments. More interestingly, Part II focuses on their sustainability in adapted niches. Overall, this book is a comprehensive exposition of microbiology of sensitive environments and provides an advantageous material for students, academicians, scientists, policy-makers, industrialists, and others who have keen interest in microbiology and biotechnology.

Dehradun, Uttarakhand, India   Raghvendra Pratap Singh

Contents

Editors and Contributors

About the Editors

Dr. Raghvendra Pratap Singh is an eminent scientist in the field of Microbial Biotechnology and is presently working at the Research and Development Department at Uttaranchal University, Dehradun, India. His research contributions largely concern the ecology of Myxobacteria, plant-microbe interaction and microbial genomics. He has received awards and honors from various scientific agencies and societies, e.g. a Young Scientist Award from the ABA-2017, SERB-DST grant, certification award from the FDA India, Chinese Postdoctoral Grant, DBT travel grant etc. In addition, Dr. Singh has published several international research articles in microbiology and biotechnology journals, numerous book chapters, and a highly successful textbook published by ICAR, New Delhi, India.

Dr. Geetanjali Manchanda is working as Head of the Botany Department at DAV University, Jalandhar, India. She has worked extensively on plant-microbe interactions in stressed and contaminated environments, with a special focus on mycorrhizae for the fortification of various crops. She has received prestigious research grants from the DST, India and IFS Sweden. She has contributed immensely to the scientific community by publishing research papers and book chapters. She recently authored a book on the use of omics technology for microbiology, published by ICAR, New Delhi, India.

Dr. Indresh Kumar Maurya received his Ph.D. in Microbiology from the School of Life Sciences (SLS), Jawaharlal Nehru University (JNU), New Delhi, India, in 2013. He has published 27 international research papers and holds one Indian patent. His teaching areas include medical microbiology, industrial microbiology and microbial biochemistry at the postgraduate level. A Life Member of four national societies, his research chiefly focuses on the design and development of bioactive molecules from synthetic as well as natural resources to combat pathogenic microorganisms.

Dr. Yunlin Wei's focus areas are environmental microbial ecology, bacterial signal transduction and small molecular recognition, as well as the development of vaccines and medicines for tumors or epidemic diseases. Prof. Wei has collaborated with numerous scientific bodies globally in the field of extremophiles and has been associated as an active member of the Japanese Biochemical Society, Yunnan Province Society of Genetics, Yunnan Province Society of Microbiology, and Yunnan Province Society of Immunology. He has received several prestigious awards, including the Japanese Government's Monbusho Scholarship, Award for Achievement in Progress of Science and Technology, Award of the Government of Yunnan Province, The Young Leading Scholar Award China, etc. He is the author of more than 50 scientific articles and 2 books, and holds 7 patents.

Contributors

Waquar Akhter Ansari National Bureau of Agriculturally Important Microorganism, Mau, Uttar Pradesh, India

Mian Nabeel Anwar State Key Laboratory of Microbial Technology, Shandong University, Jinan, People's Republic of China

Vivekanand Bahuguna Department of Biotechnology, Uttaranchal University, Dehradun, Uttarakhand, India

Dina Barman Microbial Ecology Laboratory, Department of Botany, Gauhati University, Guwahati, India

Kaushik Bhattacharjee Division of Life Sciences, Institute of Advanced Study in Science and Technology, Guwahati, Assam, India

Harleen Kaur Buttar Department of Botany and Environmental Studies, DAV University, Jalandhar, Punjab, India

Sushma Chauhan Amity Institute of Biotechnology, Amity University Chhattisgarh, Raipur, India

Rahul Dilawari CSIR-Institute of Microbial Technology (CSIR-IMTECH), Chandigarh, India

Karmveer Gautam Regional Plant Quarantine Station, Amritsar, Punjab, India

Directorate of Plant Protection Quarantine and Storage, Ministry of Agriculture and Farmer's Welfare, Govt. of India, Faridabad, Haryana, India

Chen Guo College of Food and Bioengineering, Zhengzhou University of Light Industry, Zhengzhou, Henan province, People's Republic of China

Mamta Gupta Department of Botany and Environmental Studies, DAV University, Jalandhar, Punjab, India

Priyanka Gupta Ramji Pandav Institute of Computer and Biological Studies, Nagpur, India

Yogesh Gupta Institute For Global Food Security, School of Biological Sciences, Queen's University, Belfast, UK

Dhruva K. Jha Microbial Ecology Laboratory, Department of Botany, Gauhati University, Guwahati, India

Xiuling Ji Faculty of Life Science and Technology, Kunming University of Science and Technology, Kunming, People's Republic of China

Sashi Kant Department of Immunology and Microbiology, University of Colorado School of Medicine, Anschutz Campus, Aurora, CO, USA

Ravneet Kaur Department of Botany and Environmental Studies, DAV University, Jalandhar, Punjab, India

Ram Krishna Directorate of Onion and Garlic Research (DOGR), Pune, India

Geetanjali Manchanda Department of Botany and Environmental Studies, DAV University, Jalandhar, Punjab, India

Indresh Kumar Maurya Department of Microbial Biotechnology, Panjab University, Chandigarh, India

Mohit S. Mishra Amity Institute of Biotechnology, Amity University Chhattisgarh, Raipur, India

Ashish Nayak Microbial Genomics and Diagnostics Lab., Microbiology and Plant Pathology Division, Regional Plant Resource Centre, Bhubaneswar, Odisha, India

Indrajeet Nishad Department of Plant Pathology, CSIR-Central Institute of Medicinal and Aromatic Plants, Lucknow, India

Saurabh Pandey Department of Biochemistry, School of Chemical and Life Sciences, Jamia Hamdard, New Delhi, India

Hovik Panosyan Department of Biochemistry, Microbiology and Biotechnology, Yerevan State University, Yerevan, Armenia

Shanshan Peng Collaborative Innovation Center for Food Production and Safety of Henan province, Zhengzhou, Henan Province, People's Republic of China

Arvind Saroj Department of Plant Pathology, CSIR-Central Institute of Medicinal and Aromatic Plants, Lucknow, India

Yimin Shang College of Food and Bioengineering, Zhengzhou University of Light Industry, Zhengzhou, Henan province, People's Republic of China

Department of Food and Bioengineering, Zhengzhou University of Light Industry, Zhengzhou, People's Republic of China

Nidhi Shukla Department of Translational Hematology and Oncology, Cleveland Clinic Learner Research Institute, Cleveland, OH, USA

Dipti Singh Department of Microbiology, V.B.S. Purvanchal University, Jaunpur, Uttar Pradesh, India

Raghvendra Pratap Singh Department of Research & Development, Biotechnology, Uttaranchal University, Uttarakhand, India

Ravi Kant Singh Amity Institute of Biotechnology, Amity University Chhattisgarh, Raipur, India

Savita Singh Department of Botany, Babu Shivnath Agrawal College (Affiliated to Dr BRA University, Agra), Mathura, Uttar Pradesh, India

Shailendra Singh National Bureau of Agriculturally Important Microorganism, Mau, Uttar Pradesh, India

Shashi Shekhar Singh Department of Inflammation and Immunity, Cleveland Clinic Lerner Research Institute, Cleveland, OH, USA

Atul Kumar Srivastava Department of Plant Pathology, CSIR-Central Institute of Medicinal and Aromatic Plants, Lucknow, India

Deeksha Tripathi Department of Microbiology, School of Life Sciences, Central University of Rajasthan, Ajmer, Rajasthan, India

Takshashila Tripathi Department of Neuroscience, Physiology and Pharmacology, University College London, London, UK

Yunlin Wei Faculty of Life Science and Technology, Kunming University of Science and Technology, Kunming, People's Republic of China

Akhilesh Yadav Department of Biotechnology, Indian Institute of Technology, Roorkee, India

Yingjie Yang Marine Agriculture Research Center, Tobacco Research Institute of Chinese Academy of Agricultural Sciences, Qingdao, People's Republic of China

Mohammad Tarique Zeyad Department of Agricultural Microbiology, Aligarh Muslim University, Aligarh, India

Junjie Zhang Department of Food and Bioengineering, Zhengzhou University of Light Industry, Zhengzhou, People's Republic of China

Part I

Microbes in Harsh Environments

Raghvendra Pratap Singh, Harleen Kaur Buttar, Ravneet Kaur, and Geetanjali Manchanda

Abstract

The world of microorganisms comprises a vast diversity of living organisms, each with its individual set of genes, cellular components, and metabolic reactions that interact within the cell and communicate with the environment in many different ways. Microbes perform several key ecosystem functions. They provide a number of ecological services like soil formation, nutrient cycling, plant growth, bioremediation, source for pharmaceuticals, etc. Earth may be a home for more than 10^{12} microbial species. These species are present in varied environments, many of which for other life forms are extremely hostile. Microbes have been found in varied environments ranging from the tropics to the Arctic and Antarctica, from underground mines and oil fields to the stratosphere and the top of great mountains, from deserts to the Dead Sea, from aboveground hot springs to underwater hydrothermal vents. They can survive at pressures up to 110 MPa, at extreme acid (pH 0) to extreme alkaline (pH 12.8) conditions, at temperatures as high as 122 °C to as low as −20 °C, in toxic wastes, in organic solvents, heavy metals, guts of insects, roots of plants, low oxygen conditions, etc. These microorganisms are classified according to their habitats such as thermophiles/hyperthermophiles, psychrophiles, acidophiles and alkaliphiles, barophiles, and halophiles. Studies on microbial life, their diversity, physiology, genetics, ecology, and biochemistry can reveal a lot in terms of the characteristics of biological processes, such as biochemical limits to macromolecular stability and genetic

R. P. Singh
Department of Research & Development Cell, Biotechnology, Uttaranchal University, Dehradun, India
e-mail: rpsingh@uttaranchaluniversity.ac.in

H. K. Buttar · R. Kaur · G. Manchanda (✉)
Department of Botany and Environment Studies, DAV University, Jalandhar, India
e-mail: geetanjali10194@davuniversity.org

© Springer Nature Singapore Pte Ltd. 2020
R. P. Singh et al. (eds.), *Microbial Versatility in Varied Environments*,
https://doi.org/10.1007/978-981-15-3028-9_1

instructions for constructing macromolecules. These organisms have also provided a number of clues to the origin and evolution of life. Further, a study of microbes in conflict environments has become vital in the field of research in astrobiology, since the microbes found in extreme environments may be analogous to potential life forms in other planets. In the present times of climate change, the study of microbes living at the edge of life has become even more important.

Keywords
Extremophiles · Varied environment · Genes · Ecology

1.1   Introduction

During the history of earth's evolution, microbes have played a vital role in changing and maintaining the environments on earth. An environment that was initially anoxygenic was transformed and has been maintained till today in the present state by microorganisms. Since their evolution on earth at about 3–4 million years ago, microbes have occupied every nook and corner of the earth that one can think of. The earliest microorganisms were anaerobic heterotrophic, since the atmosphere was free of oxygen. The enormous microbial biodiversity that we find today is the result of a balance between evolution, extinction, and colonization. Microbial species have been found in varied environments, many of which for other life forms are extremely hostile or not "normal," with normal being those environments where temperature is between 4 and 40 °C, pH between 5 and 8.5, and salinity between that of freshwater and that of seawater. Many microbes have the ability to survive in extreme conditions, whereas others cannot survive and die in these conditions. The study of microorganisms in varied environments allows us to get a glimpse of the environment that must have been present on the earth before the rise of eukaryotes and the role microbes must have played in making the conditions feasible for the higher life forms. Their study can also provide vital clues that can be used by humans for adapting to the changing environment.

Many microorganisms can survive in multiple extreme conditions, e.g., cyanobacterium *Chroococcidiopsis* can survive in a variety of extreme conditions such as acidity, salinity, dryness, and high and low temperatures. *Salinibacter ruber* is a red obligatory aerobic chemoorganotrophic extremely halophilic Bacterium, related to the order Cytophagales. It was isolated from saltern crystallizer ponds and requires at least 150 g l^{-1} salt for growth. Microbes, present everywhere, play an important role in the cycling of carbon, hydrogen, oxygen, nitrogen, sulfur, etc. However, archaea is the main microbial group to thrive in most of the extreme environments.

1.2   Microbes in Varied Environments

1.2.1   Microbes in High Salt Concentration

Halophiles are found in all three domains of life. Within the Bacteria, we know halophiles are found within the phyla Cyanobacteria, Proteobacteria, Firmicutes, Actinobacteria, Spirochaetes, and Bacteroidetes. Within the archaea the most salt-requiring microorganisms are found in the class *Halobacteria*. Members of the *Halobacteria* are characteristic inhabitants of salt lakes at or approaching halite saturation, saltern crystallizer ponds, and other high salt environments (i.e., the Dead Sea, the Great Salt Lake, etc.). Halophilic organisms can also be found in man-made saline environments such as salted foods and tanned hides (Antranikian 2009). Most of its relatives are obligate halophiles and require over 150–200 g L^{-1} of salts for growth and structural stability. Further, they cannot survive under lower concentrations. The halophilic species are also found within the order Methanococci. Halophiles adopt different strategies for survival in high salt concentrations and thus maintain the cytoplasmic balance with the medium. One of the widely adopted strategies is the accumulation of compatible organic osmotic solutes which do not interfere with the enzymatic activity. These compatible osmolytes include glycine betaine, glycerol, ectoine, and other amino acid derivatives, sugars and sugar alcohols, etc. These are uncharged molecules and their concentrations are adjusted to the outside salinity. Halophiles commonly contain cell membrane proteins with high ratio of acidic to basic amino acid, thus giving the surface of the proteins a negative charge (DasSarma and Arora 1997). These organisms have a highly acidic proteome. Their protein surfaces have excess of negatively charged amino acids such as aspartate and glutamate, when compared to the positively charged amino acids like lysine and arginine.

1.2.2   Microbes in Low-Temperature Environments

A number of microorganisms can grow optimally at less than 15 °C (upper limit of 20 °C) and are called psychrophiles, whereas some others are able to survive at temperatures below 0 °C and grow optimally at 20–25 °C and are called psychrotolerant organisms. The psychrophiles usually inhabit marine ecosystems, where the temperatures are permanently lesser than 5 °C. In contrast, the psychrotolerant organisms are usually isolated from terrestrial environments, which are prone to extreme temperature fluctuations (Deming 2002). Cold temperature usually limits the cell function since it has negative impact on the cell integrity, water viscosity, membrane fluidity, and macromolecular interactions. Therefore, the organisms adopt a number of adaptive strategies to maintain vital cellular functions at cold temperatures and associated stress factors, such as desiccation, radiation, excessive UV, high or low pH, high osmotic pressure, and low nutrient availability. These microorganisms contain polyunsaturated and fatty acids and cold shock proteins to help in fluidity and nutrient transportation (Feller 2013).

1.2.3   Microbes in High-Temperature Environments

The microorganisms growing optimally at temperature between 60 and 80 °C are designed as thermophiles. In contrast to this, some non-photosynthetic prokaryotic can grow at >100 °C or even more are referred as hyperthermophilic (Scambos et al. 2018). These are found in the three domains of life, Archaea, Bacteria, and Eukaryotes. There are two major types of thermophiles: the microbes that grow in geothermal sites and that grow in self-heating materials such as composts. The thermophiles contain a variety of mechanisms that allow them to survive at higher temperature no other organisms can thrive (Weigal. 2000). These microorganisms have evolved several traits including novel membrane lipid composition (saturated fatty acids with more branches), thermostable membrane proteins, and higher rates of the synthesis of various enzymes (Sterner and Liebl 2001). Apart from having thermostable membrane proteins, these microorganisms also contain stabilized proteins, DNA, and RNA (Ladenstein and Ren 2006). Further, genomic studies of thermophiles have demonstrated that the evolution of thermophiles depends on the level of heritable variation (i.e., genomic size, gene mutations, transcriptional and proteome regulations) (López-García 1999).

1.2.4   Microbes in the Atmosphere

The atmosphere of the earth is comprised of different layers, i.e., troposphere, stratosphere, mesosphere, and thermosphere. These layers are based on the temperature. The troposphere, which is the lowest layer, begins at the surface of the earth and extends up to 10 km above sea level. It is the wettest layer of the atmosphere and has almost all types of clouds and all types of weathers. The layer immediately above the troposphere is the stratosphere. The bottom of the stratosphere is at 10 km above sea level and extends up to 50 km above sea level. Ozone is abundant in this layer and heats this layer as it absorbs the energy from the incoming UV radiations. This layer is dry and thus contains very few clouds. The temperature increases as the altitude increases. The layer above the stratosphere is the mesosphere. It extends from 50 to 85 km above sea level. The temperature decreases with altitude in this layer and the coldest temperature on earth (-100 °C) can be recorded in this layer. The thermosphere extends from about 90 km to between 500 and 1000 km above our planet, and the temperatures can range from about 500 °C to 2000 °C. This is followed by the ionosphere, the part of the atmosphere ionized by solar radiation. The uppermost layer is the exosphere (up to 10,000 km), which merges with space. Diverse and viable communities of bacteria reside high in the troposphere despite the extreme cold, high UV irradiation, and thin air, suggesting adaptive mechanisms. Surprisingly, viable bacteria form up to 20% of the total number of particles found in the troposphere (DeLeon-Rodriguez et al. 2013). Some atmospheric regions have extreme conditions, but microorganisms already live under even harsher environments with extremes of pH, temperatures, and radiation

(Pikuta et al. 2007). The roles of high-altitude dwelling bacteria are not well understood. It is likely that these microbes affect meteorological events, such as cloud formation, precipitation, or atmospheric chemistry. Microbes are often considered passive inhabitants of the atmosphere, dispersing via airborne dust particles. However, recent studies suggest that many atmospheric microbes may be metabolically active (Bowers et al. 2009), even up to altitudes of 20,000 m (Griffin 2004). Some airborne microbes may alter atmospheric conditions directly by acting as cloud condensation nuclei (Bauer et al. 2003; Mohler et al. 2007) and/or ice nuclei (IN) (Pouleur et al. 1992; Mohler et al. 2007); this hypothesis is supported by the observation that most ice nuclei in snow samples are inactivated by a 95 °C heat treatment (Christner et al. 2008). However, the overall contribution of airborne microbes to atmospheric processes such as ice nucleation remains unclear.

1.2.5 Microbes in Varied pH Environments

Extremely low and high pH habitats have been observed for different ecosystems contaminated by mining waste on earth. Acidophiles thrive at low pH and come from bacteria, fungi, algae, protozoa, and archaea. Currently, the most extreme acidophiles and alkaliphiles can thrive at pH 0 and pH 12. These microorganisms thrive in hot springs, marine vents, sulfuric pools and geysers, coal mines, and metallic ores. In order to maintain the internal pH, acidophiles either actively excrete protons or use them in various metabolic reactions such as the reduction of oxygen in the membrane. Acidophiles utilize both energies derived and non-energy processes to maintain internal pH. On the other hand, the alkaliphiles thrive in environments with a pH between 10 and 12 with an optimum growth pH of about 9 (Padan et al. 2005; Singh et al. 2016; Li et al. 2017). These microorganisms are distributed worldwide like hypersaline lakes, soda lakes, industrial effluents, and alkaline soil microenvironments (Fujisawa et al. 2010). In order to survive at these extreme conditions, these microorganisms have novel adaptations to cell wall structure such as a variety of acidic compounds (i.e., phosphoric acid, aspartic acid, galacturonic acid, glutamic acid, and gluconic acid).

1.2.6 Microbes in High-Pressure Environments

These microorganisms (piezophiles/barophiles) have the capability to thrive at high-pressure area, especially higher than normal (from 0.1 MPa to 112 MPa), or they need increased pressure for their normal growth and survival. They thrive in deep sea location, hydrothermal vents, trenches, sediments, and water samples from depths (Certes 1884). *S. benthica* and *M. yayanosii* have different strains which are extremely barophilic and barophilic with optimal growth at 50–80 MPa (Kato et al. 1998). *S. violacea* and *Photobacterium profundum* and *M. japonica* strains come in the category of moderate barophilic bacteria thriving at pressures of 10–50 MPa (Kato et al. 1995). *Sporosarcina* sp. strains belong to the category of barotolerant capable of growing at 0.1 MPa pressure (Kato et al. 1995). These microbes have adapted to such extreme environments because of various modifications like secretion of polyunsaturated fatty acid (PUPA) and eicosapentaenoic acid (EPA) (Nogi et al. 1998; DeLong et al. 1997; Kato et al. 1998). Such microbes have gene expression under the control of pressure-regulated promoter sequences (Nakasone et al. 1998). There are reports of pressure-inducible proteins (Jaenicke et al. 1988) and elevated levels of heat shock protein GroES (Lin and Rye 2006). These extremophiles can be exploited for the industrial and high-pressure fermenters because genes and proteins are accustomed to high-pressure conditions; also the barophilic origin proteases and glucanases can be used for detergents and DNA polymerases in PCR amplification. The research is being extended to mesophilic piezophiles apart from commonly explored psychrophilic and thermophilic piezophiles. The yeast *S. cerevisiae* is converted into a piezophile by manipulating the genome and by introducing genes that control high-pressure growth in yeast. Tryptophan permease gene TAT2 confers high-pressure growth in *S. cerevisiae* (Abe and Horikoshi 2000). The mesophilic organisms with piezophilic applications may help industrial applications. Similarly, novel antibiotics may be produced from mutants grown under high-pressure conditions. Piezophiles/barophiles have a unique membrane composition or structure that would allow them to survive at the greatly increased pressure.

1.2.7 Microbes in Radiation Environment

These extremophiles can withstand and survive the presence of ionizing and UV radiations such as *Deinococcus radiodurans* (Sandigursky et al. 2004), *D. radiophilus* (Yun and Lee 2004), *Thermococcus piezophilus* (Jolivet et al. 2004), and Cyanobacteria like *Nostoc muscorum* and *Microcoleus vaginatus* (Singh 2018). *Acinetobacter radioresistens* (Jawad et al. 1998) are few examples of radiophiles. They thrive in various radioactive places like dry climate soil, nuclear reactors like *D. radiodurans*, and Mars Analog Antarctic Dry Valleys (Musilova et al. 2015). They have various adaptations to withstand radiation stresses like Nudix hydrolase enzyme superfamily and the homologs of plant desiccation resistance-associated proteins contributing to extreme radiation and desiccation resistance of *Deinococcus* sp. (Makarova et al. 2001). PprA protein helps in DNA ligation after DNA fragmentation due to radiation exposure reported in *D. radiodurans* (Narumi et al. 2004), increased metal concentration (i.e., Mn/Fe) ratios in protection of *D. radiodurans* cellular proteins from oxidative damage (Daly 2009), and accumulation of MnII. It helps in resistance toward gamma radiation exposure (Daly et al. 2004), production of mycosporine-like amino acids (MAAs) in cyanobacteria due to exposure of solar UV-B radiations (Sinha et al. 2001), etc. As far as applications of radiophile are concerned, they are helpful in the management of nuclear waste-polluted environments (Brim et al. 2003; Appukuttan et al. 2006).

1.2.8 Microbes in Metallic Environment

These extremophiles (metallophiles) have the capability to grow in the presence of heavy metal waste. They thrive in industrial sediments, soil, and waste effluents containing heavy metals (Mergeay et al. 2003). These metallophiles possess heavy metal resistance due to the presence of megaplasmids conferring genes for resistance by efflux mechanisms (Gomes and Steiner 2004). These microorganisms remove toxic metals via change in redox potential of metals by aiding in bioleaching of contaminants and precipitation (Lovley and Coates 1997; Wani et al. 2007; Pal and Rai 2010). These extremophiles can be exploited as bioremediation/bioleaching for different metal ores (i.e., Cu, Fe, Zn, etc.) from toxic compound removal from various industrial/mining waste effluents.

1.2.9 Microorganisms in Xerophilic Environment

These microorganisms (xerophiles) are able to thrive in low water environments and resist high desiccation, i.e., water activity below 0.8. Endolithic and halophilic microbes come under xerotolerant. Some xerophiles such as Cyanobacteria, *Nostoc commune*, were recovered from dry area after 13 years and from herbarium storage after 55 years (Shirkey et al. 2003), from dry storage conditions of herbarium after 87 years (Lipman 1941), and after 107 years from dry soil sample (Blank and Cameron 1966). There are reports of synthesis of extracellular polysaccharides (EPS) which withstand dry environment (De Philippis and Vincenzini 1998); cell component stabilization by buildup of compatible solutes like trehalose (Welsh 2000); upregulation of genes related to osmotic, salt, and low-temperature stress, osmoprotectant metabolisms, K^+ transporting system, and heat shock proteins; and downregulation of genes involved in photosynthesis, nitrogen transport, RNA polymerase, and ribosomal proteins (Katoh et al. 2004). The proteins like catalases, peroxidases, and superoxide dismutase expression are increased to neutralize ROS due to desiccation (Shirkey et al. 2000).

1.3 Future Prospective

All the three domains of life (thermophiles, halophiles, acidophiles, alkaliphiles, and piezophiles) take life to the extreme. By studying the biology of these unique microorganisms, we can gain deep insight into how life evolve on earth and even infer as to how life would be able to exist elsewhere in the universe. As for the origins of life on earth, some scientists are looking to the extremophile microbes such as *D. radiodurans* as model organisms when exploring the existence of extraterrestrial life of the solar system and beyond. This microorganism has the unique ability to survive radiation at several thousand times the lethal dose for humans.

With these groundbreaking research work and recent development in the field of extremophiles, which have been directly applicable in different branches of life sciences, our understanding about the biosphere has grown and the putative boundaries of life have expanded. However, due to the recent growth and advancement, we are just at the beginning of exploring the world of extremophiles. In this chapter, we have discussed the several aspects of these fascinating microorganisms, exploring their habitats, biodiversity, ecology, evolution, biochemistry, as well as applications.

References

Abe F, Horikoshi K (2000) Tryptophan permease gene TAT2 confers high-pressure growth in *Saccharomyces cerevisiae*. Molecul Cellula Biol 20(21):8093–8102. https://doi.org/10.1128/mcb.20.21.8093-8102.2000

Antranikian G (2009) Extremophiles and biotechnology. In eLS (ed). https://doi.org/10.1002/9780470015902.a0000391.pub2

Appukuttan D, Rao AS, Apte SK (2006) Engineering of *Deinococcus radiodurans* R1 for biopre-cipitation of uranium from dilute nuclear waste. Appl Environ Microbiol 72(12):7873–7878

Bauer H, Giebl H, Hitzenberger R, Kasper-Giebl A, Reischl G, Zibuschka F et al (2003) Airborne bacteria as cloud condensation nuclei. J Geophys Res 108:4658

Blank GB, Cameron RE (1966) Desert algae-soil crusts and diaphanous substrata as algal habitats, vol 32. Jet Propulsion Laboratory, California Institute of Technology, Pasadena, pp 1–41

Bowers RM, Lauber CL, Wiedinmyer C, Hamady M, Hallar AG et al (2009) Characterization of airborne microbial communities at a high-elevation site and their potential to act as atmospheric ice nuclei. Appl Environ Microbiol 75(15):5121–5130. https://doi.org/10.1128/AEM.00447-09

Brim H, Venkateswaran A, Kostandarithes HM, Fredrickson JK, Daly MJ (2003) Engineering *Deinococcus geothermalis* for bioremediation of high-temperature radioactive waste environments. Appl Environ Microbiol 69(8):4575–4582

Certes A (1884) Note upon the effect of high pressures on the vitality of minute fresh-water and salt-water organisms. Bull U S Fish Comm 4(28):433–435

Christner BC, Morris CE, Foreman CM, Cai R, Sands DC (2008) Ubiquity of biological ice nucleators in snowfall. Science 319:1214

Daly MJ (2009) A new perspective on radiation resistance based on *Deinococcus radiodurans*. Natur Rev Microbiol 7(3):237

Daly MJ, Gaidamakova EK, Matrosova VY, Vasilenko A, Zhai M et al (2004) Accumulation of Mn (II) in *Deinococcus radiodurans* facilitates gamma-radiation resistance. Science 306(5698):1025–1028

DasSarma S, Arora P (1997) Genetic analysis of the gas vesicle cluster in haloarchaea. FEMS Microbiol Lett 153:1–10

De Philippis R, Vincenzini M (1998) Exocellular polysaccharides from cyanobacteria and their possible applications. FEMS Microbiol Revi 22(3):151–175

DeLeon-Rodriguez N, Lathem TL, Rodriguez-RLM, Barazesh JM, Anderson BE et al (2013) Microbiome of the upper troposphere: species composition and prevalence, effects of tropical storms, and atmospheric implications. PNAS U S A 110:2575–2580

DeLong EF, Franks DG, Yayanos AA (1997) Evolutionary relationships of cultivated psychrophilic and barophilic deep-sea bacteria. Appl Environ Microbiol 63(5):2105–2108

Deming JW (2002) Psychrophiles and polar regions. Curr Opin Microbiol 5(3):301–309

Feller G (2013) Psychrophilic enzymes: from folding to function and biotechnology. Scientifica 2013:512840. https://doi.org/10.1155/2013/512840

Fujisawa M, Fackelmayer OJ, Liu J, Krulwich TA, Hicks DB (2010) The ATP synthase a-subunit of extreme alkaliphiles is a distinct variant. J Biol Chem 285(42):32105–32115

Gomes JI, Steiner W (2004) The biocatalytic potential of extremophiles and extremozymes. Food Technol Biotechnol 42(4):223–225

Griffin DW (2004) Terrestrial microorganisms at an altitude of 20,000 m in Earth's atmosphere. Aerobiologia 20:135–140

Jaenicke R, Bernhardt G, Lüdemann HD, Stetter KO (1988) Pressure-induced alterations in the protein pattern of the thermophilic archaebacterium *Methanococcus thermolithotrophicus*. Appl Environ Microbiol 54(10):2375–2380

Jawad A, Snelling AM, Heritage J, Hawkey PM (1998) Exceptional desiccation tolerance of *Acinetobacter radioresistens*. J Hospital Inf 39(3):235–240

Jolivet E, Corre E, Haridon S, Forterre P, Prieur D (2004) *Thermococcus marinus* sp. nov. and *Thermococcus radiotolerans* sp. nov., two hyperthermophilic archaea from deep-sea hydrothermal vents that resist ionizing radiation. Extremophiles 8(3):219–227

Kato C, Sato T, Horikoshi K (1995) Isolation and properties of barophilic and barotolerant bacteria from deep-sea mud samples. Biodivers Conserv 4:1–9

Kato C, Li L, Nogi Y, Nakamura Y, Tamaoka J, Horikoshi K (1998) Extremely barophilic bacteria isolated from the Mariana trench, challenger deep, at a depth of 11,000 meters. Appl Environ Microbiol 64(4):1510–1513

Katoh H, Asthana RK, Ohmori M (2004) Gene expression in the cyanobacterium *Anabaena* sp. PCC7120 under desiccation. Microbial ecol 47(2):164–174

Ladenstein R, Ren B (2006) Protein disulfides and protein disulfide oxidoreductases in hyperthermophiles. FEBS J 273(18):4170–4185

Li ZF, Li P, Jing YG, Singh RP, Li YZ (2017) Isolation and characterization of the epothilone gene cluster with flanks from high alkalotolerant strain *Sorangium cellulosum* (So0157-2). World J Microbiol Biotechnol 33(7):137. https://doi.org/10.1007/s11274-017-2301-y

Lin Z, Rye HS (2006) GroEL-mediated protein folding: making the impossible, possible. Cri Rev biochem Molecule Biol 41(4):211–239. https://doi.org/10.1080/10409230600760382

Lipman CB (1941) The successful revival of *Nostoc* commune from a herbarium specimen eighty-seven years old. Bull Torrey Bot Club 1:664–666

López-García P (1999) DNA supercoiling and temperature adaptation: a clue to early diversification of life? J Molecula Evol 49(4):439–452

Lovley DR, Coates JD (1997) Bioremediation of metal contamination. Curr Opin Biotechnol 8(3):285–289

Makarova KS, Aravind L, Wolf YI, Tatusov RL, Minton KW et al (2001) Genome of the extremely radiation-resistant bacterium *Deinococcus radiodurans* viewed from the perspective of comparative genomics. Microbiol Mol Biol Rev 65(1):44–79

Mergeay M, Monchy S, Vallaeys T, Auquier V, Benotmane A et al (2003) *Ralstonia metallidurans*, a bacterium specifically adapted to toxic metals: towards a catalogue of metal-responsive genes. FEMS Microbiol Rev 27(2–3):385–410

Mohler O, DeMott PJ, Vali G, Levin Z (2007) Microbiology and atmospheric processes: the role of biological particles in cloud physics. Biogeosciences 4:2559–2591

Musilova M, Wright G, Ward JM, Dartnell LR (2015) Isolation of radiation-resistant bacteria from Mars analog Antarctic Dry Valleys by preselection. And the correlation between radiation and desiccation resistance. Astrobiology 15(12):1076–1090

Nakasone K, Ikegami A, Kato C, Usami R, Horikoshi K (1998) Mechanisms of gene expression controlled by pressure in deep-sea microorganisms. Extremophiles 2(3):149–154

Narumi I, Satoh K, Cui S, Funayama T, Kitayama S et al (2004) PprA: a novel protein from *Deinococcus radiodurans* that stimulates DNA ligation. Mol Microbiol 54(1):278–285. https://doi.org/10.1111/j.1365-2958.2004.04272.x

Nogi Y, Kato C, Horikoshi K (1998) Taxonomic studies of deep-sea barophilic *Shewanella* strains and description of *Shewanella violacea* sp. Nov. Arch Microbiol 170(5):331–338

Padan E, Bibi E, Ito M, Krulwich TA (2005) Alkaline pH homeostasis in bacteria. New insights. Biochim Biophys Acta 1717(2):67–88

Pal R, Rai JPN (2010) Phytochelatins: peptides involved in heavy metal detoxification. Appl Biochem Biotechnol 160(3):945–963

Pikuta EV, Hoover RB, Tang J (2007) Microbial extremophiles at the limits of life. Crit Rev Microbiol 33:183–209

Pouleur S, Richard C, Martin J-G, Antoun H (1992) Ice nucleation activity in *Fusarium acuminatum* and *Fusarium avenaceum*. Appl Environ Microbiol 58:2960–2964

Sandigursky M, Sandigursky S, Sonati P, Daly MJ, Franklin WA (2004) Multiple uracil-DNA glycosylase activities in *Deinococcus radiodurans*. DNA Repair 3(2):163–169

Scambos TA, Camppbell GG, Pope A, Haran T, Muto A et al (2018) Ultralow surface temperature in east Antarctica from satellite thermal infra-red mapping: the coldest place on earth. Geophys Res Lett 45(12):6124–6133

Shirkey B, Kovarcik DP, Wright DJ, Wilmoth G, Prickett TF et al (2000) Active Fe-containing superoxide dismutase and abundant sodF mRNA in *Nostoc* commune (cyanobacteria) after years of desiccation. J Bacteriol 182(1):189–197

Shirkey B, McMaster NJ, Smith SC, Wright DJ, Rodriguez H, Jaruga P, Potts M (2003) Genomic DNA of Nostoc commune (Cyanobacteria) becomes covalently modified during long-term (decades) desiccation but is protected from oxidative damage and degradation. Nucleic Acids Res 31(12):2995–3005

Singh H (2018) Desiccation and radiation stress tolerance in cyanobacteria. J Basic Microbiol 58(10):813–826

Singh RP, Manchanda G, Singh RN, Srivastava AK, Dubey RC (2016) Selection of alkalotolerant and symbiotically efficient chickpea nodulating rhizobia from North-West Indo Gangetic Plains. J Basic Microbiol 56:4–25

Sinha RP, Klisch M, Gröniger A, Hader DP (2001) Responses of aquatic algae and cyanobacteria to solar UV-B. In: Responses of plants to UV-B radiation. Springer, Dordrecht, pp 219–236

Sterner RH, Liebl W (2001) Thermophilic adaptation of proteins. Crit Rev Biochem Mol Biol 36(1):39–106

Wani PA, Khan MS, Zaidi A (2007) Effect of metal tolerant plant growth promoting *Bradyrhizobium* sp. (*Vigna*) on growth, symbiosis, seed yield and metal uptake by green-gram plants. Chemosphere 70(1):36–45

Welsh DT (2000) Ecological significance of compatible solute accumulation by micro-organisms: from single cells to global climate. FEMS Microbiol Rev 24(3):263–290

Yun YS, Lee YN (2004) Extremophiles 8(3):237–242. https://doi.org/10.1007/s00792-004-0383-6

Thermophilic and Halophilic Prokaryotes Isolated from Extreme Environments of Armenia and Their Biotechnological Potential

2

Hovik Panosyan

Abstract

Numerous geothermal springs of different geotectonic origins and with different physicochemical properties as well as various saline and hypersaline environments with a broad microbial diversity and opportunities for newly isolated microorganisms for many industrial applications are found in the territory of Armenia. Despite intensive microbiological studies on terrestrial geothermal springs and hypersaline environments in various regions of the globe, very little is known about the microbial diversity of similar ecosystems in Armenia. During the past decades, the phylogenetic diversity of microbial community thriving in geothermal springs and hypersaline environments located on the territory of Armenia has been explored following both cultivation-based and culture-independent approaches. The thermophilic bacterial members studied were microbes belonging to the *Bacillus*, *Geobacillus*, *Anoxybacillus*, *Paenibacillus*, *Sporosarcina*, *Thermoactinomyces*, *Rhodobacter*, *Methylocaldum*, *Arcobacter*, *Desulfomicrobium*, *Desulfovibrio*, *Treponema*, and archaeal genus *Methanoculleus*, whereas the isolated halophilic microorganisms were mainly found to be the members of bacterial phylum Firmicutes and archaeal family *Halobacteriaceae*. This chapter contains a review of the results of microbial diversity analyses of geothermal springs, saline–alkaline soils, and subterranean salt deposits of Armenia with special emphasis to its distribution, ecological significance, and biotechnological potential.

Keywords

Thermophiles · Halophiles · Geothermal springs · Subterranean salt deposit · Saline · Alkaline soils · Biotechnological applications

H. Panosyan (✉)
Department of Biochemistry, Microbiology and Biotechnology, Yerevan State University, Yerevan, Armenia
e-mail: hpanosyan@ysu.am

© Springer Nature Singapore Pte Ltd. 2020
R. P. Singh et al. (eds.), *Microbial Versatility in Varied Environments*,
https://doi.org/10.1007/978-981-15-3028-9_2

13

2.1   Introduction

Extremophilic microorganisms are adapted to survive in ecological niches with extreme conditions, such as at high or low temperatures, extreme values of pH, high salt concentrations, and high pressure. These microorganisms produce unique bio-molecules and biocatalysts that function under extreme conditions (Antranikian and Egorova 2007; Kumar et al. 2011; Rampelotto 2013; Singh et al. 2016; Orellana et al. 2018). To date, some of the enzymes (extremozymes) produced by extremo-philes are currently being assessed for their industrial applications (Raddadi et al. 2015; Dumorné et al. 2017). Due to stability at high temperatures and under high salt concentrations, the demand for enzymes produced by thermophiles and halo-philes, respectively, has increased considerably (DasSarma and DasSarma 2015; Littlechild 2015; Edbeib et al. 2016; Atalah et al. 2019). Despite benefits of these enzymes, their potentials remain largely unexplored (DeCastro et al. 2016; Khan and Sathya 2018).

Natural geothermal springs, including terrestrial hot springs, are widely distrib-uted in various regions of our planet and offer a new source of fascinating microor-ganisms with unique properties (Hreggvidsson et al. 2012; Deepika and Satyanarayana 2013; Saxena et al. 2017).

Halophiles are found in all three life domains and are distributed in habitats cov-ering a wide range of salinities. Hypersaline environments are widely distributed and mainly represented by saline water systems, including saline lakes, saline–alka-line soils, solar salterns, and subterranean salt deposits (Oren 2013; Canfora et al. 2014; Ventosa et al. 2015; Vera-Gargallo et al. 2019). Salinity is usually associated with alkalinity, and many saline or hypersaline environments are characterized by alkaline or extremely alkaline pH values (Marquez et al. 2011). Saline–alkaline soils differ from each other in terms of the salt concentrations caused by primary or secondary salinization processes and their chemical compositions (Ventosa 2006; Canfora et al. 2014).

The scientific interest in the microbial diversity of these exotic niches has increased during the last decades. The use of a combination of several approaches of traditional microbiology with state-of-the-art molecular biology techniques has substantially increased our understanding of the structural and functional diversity of the microbial communities of hot springs, saline–alkaline soils, and subterranean salt deposits (López-López et al. 2013; Vera-Gargallo et al. 2019).

Despite the relatively small territory, Armenia has many different extreme eco-logical areas, which possess a broad microbial diversity. The geology of the region where Armenia is situated is complex, owing to the accretion of terrains through plate-tectonic processes, ongoing tectonic activity, and volcanism (Badalyan 2000; Henneberger et al. 2000). Numerous geothermal springs with different geochemical properties have been found on the territory of Armenia (Mkrtchyan 1969). Hydromorphic saline–alkaline soils (occupying more than 29,000 ha of the Ararat

Plain) and hypersaline subterranean salt deposits were also found in Armenia (Baghdasaryan 1971).

Thermal springs and saline–hypersaline environments located on the territory of Armenia still represent a challenge for exploring biodiversity and searching of undescribed biotechnological resources. During the last decades, many thermophilic and halophilic microbes have been isolated from Armenian geothermal springs and saline–hypersaline environments, respectively (Panosyan et al. 2018a, b). The thermophiles studied were microbes belonging to bacterial genera *Bacillus*, *Geobacillus*, *Anoxybacillus*, *Paenibacillus*, *Sporosarcina*, *Thermoactinomyces*, *Rhodobacter*, *Methylocaldum*, *Arcobacter*, *Desulfomicrobium*, *Desulfovibrio*, *Treponema*, and archaeal genus *Methanoculleus*, whereas the isolated halophilic microorganisms were mainly found to be members of bacterial phylum Firmicutes and archaeal family *Halobacteriaceae*. In this chapter, an attempt has been made to review the microbes isolated and identified in extreme conditions of Armenia like geothermal springs, saline–alkaline soils, and subterranean salt deposits, as well as on the potential biotechnological applications of their extremozymes and novel biomolecules.

2.2 Thermal Springs Studied in Armenia

The distribution of natural geothermal springs is primarily associated with tectonically active zones in areas where the Earth's crust is relatively thin. On the territory of Armenia, where traces of recently active volcanic processes are still noticeable, many geothermal springs with different geotectonic origins are found (Mkrtchyan 1969; Henneberger et al. 2000). Most of the studies were focused on the hot springs at higher altitude. Hot springs in Armenia, where temperature varies from 25.8 to ≥ 53 °C, are neutral, moderately alkaline, or alkaline in nature and have mainly mixed cation–mixed anion compositions. As is typically the case, the hotter and more saline samples tend to have higher ratios of (Na^+ and K^+) / (Ca^{2+} and Mg^{2+}) and relatively high ratios of chloride to bicarbonate (Cl^- / HCO_3^-) or sulfate to bicarbonate (SO_4^{2-} /HCO_3^-). The cooler waters tend to be higher in Ca^{2+} and Mg^{2+} and bicarbonate (Mkrtchyan 1969; Henneberger et al. 2000). All studied springs are rich in heavy metals. Some of the springs contain gasses such as hydrogen sulfide, methane, nitrogen, and carbon dioxide (Mkrtchyan 1969). A majority of the hot springs found in Armenia are anthropogenically influenced and often used by tourists and local people for bath. Some of the geothermal springs are used for balneology for curing rheumatismal and dermatological as well as digestive disorders and physical exhaustion (Mkrtchyan 1969; Panosyan et al. 2018a).

The geographical locations and physicochemical profiling of some geothermal springs distributed on the territory of Armenia are summarized in Table 2.1. Figure 2.1 shows photographs of some Armenian geothermal springs studied and documented within this review.

Table 2.1 Geographical location and physicochemical profiling of main geothermal (mesothermal) springs distributed on the territory of Armenia

Geothermal spring	Spring GPS location	Altitude, m, above sea level	pH	Conductivity, $\mu S/cm$	Outlet temperature T, °C
Akhurik	40°44′34.04″N 43°46′53.95″E	1490	6.5	2490	30
Arzakan	40°27′36.10″N 44°36′17.76″E	1490	7.2	4378.3	44
Bjni	40º45'94.44″N 44º64'86.11″E	1610	6.2–7.0	4138.3	30–37
Hankavan	40º63′26.50″N 44º48′46. 00″E	1900	7.0–7.2	6722.9	42–44
Jermuk	39°96′63.90″N 45°68′52.80″E	2080	7.5	4340	>53
Tatev	39°23′76.00″ N 46°15′48.00″ E	960	6.0	1920	27.5
Uytc (Uz)	39°31′00″ N 46°03′09″ E	1600	6.23	2700	25.8

Fig. 2.1 Some microbiologically explored terrestrial geothermal springs in Armenia: (**a**) Akhurik, (**b**) Arzakan, (**c**) Bjni, (**d**) Hankavan, (**e**) Jermuk, (**f**) Tatev

2.3   Thermophilic Microbes Isolated and Identified from Armenian Hot Springs and Their Biotechnological Importance

Cultivable approaches have been used for the analysis of microbial diversity associated with hot springs. Several studies have been performed on the description of novel genera, species, and strains, characterization of different bioresources, and whole genome analysis of a few isolates from geothermal springs in Armenia. Many thermostable enzymes, including lipase, protease, amylase, aspartase, aminoacylase, glucose isomerase, and inulinase, producers of EPS, protein and vitamins, and other biomolecule producers with potential biotechnological applications have been reported by several authors (Table 2.2).

Panosyan (2010) studied the distribution of thermophilic aerobic endospore-forming bacteria in two Armenian terrestrial geothermal springs (Arzakan and Akhurik). The phylogenetic diversity of 19 isolates was studied by the 16S rRNA gene analysis. The comparison of generated 16S rDNA sequences of the isolates with the ones available in GenBank database indicates their relation to genera *Bacillus*, *Geobacillus*, *Sporosarcina*, *Paenibacillus*, and *Thermoactinomyces*. In total of 14 (ArzA-2 to ArzA-11, ArzA-13, ArzA-13A, ArzA-33, and ArzA-33A) and 5 (AkhA-1, AkhA-12, AkhA-14, AkhA-14a, and AkhA-15) aerobic thermophilic bacilli, strains were isolated from Arzakan and Akhurik hot springs, respectively. All strains were Gram-positive, endospore-forming, and catalase-positive bacteria. Most of the isolates were able to optimally grow at temperatures 37–70 °C (optimum 50–65 °C), pH range 6.5–8.5 and 0–5% NaCl. *Geobacillus* (seven isolates) and *Bacillus* (five isolates) were the most predominant recovered genera.

Recently Panosyan and Trchounian (2019) analyzed polar lipid patterns and fatty acid (FA) compositions of the membrane of four *Geobacillus* strains (*G. caldoxylosilyticus* ArzA-3, *G. thermodenitrificans* ArzA-6, *G. toebii* ArzA-8, and *G. toebii* ArzA-33a) isolated from Arzakan geothermal spring. Phospholipids were found as the main polar lipids. The branched chains of *iso-* and *anteiso-*FAs were predominant, ranging from 80% to 89% of total FAs. The branched *iso-*family FAs (*i*C15:0, *i*C17:0) were the most abundant components (73–76.6%). Unsaturated FAs were not detected. Temperature-induced changes in polar lipids and FA composition were analyzed. With increasing temperature, a decrease of aminophospholipids and short straight (*n*C14) chains of FAs and an increase of phosphoglycolipids and long straight (*n*C15–*n*C17) chains of FAs were observed. The ratio of *i*C15:0/*i*C17:0 was decreased along with the growth temperature rise.

Ghazaryan et al. (2015) studied redox stress and membrane-associated response mechanisms in *G. toebii* ArzA-8 strain. It was shown that *G. toebii* ArzA-8 likely presents specific membrane-associated response mechanisms involving F_0F_1-ATPase to overcome redox stress and survive.

Bacilli isolated from Armenian geothermal springs were characterized based on tRNA modification profiles (Panosyan et al. 2014a). LC-MS was applied for the quantification of modified nucleosides by using isotopically labeled standards. 16 tRNA

Table 2.2 Some of the thermophilic microbes isolated from geothermal springs of Armenia and their potential in biotechnology

Strains	GenBank accession no.	Isolation source	Biotechnological application	References
Bacillus sp. ArzA-9	JQ929023	Arzakan	Amylase	Panosyan and Birkeland (2014)
B. licheniformis Akhurik 107	KY203975	Akhurik	Lipase	Shahinyan et al. (2017)
B. licheniformis ArzA-4	JQ929018	Arzakan	Amylase	Panosyan and Birkeland (2014)
B. simplex ArzA-2	JQ929014	Arzakan	Amylase	Panosyan and Birkeland (2014)
B. simplex ArzA-13A	JQ929012	Arzakan	Amylase	Panosyan and Birkeland (2014)
B. simplex ArzA-10	JQ929010	Arzakan	Amylase	Panosyan and Birkeland (2014)
Paenibacillus sp. ArzA-5	JQ929019	Arzakan	Amylase	Panosyan and Birkeland (2014)
Geobacillus sp. Tatev 4	KY203974	Tatev	Lipase	Shahinyan et al. (2017)
Geobacillus sp. ArzA-7	JQ929021	Arzakan	Protease	Panosyan and Birkeland (2014)
G. toebii ArzA-8	JQ929022	Arzakan	EPS	Panosyan et al. (2018c)
G. toebii ArzA-33a	JQ929015		EPS	Panosyan et al. (2014b)
G. toebii ArzA-33	JQ929016		EPS	Panosyan et al. (2014b)
G. stearothermophilus ArzA-11	JQ929011	Arzakan	Amylase	Panosyan and Birkeland (2014)
G. caldoxylosilyticus ArzA-3	JQ929017	Arzakan	Amylase	Panosyan and Birkeland (2014)
G. thermodenitrificans ArzA-6	JQ929020	Arzakan	EPS	Panosyan et al. (2018c)
Anoxybacillus sp. LF_2	KX018621	Jermuk	Amylase	Poghosyan (2015)
Thermoactinomyces sp. AkhA-12	MK418253	Akhurik	Protease	Panosyan et al. (2018a)

(continued)

Table 2.2 (continued)

Strains	GenBank accession no.	Isolation source	Biotechnological application	References
Sporosarcina sp. ArzA-13	JQ929013	Arzakan	Amylase	Panosyan and Birkeland (2014)
R. palustris D-6	–	Jermuk	Aspartase	Paronyan (2003, 2007)
Methylocaldum sp. AK-K6	KP272135	Akhurik	–	Islam et al. (2015)
Methylocaldum sp. Arz-AM-1	JQ929024	Arzakan	–	Panosyan and Birkeland (2014)
Arcobacter sp. Arz-ANA-2	JQ929025	Arzakan	Hydrolases	Panosyan and Birkeland (2014)
D. thermophilum SRB_21	KX018622	Jermuk	–	Poghosyan et al. (2014)
D. psychrotolerans SRB_141	–	Jermuk	–	Poghosyan et al. (2014)
Treponema sp. J_25	MG970326	Jermuk	–	Poghosyan (2015)
Methanoculleus sp. Arz-ArchMG-1	JQ929040	Arzakan	–	Panosyan and Birkeland (2014)

modifications (m^1A, m^6A, Am, t^6A, i^6A, ms^2i^6A, Q, m^1G, m^6_2A, m^2G, m^2_2G, Gm, m^2A, m^7G, m^5C, Cm) in both their natural and isotopically labeled forms were synthesized for the parallel quantification. Results obtained confirmed the presence of m^6A, i^6A, m^6_2A, and Cm modifications at low levels in all the species studied. In contrast, m^2A and m^7G modifications were present at high levels in all species indicating their oldest origin. Relatively high level of i^6A modification was observed for *Paenibacillus*. The lowest level of Cm modification was found in *Bacillus*. The highest level of m^5C modification was presented in *Ureibacillus*. The ms^2i^6A modification was found in all bacteria except *Ureibacillus*, while modifications of Am and m^2_2G were observed only among the representatives of *Ureibacillus*. In *Brevibacillus*, m^1G modification was absent. Large quantities of ms^2i^6A and Gm modification were detected only within representatives of genera *Geobacillus* and *Anoxybacillus*. The presented results offer a deeper insight into the evolution of tRNA modifications and show that they characterize species at a very fine level and are linked to phylogenetic variation of endospore-forming bacteria.

Poghosyan (2015) isolated from Jermuk geothermal spring a thermophilic *Anoxybacillus* strain designated as *Anoxybacillus* sp. LF_2, which was able to use dextrin and lactate as carbon sources. The partial 16S rRNA gene sequence analysis confirmed its belonging to *A. mongoliensis* with 99% similarity. The analysis of strain LF_2 by electron microscopy indicated rod-shaped cells about 2.5–2.8 µm in length and 0.4–0.5 µm in diameter. The strain *Anoxybacillus* sp. LF_2 (KX018621) has been deposited in the German Collection of Microorganisms and Cell Cultures (DSMZ) under the number DSM101951.

Recently, identification based on phenotypic and phylogenetic characteristics of the thermophilic bacilli isolated from Akhurik, Arzakan, Bjni, Hankavan, Jermuk, Tatev, and Uyts geothermal springs was carried out by Panosyan (2017). In total 135 thermophilic and thermotolerant bacilli strains were isolated under aerobic conditions at 55–65 °C and identified based on 16S rRNA gene sequence analysis as representatives of genera *Anoxybacillus*, *Bacillus*, *Brevibacillus*, *Geobacillus*, *Paenibacillus*, *Sporosarcina*, *Ureibacillus*, and *Thermoactinomyces*. While these results are important for further taxonomic work, positive results on hydrolytic activities are indicative of potential application of these bacterial cultures. All isolated thermophilic bacilli were tested for enzyme production capacities such as lipase, protease, and amylases, and a number of biotechnologically valuable enzyme producers were selected (Panosyan 2017).

Hovhannisyan et al. (2016) isolated several thermophilic amylase-producing bacilli from Arzakan and Jermuk geothermal springs from which five active producers (Arzakan 2, Arzakan 3, Arzakan 4, Arzakan 5, Jermuk 6) were selected. The isolates were identified as species of the genera *Anoxybacillus* and *Geobacillus* based on the 16S rRNA gene analysis. It was shown that at optimum growth conditions (65 °C, pH 7.0), the amylase production of most active amylase producer *A. rupiences* strain Arzakan 2 started in the early log phase and reached a maximum in the late exponential phase.

Vardanyan et al. (2015) studied the distribution of lipase-producing bacilli in Tatev geothermal spring. Two thermophilic lipase-producing bacilli strains were identified based on phenotypic characteristics and 16S rRNA gene analysis as *G. toebii* Tatev 5 and Tatev 6. The lipase activities of strains at optimal growth temperature and pH (65 °C, pH 7) were 70.3 and 80.7 U/ml, correspondingly. Shahinyan et al. (2015, 2017) studied the ability to produce lipases by 72 bacilli strains previously isolated from different geothermal springs of Armenia. Two most active lipase-producing isolates were identified as *B. licheniformis* Akhurik 107 (KY203975) and *Geobacillus* sp. Tatev 4 (KY203974). The activity of crude lipases of strains was carried out in different pH (5–11) and temperatures (25–75 °C). Lipase activity was determined spectrophotometrically using p-nitrophenyl palmitate as substrate. The highest lipase activity (0.89 U/ml) of *B. licheniformis* Akhurik 107 strain was observed at pH 6–7 and 55 °C temperature. *Geobacillus* sp. Tatev 4 strain showed high lipase activity (1.5–3.4 U/ml) at pH 6–7 and 65 °C.

The lipase genes of the two studied active lipase producers were sequenced by using initially designed primer sets. The PCR amplification revealed a presence of 1100 bp and 600 bp sized genes in *Geobacillus* sp. Tatev 4 and *B. licheniformis* Akhurik 107 strains, respectively. Multiple alignments generated from primary structures of the lipase proteins and annotated lipase protein sequences, conserved region analysis, and amino acid composition have illustrated the similarity (98–99%) of the lipases with true lipases (family I) and GDSL esterase family (family II). It was shown also that lipases contain Zn^{2+} and Ca^{2+} as ligands (Shahinyan et al. 2017).

The strains, *B. licheniformis* Akhurik 107 and *Geobacillus* sp. Tatev 4, were deposited at the Microbial Depository Center of Armenia under accession numbers MDC11855 and MDC11856, respectively.

Among several thermophilic bacilli isolated from sediment sampled from Arzakan geothermal spring, two best extracellular polysaccharide (EPS) producers identified as *G. thermodenitrificans* ArzA-6 and *G. toebii* ArzA-8 were selected (Panosyan et al. 2014b, c). EPS production was investigated under different time, temperature, and culture media's composition. The highest specific EPS production yield (0.27 g g^{-1} dry cells and 0.22 g g^{-1} dry cells for strains *G. thermodenitrificans* ArzA-6 and *G. toebii* ArzA-8, respectively) was observed after 24 h when fructose was used as the sole carbon source at 65 °C and pH 7.0. Purified EPSs displayed a high molecular mass: 5 × 105 Da for *G. thermodenitrificans* ArzA-6 and 6 × 105 Da for *G. toebii* ArzA-8. Chemical composition and structure of the biopolymers, determined by GC–MS, HPAE-PAD, and NMR, showed that both the EPSs are heteropolymers composed by mannose as the major monomer unit. Optical rotation values $[\alpha]_D^{25\,°C}$ of the two EPSs (2 mg ml^{-1} H$_2$O) were − 142,135 and − 128,645 for *G. thermodenitrificans* ArzA-6 and *G. toebii* ArzA-8, respectively. The strains ArzA-6 and ArzA-8 were deposited at the Microbial Depository Center of Armenia under accession numbers MDC11858 and MDC11859, respectively.

In a recent study, Islam et al. (2015) isolated a methanotrophic bacteria from Akhurik geothermal spring. Strain identified and designated as *Methylocaldum* sp. AK-K6 (accession number KP272135) had a temperature range for growth of 8–35 °C (optimal 25–28 °C) and pH range of 5.0–7.5 (optimal 6.4–7.0). 16S rRNA gene sequences showed that it was a new gammaproteobacterial methanotroph, which forms a separate clade in the family Methylococcaceae. It fell into a cluster with thermotolerant and mesophilic growth tendency, comprising the genera *Methylocaldum-Methylococcus-Methyloparacoccus-Methylogaea*. The strains possessed type I intracytoplasmic membranes. The genes *pmoA*, *mxaF*, *cbbL*, *nifH*, but no *mmoX* were detected in the genome of *Methylocaldum* sp. AK-K6.

Panosyan and Birkeland (2014) also isolated a methanotrophic isolate from Arzakan geothermal spring. The isolate designated as Arz-AM-1 was identified based on 16S rRNA gene sequences as *Methylocaldum* sp. (96% similarity). Methanotrophs, as a group of chemoorgano-autotrophic bacteria in geothermal spring, support the primary production of the ecosystem. Another isolate designated as Arz-ANA-2 was anaerobic. The analysis of the amplified 16S rRNA gene revealed a 99% phylogenic relationship to *Arcobacter* sp. (HM584709), an Epsilonproteobacteria (Panosyan and Birkeland 2014).

Two sulfate-reducing thermophilic bacteria optimally growing at 55 °C and designated as SRB_21 and SRB_141 were isolated from Jermuk hot spring using lactate as the sole carbon source (Poghosyan et al. 2014; Poghosyan 2015). Sequence analysis of 16S rRNA gene revealed 99% similarity of SRB_21 with *Desulfomicrobium thermophilum*. Strain SRB_21 (KX018622) grew lithotrophically with hydrogen as electron donor and organotrophically with lactate, ethanol, propionate, acetate, or propanol in the presence of sulfate as the terminal electron acceptor. Growth was not detected on formate or methanol. Morphological analysis by light and electron microscopy revealed the presence of 2.5–3.5 μm long non-sporulating motile rods. The 16S rRNA gene analysis of the strain SRB_141 revealed its 99% similarity with *Desulfovibrio psychrotolerans*. Electron micros-copy confirmed the presence of 2.5–3.0 μm vibrio-shaped cells. Strain SRB_141 grew on hydrogen, lactate, and ethanol in the presence of sulfate as the terminal electron acceptor.

A novel strictly anaerobic and moderately thermophilic spirochete, designated as *Treponema* sp. J25, was isolated from Jermuk hot spring. The strain has a fermenta-tive sugar-based metabolism using xylan as carbon and energy source. Strain uti-lizes D-galactose, D-fructose, and D-ribose as growth substrates in addition to xylan, but does not grow on D-glucose, L-arabinose, D-mannose, cellulose, cellobi-ose, lactose, maltose, or peptone. It grows optimally around 55 °C, with an upper limit at 60 °C, and at circumneutral pH. The cells of strain J25 are long and helical, with two periplasmic flagella. Strain J25 shares 95.1% 16S rRNA sequence identity with its closest relative, *Treponema caldarium* H1^T, suggesting that it represents a separate and novel species (Poghosyan 2015). The strain *Treponema* sp. J25 (acces-sion number MG970326) has been deposited in DSMZ under the number DSM100394.

Paronyan (2003, 2007) isolated a purple nonsulfur bacterial strain designated as *Rhodopseudomonas palustris* D-6 from water/sediment samples of Jermuk hot spring. The strain was mesophilic but was able to grow up to 45 °C (moderate thermophilic) using organic compounds as carbon and nitrogen organic source. The strain *R. palustris* D-6 produces aspartase (with activity 33.05% per 100 mg dry biomass, at 37 °C, pH 6.0–9.0). The strain was deposited at the Microbial Depository Center of Armenia under accession number ИНМИА B-6506. A high acylase activity of *R. palustris* D-6 was shown as well.

Several phototrophic bacteria belonging to genera *Rhodobacter* (*R. capsulatus*, *R. sulfidophilus*, *R. sphaeroides*), *Rhodopseudomonas* (*R. palustris*), *Thiospirillum* (*T. jenense*), and *Thiocapsa* (*T. roseopersicina*) have been isolated from Arzakan, Jermuk, and Akhurik geothermal springs. Some of those isolates were good produc-ers of enzymes such as aspartase, β-carotene, aminoacylase, glucose isomerase, and inulinase, as well as sources of protein, carbohydrates, and vitamins (Paronyan 2003). All isolated phototrophic bacteria were mesophiles unable to grow at tem-peratures above 40 °C.

Edwards and coauthors (2013) successfully obtained nitrite-oxidizing bacteria (NOB) enrichment from sediment samples of Arzakan and Jermuk geothermal springs. All successful enrichments contained organisms with high 16S rRNA gene sequence identity ($\geq 97\%$) with *Nitrospira calida* and *N. moscoviensis*. Physiological properties of all enrichments were similar, with an optimum temperature of 45–50 °C, yielding nitrite oxidation rates of 7.53 ± 1.20 to 23.0 ± 2.73 fmoles cell^{-1} h^{-1} and an upper temperature limit between 60 and 65 °C. The highest rates of NOB activity occurred with initial NO^{-2} concentrations of 0.5–0.75 mM; however, lower initial nitrite concentrations resulted in shorter lag times.

Hedlund et al. (2013) obtained active methanogenic enrichments on acetate and H$_2$/CO$_2$ (acetoclastic and hydrogenotrophic methanogenesis) at 45 and 55 °C, but not 65 °C using sediments sampled from Jermuk and Arzakan geothermal springs. Methylotrophic enrichments were only successful with samples from Jermuk.

Only archaeal strain isolated so far from Armenian geothermal springs was reported by Panosyan and Birkeland (2014). A hydrogenotrophic methanogenic archaea designated as *Methanoculleus* sp. Arz-ArchMG-1 (JQ929040) was successfully isolated from Arzakan hot spring. Based on 16S rRNA gene sequence, it shared 97% similarity with *Methanoculleus* sp. LH2 (DQ987521).

Prospective microbes from hot springs offer a major advantage of preserving those strains for future studies and exploring them in due course for potential biotechnological applications in medical, industrial, and agricultural processes.

2.4   Saline and Hypersaline Environments in Armenia

Halophiles are distributed in habitats covering a wide range of salinities, such as saline lakes, saline and saline–alkaline soils, solar salterns, and salt mines (Ma et al. 2010; Ventosa et al. 2015; Edbeib et al. 2016; Paul and Mormile 2017; Vera-Gargallo et al. 2019). Hydromorphic saline–alkaline soils (occupying more than 29,000 ha) with up to 3% salinity and highly alkaline conditions (pH 9–11) are also found in the territory of Ararat Plain, Armenia (Baghdasaryan 1971; Panosyan et al. 2018b). These soils are found at 850–900 m above sea level and have been formed on alluvial–proluvial stratified sediments from the Araks River. They are characterized by strong salinity (total soil content 1–3%), considerable carbonization, low humus content (< 1.0%), and a high absorbed sodium content. Hydromorphic saline–alkaline soils are developed in the areas of Ararat Plain, where subsoil water is mineralized and located close to the surface (1–2 m). The formation of salt crusts on the soil surface selectively supports the development of halophilic (mainly halotolerant) microbes (Panosyan 2007; Panosyan et al. 2018b).

Subterranean salt deposits are the remains of ancient hypersaline waters that presumably supported dense populations of halophilic archaea and bacteria. A hypersaline subterranean salt deposit, developed after the closing of the Neo-Tethys Ocean during the Pyrenees land forming period, was found in Armenia. It is located on the eastern part of Yerevan and currently is called Avan subterranean salt deposit (40°13′30.15″ N, 44°33′50.32″ E). The major chemical components of the salt stones from deposit are NaCl (95.5%), $CaSO_4$ (1.48%), $CaCl_2$ (0.14%), and $MgCl_2$ (0.08%) and insoluble residues constitute 2.7% (Poghosyan 2008). Pliocene and partially Pleistocene dolomite and basaltic andesite having 220 m thickness are involved in the geological structure of deposit. Miocene salt clay and salt stone layers with 700 m thickness are located under the basalts. Currently based on salt deposit, the Avan Salt Plant is exploited. The Plant is the only salt-producing company in the region and produces two main types of production: iodized, high-quality common food salt and the mining complex of rock salt. The Republican Speleological Therapeutic Center established in Avan salt deposit is used for curing patients suffering from bronchial asthma and respiratory and some other diseases. This original clinic is located in a man-made cave, at a depth of 235 m from a surface of the ground.

2.5   Halophilic Microbes Isolated from Saline and Hypersaline Environments in Armenia and Their Biotechnological Importance

High salinity represents an extreme environment to which relatively few organisms have been able to adapt and occupy. Halophiles, salt-loving organisms that flourish in saline environments, are classified as slight, moderate, or extreme, depending on their requirement for sodium chloride. The world of halophilic microorganisms is highly diverse. Microbes adapted to life at high salt concentrations are found in all three domains of life: *Archaea*, *Bacteria*, and *Eucarya* (Ma et al. 2010). A large number of members from these domains are found growing in diverse habitats including both thalassohaline and athalassohaline environments. The halophiles of all three domains have been relatively little exploited in biotechnological processes, with notable exceptions of β-carotene from *Dunaliella*, bacteriorhodopsin from *Halobacterium*, and ectoine from *Halomonas* (Oren 2010).

Halophilic microbial communities of saline–alkaline soils have been studied recently in Columbia (Díaz-Cárdenas et al. 2017), India (Keshri et al. 2013), Mexico (Delgado-García et al. 2018), Turkey (Orhan and Gulluce 2014), Italy (Canfora et al. 2014), and several other regions of the world (Ventosa et al. 2008; Vera-Gargallo et al. 2019). Microbial diversity of saline–alkaline soils in Ararat Plain (Armenia) as well as the microbial processes and key microbes involved in soil formation and biogeochemical cycles has been studied in the last years (Panosyan et al. 2018b). Figure 2.2 shows photographs of saline–alkaline soils located in Ararat Plain.

Khachaturian et al. (1995) studied the distribution of various extremophilic bacilli in all main types of soils, including saline–alkaline soils of Armenia. Several halophilic bacilli strains belonging to *B. alcalophilus* and *B. alcalophilus* subsp.

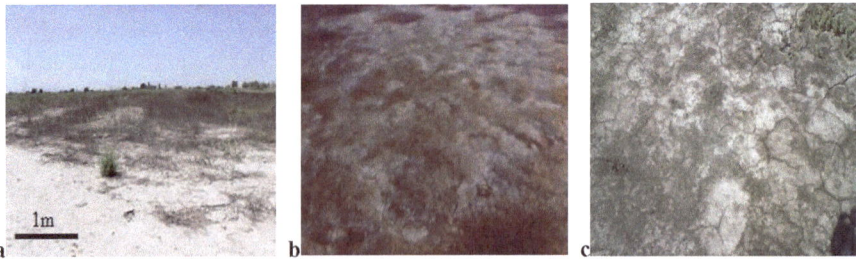

Fig. 2.2 Saline-alkaline soils located in Metsamor (40°04′32.07″ N; 44°14′30.02″ E) (**a**) and crusted soil surfaces in Artashar (located at 40°11′65.85″ N; 44°20′38.60″ E) (**b**) and Jrarat (40°07′18.03″ N; 44°33′49.80″ E) (**c**) of Ararat Plain

halodurans were isolated. Optimum growth was observed at 37 °C, at pH 8.6–10.1, and in the presence of 15–17% NaCl.

Recently the bacterial community composition in the A horizon of natural saline–alkaline soils of Ararat Plain was studied using molecular and culture-based methods (Panosyan et al. 2018b). The sequence analysis of a 16S rRNA gene clone library and denaturing gradient gel electrophoresis (DGGE) profiles indicated the dominance of Firmicutes populations. The majority of the sequences of the bacterial 16S rRNA gene library were close relatives of representatives belonging to the genera *Halobacillus* (41.2%), *Piscibacillus* (23.5%), *Bacillus* (23.5%), and *Virgibacillus* (11.8%). Eight novel moderately halophilic bacilli isolates were successfully obtained from the enriched cultures of the saline–alkaline soil samples and characterized by phenotypic and phylogenetic techniques.

16S rRNA gene sequence analyses of isolates revealed their affiliation (97.7–99.7% similarity) to representatives of the genera *Bacillus* (4 isolates), *Piscibacillus* (2 isolates), and *Halobacillus* (2 isolates). All isolates grew well in a range of 1–25% NaCl with an optimum at 9–18% NaCl, and they were considered to be moderate halophiles. Most of the isolates were able to grow in the pH range 6.0–11.5 (optimum 7.5–8.5) and at 10–55 °C (optimum 35–37 °C). Earlier it was shown that the amount of haloalkaliphilic bacilli (belonging to *B. subtilis*, *B. licheniformis*, *B. alcalophilus*, and *B. circulans*) was up to 10^5 CFU/g soil (Panosyan 2007).

While these results will be important for further taxonomic work, positive results for hydrolytic activities among the isolates and an ability to tolerate not only high concentrations of NaCl but also highly alkaline pH values indicated the biotechnological potential of these cultures. They may also be explored further for gene cataloging and metabolic profiling.

As key microbes in bacterial saprotrophic complexes, these bacilli are involved in the food webs and biogeochemical cycles of saline–alkaline soils. A detailed knowledge of the microbial diversity in these environments would be helpful for the ecological restoration of areas affected by salinization processes and recovery of arid and saline soils for agriculture (Panosyan et al. 2018b).

Another recent study to identify bacterial diversity of saline–alkaline soils of Ararat Plain revealed the presence of halophilic actinobacteria. In total five moderately haloalkaliphilic *Streptomyces* strains have been isolated and identified based on phenotypic characteristics and 16S rRNA gene sequence analysis (Hakobyan and Panosyan 2012; Hakobyan et al. 2013). The moderately haloalkaliphilic isolate designated as *Streptomyces* sp. A3 (KU681239) exhibited optimal growth at 5% NaCl and pH 9 at 37 °C. Based on phenotypic characteristics, this isolate has been primarily identified as *S. roseosporus* (Hakobyan and Panosyan 2012), while 16S rRNA gene sequence sheared 98 and 97% similarity with *Streptomyces* sp. YIM 75712 (JQ808019), isolated from Limestone Quarries of South India, and *S. fradiae* strain NIOT-Cu-51 (KJ575069), isolated from deep-sea sediments of the Bay of Bengal and Andaman Sea, respectively.

The strain *Streptomyces* sp. A3 was examined for their antimicrobial activity using *Saccharomyces cerevisiae*, *B. subtilis*, *Staphylococcus* sp., *Enterococcus faecalis*, and *Escherichia coli* as test organisms. High antimicrobial activity against

Gram-positive bacteria and yeasts was observed. It was shown that antimicrobial compound of *Streptomyces* sp. A3 was extensively synthesized at the stationary stage of growth and had high stability against proteinase K (Hakobyan and Panosyan 2012).

Another actinobacterial isolate has been identified based on phenotypic characteristics and designated as *S. griseus* A5 (Hakobyan et al. 2013). 16S rRNA gene sequence similarities between strain *S. griseus* A5 and related species showed close relation (98% similarity) to *S. albogriseolus* 71.1 (KX454152) and *S. labedae* NBRC 15864 (AB915631). Further investigation to confirm the affiliation of the strain is needed. Isolates were able to optimally grow at 5% NaCl concentration with pH 8 and at 37 °C. The strains utilized glucose, mannitol, xylose, and arabinose as the carbon source along with acid production. The utilization of starch and casein showed that these isolates could also produce the extracellular enzymes as amylase and protease to metabolize the polymeric components.

Aiming to select active cellulase producers, actinobacterial strains were incubated on solid medium containing carboxymethylcellulose and then flooding by 0.1% Congo red. Cellulase activity (3000–3300 ± 2 U/mg) determined by spectrophotometric method was observed at pH 8 and 37 °C in a medium supplemented with 2% and 5% NaCl for *Streptomyces* sp. A3 and *S. griseus* A5, respectively (Hakobyan et al. 2013).

Adamyan (2004) isolated two halophilic bacilli strains designated as *Bacillus* sp. 3849 and *B. sphaericus* 3863 from saline–alkaline soils of Ararat Plain. They have been identified based on phenotypic properties. Both strains were able to grow in wide range of salt and showed optimal growth at 12–15% NaCl. The strain *Bacillus* sp. 3849 was an active producer of cyclodextrin glucanotransferase (CGTase) with 71 kDa molecular mass. pH optimum of CGTase was 7.0, but the enzyme kept the stability at pH 6.5–8.0. It was thermostable and was active at 50 °C temperature. The main product was β-CD.

Several alkaliphilic and moderate halophilic phototrophic bacteria able to grow optimally at 6–8% NaCl and pH 9.0–10.5 and belonging to genera *Ectothiorhodospira* (*E. mobilis*), *Thiospirillum*, *Chromatium* (*C. vinosum*), *Thiocapsa* (*T. roseopersicina*, *T. jenense*), and *Lamprocystis* have been isolated from the saline–alkaline soils

Fig. 2.3 Photographs of Avan subterranean salt deposit (located 40°13′30.15" N; 44°33′50.32" E) (**a**) and examples of salt stones

of Ararat Plain (Paronyan 2002, 2003). Some of them have been used for studies of aspartase, aminoacylase, glucose isomerase, and inulinase activities, as well as sources of protein, carbohydrates, and vitamins.

Halophilic bacteria and extremely halophilic archaea are well known to survive in the hypersaline environments like salt mines or subterranean salt deposits. A study was also conducted on microbial diversity in subterranean salt deposit located in Avan, Armenia (Fig. 2.3). Both culture-based and not culture-independent approaches have been used for addressing microbial diversity associated with salt mine. It has been reported that Avan subterranean salt deposits are inhabited by a variety of microbes belonging to the Bacteria and Archaea domains that tolerate environmental extremes and could have some yet undescribed biotechnological potential (Hakobyan et al. 2014a, b, 2015a).

Studies based on sequence analysis of 16S rRNA gene clone libraries and from sampled salt stones have been done recently (Hakobyan et al. 2015b). 16S rRNA gene libraries were generated from total DNA extracts using universal archaeal and bacterial oligonucleotide primer sets. The PCR-DGGE fingerprinting method also was applied to obtain information about the presence of the dominant bacterial and archaeal populations. It was the first microbiological investigation on salt mine in Armenia.

The study of bacterial clone library indicated abundance of representatives belonging to genera *Bacillus*, *Virgibacillus*, *Halobacillus*, *Filobacillus*, *Anoxybacillus*, and *Streptomyces*. Several sequences of clone library were affiliated with uncultured bacteria. DGGE profile was in good agreement with the clone library results indicating predominance of halophilic bacilli (unpublished data).

Sequence analysis of clones in the archaeal library and DGGE profiles indicated that they originated from *Euryarchaeota* and were mainly affiliated with the genera *Haloarcula*, *Halobacterium*, *Halarchaeum*, and *Natronomonas* (Hakobyan et al. 2015b). While the molecular data indicated *Halobacterium* as one of the major groups in the microbial community in the Avan salt mine, no isolates of this group were obtained.

Several aerobic chemoorganotrophic endospore-forming bacteria were isolated from the salt samples and based on 16S rRNA sequence analyses identified as representatives of the genera *Halobacillus*, *Piscibacillus*, and *Bacillus* (Hakobyan et al. 2011).

Hakobyan et al. (2014a) reported on the diversity of archaeal strains isolated from salt stone sample. Five halophilic archaeal strains were isolated and identified based on 16S rRNA gene sequence analysis. Four of the archaeal strains were most closely related to members of the genus *Haloarcula* (97–99% similarity) and one strain most closely to the genus *Halarchaeum* (< 97% similarity), indicating that the Avan salt mine harbors a unique community of possible novel species. The strains required at least 1.5–2.0 M NaCl (optimal growth at 3.5–4 M NaCl) and 10 mM Mg^{2+} for growth and were able to grow at pH 5.5–10 (optimum, pH 7–8) and at 21–45 °C (optimum 37 °C). Most of the strains grew in the presence of casamino acids, Na-glutamate, D-glucose, L-propionate, L-arginine, and D-cellobiose as carbon sources. Chemotaxonomic studies revealed the presence of the dieter phospholipid phosphatidylglycerol and phosphatidylglycerol phosphate methyl ester derived from C20C20 archaeol.

The *Haloarcula* isolates were studied for 16S rRNA gene sequence polymorphism using specially designed 16S rRNA type-specific primers. They harbored at least two different 16S rRNA gene copies organized into three operons designated as *rrnA*, *rrnB*, and *rrnC*. Phylogenetic analysis based on 16S rRNA genes indicated that these genes fall into two distinct clusters, type I (*rrnA*) and type II (*rrnB* or *rrnC*), differing 6–7% in sequence similarity. This underlines the importance of polyphasic approaches for taxonomic analyses of the *Haloarcula* group (Hakobyan et al. 2014a, b; Azaryan et al. 2018a).

The strain designated as *Haloarcula* sp. salt stone-1 shared 98% 16S rRNA sequence identity with strain *Haloarcula marismortui* ATCC 43049 isolated from the Dead Sea. The draft genome of the strain was sequenced using PacBio RS technology and assembled using the Celera Assembler GATC Biotech, Germany (http://www.gatc-biotech.com), resulting in 4 Mb of unique sequence data distributed into 48 contigs constituting a total of 4,331,370 bp. The G + C content is 61.4%. Gene prediction carried out with the NCBI Prokaryotic Genome Annotation Pipeline, as well as the RAST server (http://rast.nmpdr.org/rast.cgi), identified a total of 2,529 genes, including 4,261 coding DNA sequences, 68 sets of rRNA genes (Azaryan 2018).

Carotenoids are of great interest in many scientific disciplines because of their wide distribution, diverse functions, and interesting properties. Extremely halophilic archaea generating red-colored colonies produce carotenoids which are valuable compounds for the pharmaceutical, chemical, food, and feed industries, not only because they can act as vitamin A precursors but also for their coloring, antioxidant, and possible anticancer activities (Yatsunami et al. 2014; Rodrigo-Bañon et al. 2015).

The optimization of culture conditions for both cell growth and total carotenoid production of the halophilic *Haloarcula* sp. salt stone-1 and *H. japonica* A2 strains was carried out as well. The effects of various factors including temperature, pH, light, salinity, and carbon sources have been studied too. The optimum conditions for cell growth and total carotenoid production were observed at 30 °C, pH 7.2, 20% NaCl (w/v), using casamino acids as carbon source. The production of biomass ranged up to 0.7 and the total carotenoids up to 10.2 mg/l. The maximal productivity of carotenoids at the late stationary phase of growth was 14.6 mg g^{-1}. Under 24 hours of light conditions, the biomass increased 1.16-fold, while the production of carotenoids increased 1.4-fold, demonstrating the need of light for optimal carotenoid production (Azaryan et al. 2017, 2018b).

Biotechnological potential of isolates from Armenian saline and hypersaline environments is summarized in Table 2.3. Despite this progress, performing on halophilic microorganisms in different saline and hypersaline ecosystems of Armenia still needs special interest in terms of isolation and evaluation of their biotechnological potential and industrial demands.

Table 2.3 Summary of halophilic and halotolerant microbes isolated from saline and hypersaline environments of Armenia and their potential in biotechnology

Microbial species	Strain code	GenBank accession number	Isolation source	Comments	References
Bacillus sp.	3849	–	Saline–alkaline soil	CGTase, β-CD	Adamyan (2004)
Bacillus sp.	Avan-3	–	Avan salt deposit	Hydrolases	Hakobyan et al. (2011)
B. sphaericus	3863	–	Saline–alkaline soil	CGTase, β-CD	Adamyan (2004)
B. alcalophilus	–	–	Saline–alkaline soil	Hydrolases, CGTase	Khachaturian et al. (1995)
B. alcalophilus subsp. halodurans)	–	–	Saline–alkaline soil	Hydrolases	Khachaturian et al. (1995)
B. licheniformis	Art 4	KU681232	Saline–alkaline soil	Hydrolase	Panosyan et al. (2018b)
B. pumilus	Ech 2.9	KU681233	Saline–alkaline soil	Protease	Panosyan et al. (2018b)
B. sonorensis	Art 5	KU681234	Saline–alkaline soil	Hydrolases	Panosyan et al. (2018b)
B. halodurans	Sov 4	KU681231	Saline–alkaline soil	Amylase, protease	Panosyan et al. (2018b)
Halobacillus sp.	Ech 2.1	KU681236	Saline–alkaline soil	Amylase Protease	Panosyan et al. (2018b)
Halobacillus sp.	Avan-1	–	Avan salt deposit	Hydrolase	Hakobyan et al. (2011)
H. karajensis	Sov 2.2	KU681235	Saline–alkaline soil	Amylase Protease	Panosyan et al. (2018b)
Piscibacillus sp.	Sov 2.1	KU681238	Saline–alkaline soil	Lipases, protease	Panosyan et al. (2018b)
Piscibacillus sp.	Art 1	KU681237	Saline–alkaline soil	Lipases, protease	Panosyan et al. (2018b)
Piscibacillus sp.	Avan-2	–	Avan salt deposit	Hydrolase	Hakobyan et al. (2011)

(continued)

Table 2.3 (continued)

Microbial species	Strain code	GenBank accession number	Isolation source	Comments	References
Streptomyces sp.	A3	KU681239	Saline–alkaline soil	Cellulase antibacterial agents	Hakobyan and Panosyan (2012), Hakobyan et al. (2013)
S. griseus	A5	–	Saline–alkaline soil	Cellulase antibacterial agents	Hakobyan et al. (2013)
Haloarcula sp.	Salt stone-1	–	Avan salt deposit	Carotenoids	Azaryan et al. (2018b)
H. japonica	A2	–	Avan salt deposit	Carotenoids	Azaryan et al. (2017)

2.6 Conclusion

The isolation and study of extremophiles from different extreme environments are important for the understanding of their diversity, microbe-mediated biogeochemical cycles, and ecosystem, functioning and exploring their biotechnological potency. Armenia is rich in many geothermal springs and (hyper)saline environments. Both culture-dependent and molecular studies showed that those extreme environments are diverse with respect to their content and microbial diversity. Many new thermophilic microbes mainly belonging to the *Bacillus* and related genera have been isolated, identified, and evaluated, taking into account their biotechnological potency. Only few reports have been described in studying halophilic microbes in Armenia. In this context, we summarized here information on thermophilic and halophilic bacterial and archaeal isolates from Armenia and their potential use in various biotechnologies obtaining to date. The results summarized in this chapter suggest that Armenian hot springs, saline–alkaline soils, and subterranean salt deposits are inhabited by abundant and diverse thermophilic, halotolerant, and halophilic microbes, respectively, with great deal of opportunities for their use in biotechnological applications. This is the first comprehensive census of thermophilic and halophilic microbes thriving in geothermal springs and hyper(saline) ecosystems of Armenia.

Acknowledgments This work was supported by grants from the Eurasia Programme of the Norwegian Center for International Cooperation in Education (CPEA-2011/10081, CPEA-LT-2016/10095) and partially supported by the RA MES State Committee of Science, in the frames of the research projects №15 T-1F399 and № 18 T-1F261, Armenian National Science and Education Fund based in New York, USA, to HP (ANSEF-NS-microbio 2493, 3362 and 4676).

References

Adamyan MO (2004) Ekologia i biosinteticheskie osobennosti galofilnikh sporoobrazujush-chikh bakterii (Ecology and biosynthetic pecularities of halophilic spore-forming bacteria). Dissertation, Institute of Microbiology NAS of RA

Antranikian G, Egorova K (2007) Extremophiles, a unique resource of biocatalysts for industry. In: Gerday C, Glansdorff N (eds) Physiology and biochemistry of extremophiles. ASM Press, Washington, DC, pp 361–406

Atalah J, Cáceres-Moreno P, Espina G, Blamey JM (2019) Thermophiles and the applications of their enzymes as new biocatalysts. Bioresour Technol 280:478–488. https://doi.org/10.1016/j.biortech.2019.02.008

Azaryan A (2018) Study of whole genome and some biochemical peculiarities of *Haloarcula* sp. Salt stone-1 archaeal strain isolated from Avan salt deposit. Master thesis, Yerevan State University (in Armenian)

Azaryan AS, Gabrielyan LS, Panosyan HH, Trchounian AH (2017) Effect of growth conditions on total carotenoid production by *Haloarcula japonica* A-2 isolated from Avan (Armenia) sub-terranean deposit. In: Book of abstracts and papers of the conference. 2–3 November, 2017, Yerevan, Lusabac, pp 25–33

Azaryan A, Panosyan H, Trchounian A, Birkeland NK (2018a) Thermophiles and halophiles in Armenian extreme ecosystems: diversity and biotechnology. In: Abstract book of extremo-philes: from biology to biotechnology, 19–25 August, 2018, Tashkent, Uzbekistan, p 15

Azaryan A, Trchounian A, Birkeland NK (2018b) Optimization of the culture conditions for cell growth and total carotenoid production of *Haloarcula japonica* A2 isolated from the Avan subterranean salt deposit in Armenia. In: Abstract book of 12th International Congress of Extremophiles, 16–20 September, 2018, Ischia, Naples, Italy, p 63

Badalyan M (2000) Geothermal features of Armenia: a country update. In: Proceedings World Geothermal Congress, 28 May–10 June 2000, Kyushu-Tohoku, Japan, pp 71–75

Baghdasaryan AB (1971) Physical geography of Armenian SSR. Publishing house of NA of ASSR, Yerevan. (in Armenian)

Canfora L, Bacci G, Pinzari F, Lo Papa G, Dazzi C et al (2014) Salinity and bacterial diversity: to what extent does the concentration of salt affect the bacterial community in a saline soil? PLoS One 9(9):e106662. https://doi.org/10.1371/journal.pone.0106662

DasSarma S, DasSarma P (2015) Halophiles and their enzymes: negativity put to good use. Curr Opin Microbiol 25:120–126. https://doi.org/10.1016/j.mib.2015.05.009

DeCastro ME, Rodríguez-Belmonte E, González-Siso MI (2016) Metagenomics of thermo-philes with a focus on discovery of novel thermozymes. Front Microbiol 7:1521. https://doi.org/10.3389/fmicb.2016.01521

Deepika M, Satyanarayana T (2013) Diversity of hot environments and thermophilic microbes. In: Satyanarayana T, Littlechild J, Kawarabayasi Y (eds) Thermophilic microbes in environmental and industrial biotechnology. Springer, Dordrecht/Heidelberg/New York/London, pp 3–60

Delgado-García M, Contreras-Ramos SM, Rodríguez JA, Mateos-Díaz JC, Aguilar CN et al (2018) Isolation of halophilic bacteria associated with saline and alkaline-sodic soils by culture dependent approach. Heliyon 4(11):e00954. https://doi.org/10.1016/j.heliyon.2018.e00954

Díaz-Cárdenas C, Cantillo A, Rojas LY, Sandoval T, Fiorentino S et al (2017) Microbial diver-sity of saline environments: searching for cytotoxic activities. AMB Express 7:223. https://doi.org/10.1186/s13568-017-0527-6

Dumorné K, Córdova DC, Astorga-Eló M, Renganathan P (2017) Extremozymes: a poten-tial source for industrial applications. J Microbiol Biotechnol 27(4):649–659. https://doi.org/10.4014/jmb.1611.11006

Edbeib MF, Wahab RA, Huyop F (2016) Halophiles: biology, adaptation, and their role in decon-tamination of hypersaline environments. World J Microbiol Biotechnol 32:135. https://doi.org/10.1007/s11274-016-2081-9

Edwards TA, Calica NA, Huang DA, Manoharan N, Hou W et al (2013) Cultivation and characterization of thermophilic *Nitrospira* species from geothermal springs in the U.S. Great Basin, China, and Armenia. FEMS Microbiol Ecol 85(2):283–292

Ghazaryan A, Blbulyan S, Poladyan A, Trchounian A (2015) Redox stress in geobacilli from geothermal springs: phenomenon and membrane-associated response mechanisms. Bioelectrochemistry 105:1–6. https://doi.org/10.1016/j.bioelechem.2015.04.007

Hakobyan A, Panosyan H (2012) Antimicrobial activity of moderately haloalkaliphilic *Streptomyces roseosporus* A3 isolated from saline-alkaline soils of Ararat Plain, Armenia. In: Proceedings of international young scientist's conference "Prospectives for development of molecular and cellular biology-3", 26–29 September, 2012, Yerevan, Armenia, pp 89–95

Hakobyan A, Margaryan A, Panosyan H (2011) Halophilic aerobic endospore-forming bacteria of alkali-saline soils and salt mine of Armenia. In: Abstracts' book of XV Pushchinian international school-conference of youth scientists "Biolog-Science of XXI century". 21–26 April, 2011, Pushchino, Russia, p 361 (in Russian)

Hakobyan A, Panosyan H, Trchounian A (2013) Production of cellulose by the haloalkaliphilic strains of *Streptomyces* isolated from saline-alkaline soils of Ararat plain, Armenia. Electron J NAS RA Nat Sci Biotechnol 21(2):44–46

Hakobyan A, Panosyan H, Trchounian A, Birkeland NK (2014a) Identification of halophilic archaea from the Avan salt mine in Armenia, using a polyphasic approach including 16S rRNA gene sequence polymorphism. Abstracts book of international scientific workshop on "Trends in microbiology and microbial biotechnology", Yerevan, Armenia, 5–8 October 2014, YSU press, p 37

Hakobyan A, Panosyan H, Trchounian A, Birkeland NK (2014b) Microbial diversity analysis of an Armenian subterranean salt deposit using molecular and culture-based methods. In: Book of abstracts 10th international congress on extremophiles, September 7–11, 2014, Saint Petersburg, Russia, p 144

Hakobyan A, Panosyan H, Trchounian A (2015a) Isolation of halophilic archaea from Avan salt mine in Armenia and their identification using polyphasic approaches. In: Book of Abstracts 3rd international scientific conference "Dialogues on science", June 23–26, 2017, Yerevan, Armenia, p 27

Hakobyan A, Panosyan H, Trchounian A Birkeland NK (2015b) Microbial diversity in Armenian saline-alkaline soils and subterranean salt deposits analyzed by molecular and culture-based methods. The FEMS 2015, 6th Congress of European Microbiologists, Maastricht, The Netherlands, p 279

Hedlund BP, Dodsworth JA, Cole JK, Panosyan HH (2013) An integrated study reveals diverse methanogens, Thaumarchaeota, and yet-uncultivated archaeal lineages in Armenian hot springs. Anton Leeuw 104(1):71–82

Henneberger R, Cooksley D, Hallberg J. (2000) Geothermal resources of Armenia: In: Proceedings world geothermal congress, Kyushu-Tohoku, Japan, 28 May–10 June 2000, pp 1217–22

Hovhannisyan P, Turabyan A, Panosyan H, Trchounian A (2016) Thermostable amylase production bacilli isolated from Armenian geothermal springs. Biol J Armenia 68:6–15

Hreggvidsson GO, Petursdottir SK, Bjornsdottir SH, Fridjonsson OH (2012) Microbial speciation in the geothermal ecosystem. In: Stan H, Fendrihan LS (eds) Adaption of microbial life to environmental extremes: novel research results and application. Springer, Wien, pp 37–68

Islam T, Larsen Ø, Torsvik V, Øvreås L, Panosyan H, Murrell C, Birkeland NK, Bodrossy L (2015) Novel methanotrophs of the family *Methylococcaceae* from different geographical regions and habitats. Microorganisms 3:484–499

Keshri J, Mishra A, Jha B (2013) Microbial population index and community structure in saline-alkaline soil using gene targeted metagenomics. Microbiol Res 168:165–173

Khacaturian AA, Kazanchian NL, Khacaturian NS, Adamian MO, Khachikian LO (1995) About the ecology of extremophilic forms of bacilli in the main types of soils of Armenia. Biol J Armenia 48(1):12–18. (in Russian)

Khan M, Sathya TA (2018) Extremozymes from metagenome: potential applications in food processing. Crit Rev Food Sci Nutr 58(12):2017–2025. https://doi.org/10.1080/1040839 8.2017.1296408

Kumar L, Awasthi G, Singh B (2011) Extremophiles: a novel source of industrially important enzymes. Biotechnology 10:121–135. https://doi.org/10.3923/biotech.2011.121.135

Littlechild JA (2015) Enzymes from extreme environments and their industrial applications. Front Bioeng Biotechnol 3:161. https://doi.org/10.3389/fbioe.2015.00161

López-López O, Cerdán ME, González-Siso MI (2013) Hot spring metagenomics. Life 3(2):308–320

Ma Y, Galinski EA, Grant WD, Oren A, Ventosa A (2010) Halophiles 2010: life in saline environments. Appl Environ Microbiol 76(21):6971–6698. https://doi.org/10.1128/AEM.01868-10

Marquez MC, Sanchez-Porro C, Ventosa A (2011) Halophilic and haloalkaliphilic, aerobic endospore-forming bacteria in soil. In: Logan NA, De Vos P (eds) Endospore-forming soil bacteria. Springer, Berlin/Heidelberg, pp 309–339

Mkrtchyan S (ed) (1969) Geology of Armenian SSR. Publishing house of AS of ASSR, Yerevan. (in Russian)

Orellana R, Macaya C, Bravo G, Dorochesi F, Cumsille A, Valencia R, Rojas C, Seeger M (2018) Living at the frontiers of life: extremophiles in Chile and their potential for bioremediation. Front Microbiol 9:2309. https://doi.org/10.3389/fmicb.2018.02309

Oren A (2010) Industrial and environmental applications of halophilic microorganisms. Environ Technol 31:825–834

Oren A (2013) Strategies for the isolation and cultivation of halophilic microorganisms. In: Singh OV (ed) Extremophiles sustainable resources and biotechnological implications. Wiley-Blackwell/Wiley, Hoboken, pp 75–94

Orhan F, Gulluce M (2014) Isolation and characterization of salt-tolerant bacterial strains in salt-affected soils of east Anatolian region. Geomicrobiol J 32(1):10–16. https://doi.org/10.108 0/01490451.2014.917743

Panosyan H (2007) Bacterial population of alkali-saline soils of Ararat plain. Bull State Agrarian Univ Armenia 20(4):20–22. (in Russian)

Panosyan H (2010) Phylogenetic diversity based on 16S rRNA gene sequence analysis of aerobic thermophilic endospore-forming bacteria isolated from geothermal springs in Armenia. Biol J Armenia 62(4):73–80

Panosyan HH (2017) Thermophilic bacilli isolated from Armenian geothermal springs and their potential for production of hydrolytic enzymes. Int J Biotech Bioeng 3(8):239–244

Panosyan H, Birkeland NK (2014) Microbial diversity in an Armenian geothermal spring assessed by molecular and culture-based methods. J Basic Microbiol 54(11):1240–1250

Panosyan H, Trchounian A (2019) Polar lipid pattern and fatty acid composition of geobacilli and their temperature induced changes. RepNAS Armenia 119(1):86–94

Panosyan H, Wagner M, Brandmayr C, Carell T (2014a) Differentiation of bacilli on genera level based on tRNA modification profiles. FEBS J 281(1):655–556

Panosyan H, Anzelmo G, Nicolaus B (2014b) Production and characterization of exopolysaccharides synthesized by geobacilli isolated from an Armenian geothermal spring. FEBS J 281(1):667

Panosyan H, Margaryan A, Poghosyan L, Saghatelyan A, Gabashvili E, Jaiani E, Birkeland NK (2018a) Microbial diversity of terrestrial geothermal springs in lesser Caucasus. In: Egamberdieva D, Birkeland NK, Panosyan H, Li WJ (eds) Extremophiles in Eurasian ecosystems: ecology, diversity, and applications. Springer, Singapore, pp 81–117

Panosyan H, Hakobyan A, Birkeland NK, Trchounian A (2018b) Bacilli community of saline–alkaline soils from the Ararat plain (Armenia) assessed by molecular and culture-based methods. Syst Appl Microbiol 41:232–240

Panosyan H, Di Donato P, Poli A, Nicolaus B (2018c) Production and characterization of exopolysaccharides by *Geobacillus thermodenitrificans* ArzA-6 and *Geobacillus toebii* ArzA-8 strains isolated from an Armenian geothermal spring. Extremophiles 22:725–737. https://doi. org/10.1007/s00792-018-1032-9

Paronyan AK (2002) Ecology and biodiversity of phototrophic bacteria of various ecosystems of Armenia. Biol J Armenia 54(1-2):91–98

Paronyan A (2003) Ekologia, biologicheskie osobennosti fototrofnikh bakterii Armenii i perspe-
    ktiwi ikh ispolzovanija. (Ecology, biological peculiarities of phototrophic bacteria of Armenia
    and prospects its application). Dissertation, Institute of Microbiology NAS of RA
Paronyan A (2007) Ecophysiological characteristics of phototrophic bacteria *Rhodopseudomonas
    palustris* isolated from mineral geothermal Jermuk. Biol J Armenia 59(1–2):73–77. (in Russian)
Paul VG, Mormile MR (2017) A case for the protection of saline and hypersaline environ-
    ments: a microbiological perspective. FEMS Microbiol Ecol 93(8). https://doi.org/10.1093/
    femsec/fix091
Poghosyan DA (2008) Landscape of the Armenian Highland and physico-geographical regions.
    YSU press, Yerevan
Poghosyan L (2015) Prokaryotic diversity in an Armenian geothermal spring using metagenomics,
    anaerobic cultivation and genome sequencing. Master thesis, University of Bergen
Poghosyan L, Birkeland NK, Panosyan H (2014) Diversity of thermophilic anaerobes in the geo-
    thermal spring Jermuk in Armenia. In: Abstracts book of international scientific workshop on
    "Trends in microbiology and microbial biotechnology", 5–8 October, 2014, Yerevan, Armenia,
    YSU press, p 83
Raddadi N, Cherif A, Daffonchio D, Neifar M, Fava F (2015) Biotechnological applica-
    tions of extremophiles, extremozymes and extremolytes. Appl Microbiol Biotechnol
    99(19):7907–7913
Rampelotto PH (2013) Extremophiles and extreme environments. Life 3:482–485. https://doi.
    org/10.3390/life3030482
Rodrigo-Bañon M, Garbayo I, Vílchez C, José Bonete M, Martínez-Espinosa M (2015) Carotenoids
    from Haloarchaea and their potential in biotechnology. Mar Drug 13:5508–5532
Saxena R, Dhakan DB, Mittal P, Waiker P, Chowdhury A et al (2017) Metagenomic analysis of
    hot springs in Central India reveals hydrocarbon degrading thermophiles and pathways essen-
    tial for survival in extreme environments. Front Microbiol 7:2123. https://doi.org/10.3389/
    fmicb.2016.02123
Shahinyan GS, Margaryan AA, Panosyan HH, Trchounian AH (2015) Isolation and character-
    ization of lipase-producing thermophilic bacilli from geothermal springs in Armenia and
    Nagorno-Karabakh. Biol J Armenia 67(2):6–15
Shahinyan G, Margaryan AA, Panosyan HH, Trchounian AH (2017) Identification and sequence
    analysis of novel lipase encoding novel thermophilic bacilli isolated from Armenian geothermal
    springs. BMC Microbiol 17:103. https://doi.org/10.1186/s12866-017-1016-4
Singh RP, Manchanda G, Singh RN, Srivastava AK, Dubey RC (2016) Selection of alkalotoler-
    ant and symbiotically efficient chickpea nodulating rhizobia from North-West Indo Gangetic
    Plains. J Basic Microbiol 56:4–25
Vardanyan G, Margaryan A, Panosyan H (2015) Isolation and characterization of lipase-producing
    bacilli from Tatev geothermal spring (Armenia). In: Collection of scientific articles of YSU
    SSS: Materials of the scientific session dedicated to the 95th anniversary of YSU, 27-30 April,
    2015, Yerevan State University, Yerevan, pp 33–36
Ventosa A (2006) Unusual microorganisms from unusual habitats: hypersaline environments.
    In: Logan NA, Lappin-Scott HM, Oyston PCF (eds) Prokaryotic diversity: mechanisms and
    significance. Cambridge University Press, Cambridge, pp 223–253
Ventosa A, Mellado E, Sanchez-Porro C, Marquez MC (2008) Halophilic and halotolerant
    microorganisms from soils. In: Dion P, Nautiyal CS (eds) Microbiology of extreme soils.
    Springer-Verlag, Berlin/Heidelberg, pp 87–116
Ventosa A, de la Haba RR, Sánchez-Porro C, Papke RT (2015) Microbial diversity of hypersa-
    line environments: a metagenomic approach. Curr Opin Microbiol 25:80–87. https://doi.
    org/10.1016/j.mib.2015.05.002
Vera-Gargallo B, Chowdhury TR, Brown J, Fansler SJ, Durán-Viseras A et al (2019) Spatial
    distribution of prokaryotic communities in hypersaline soils. Sci Rep 9:1769. https://doi.
    org/10.1038/s41598-018-38339-z
Yatsunami R, Ando A, Yang Y, Takaichi S, Kohno M et al (2014) Identification of carotenoids from
    the extremely halophilic archaeon *Haloarcula japonica*. Front Microbiol 5:100. https://doi.
    org/10.3389/fmicb.2014.00100

Savita Singh, Mamta Gupta, and Yogesh Gupta

Abstract

Microbial life attracts much attention due to its extraordinary diversity. Microbial populations inhabiting in extremes of environmental factors like temperature, pH, salinity, pressure, and radiation present a fascinating world of life. The present chapter attempts to explore the uniqueness of microbial life thriving in extremes of environmental condition particularly salinity. The biochemical and molecular mechanisms responsible for their survival under salinity extremes have been discussed. Moreover, the relevance and application of these microbial populations have been realized in terms of their potential to contribute to agriculture, industry, and research relating extra-terrestrial life on other planets such as Mars which present an exciting field of knowledge with immense potential. The chapter aims to present a systematic and upgraded findings made to this topic.

Keywords

Bacteria · Salinity · Mechanisms · Genes · Adaptability

S. Singh
Department of Botany, Babu Shivnath Agrawal College (Affiliated to Dr BRA University, Agra), Mathura, Uttar Pradesh, India

M. Gupta
Department of Botany and Environment Studies, DAV University, Jalandhar, India

Y. Gupta (✉)
Institute For Global Food Security, School of Biological Sciences, Queen's University, Belfast, UK
e-mail: Y.Gupta@qub.ac.uk

© Springer Nature Singapore Pte Ltd. 2020
R. P. Singh et al. (eds.), *Microbial Versatility in Varied Environments*,
https://doi.org/10.1007/978-981-15-3028-9_3

3.1   Introduction

Extreme environments and the life thriving under such conditions have always been an intriguing area of research. Extreme condition refers to harsh environmental condition which is beyond the optimal surviving condition of an organism (Vorgias and Antranikian 2004). Microbial populations inhabiting in extremes of environmental factors like temperature, pH, salinity, pressure, and radiation present a fascinating world of life. Among these conditions, salinity is a much widespread but still less explored topic. Salt water-containing environments are worldwide spread throughout Earth (Oren 2002), and only few microbial communities thrive in such extreme habitat (Oren 1999a; Yang et al. 2016). These salt-containing environments constitute the favorable systems where the native microbial consortia survive via various molecular mechanisms of adaptation in elevated concentrations of NaCl from saline to hypersaline stages. However, the salt-containing oceans are the biggest saline water habitat; generally the 3.5% total dissolved salt-containing waters are known as hypersaline environment (DasSarma and Arora 2001; Benlloch et al. 2002). Hypersaline lakes may be of two types: thalassohaline and athalassohaline (Banciu et al. 2004; Banciu and Sorokin 2013). Thalassohaline lakes originate by evaporation of seawater; consequently, Na^+ and Cl^- are the predominant ions. Athalassohaline lakes are another classical example of extreme salt niche different from the seawater (e.g., Red Sea). Investigations of microbial life have always been fascinating due to reasons like novel potential of hypersaline microbiota in biotechnology and industry; terrestrial life forms of saline environment mimic the conditions of life forms thriving on Mars; hence, their study becomes exciting; and moreover, it's a way to study evolution of biosphere of Earth (Dong 2008).

Extremophiles occur both in ocean and terrestrial environments globally and especially in unusual ecological habitat like solfataric fields, soda lakes, abyssal hypothermal vents, and polar regions (Vorgias and Antranikian 2004). Most of the bacteria have been identified from the phylogenetic group of *Alphaproteobacteria*, *Betaproteobacteria*, and *Gammaproteobacteria* as well as *Cytophaga-Flavobacterium-Bacteroides*. Halophiles are a category of extremophiles which inhabit at high salinity and require sodium chloride to flourish well by employing the "salt-in-cytoplasm" mechanism. Moreover, they adapted the compatible solute strategies to sustain the osmotic balance between surrounding medium and inside the cytoplasm. During the enhancement of ionic concentration inside the cell cytoplasm, the enzymatic machinery of microbial cell has significantly changed, and the high acidic amino acids such as aspartate (Asp) have been overrepresented, while the hydrophobic residues are low in proportion (Rhodes et al. 2010). While the use of compatible solutes or osmoprotectants which haven't obstruct during the biochemical process of cell was applicable for the compatible solute strategy, compatible solute strategy is generally prevalent in several genera and species in comparison to "salt-in" strategy.

Halophiles are present in all three domains of life (Oren 2002). The Archaea consist the majority of salt-loving microbes comprising the class Halobacteria. Most of its members need more than $100–150 gl^{-1}$ of salt for normal metabolism and structural stability. Halophiles can be classified based on their survival and growth ability in various salt environments. There are various types of halophiles such as non-tolerant (1%w/v), slightly tolerant (6–8%w/v), moderately tolerant (18–20%w/v), and extremely tolerant (salt concentrations from 0 < saturation) to salt concentration (Yakimov et al. 2013; Sorokin et al. 2014; Prajakta et al. 2019). Some salt-loving microorganisms are also recognized without the microscope. For example, Archaea (*Haloquadratum*), *Bacteria* (*Salinibacter*), and Eukarya (*Dunaliella salina*) from saltern crystallizer pond brines adapted a pink-red morphology worldwide (Yanhe et al. 2010).

Some specific lipid molecules (phosphatidylglycerol or glycolipids) and physiological mechanisms present in the halophiles to counter with the environmental stress conditions such as high osmolarity and ionic strength of salt habitat (Van Vossenberg et al. 1999; Konings et al. 2002).

Hence, the current chapter focuses on the microbial life in extreme of salinity, the mechanisms adopted to survive under such conditions, and the genes involved in this potential applied aspect of these salt extremophiles in the field of biotech industry, study of evolution of life, and potential of life on other planets.

3.2 Basic Adaptation Strategies Under Extremes of Salinity

Advances in molecular biology have opened the new era of information which is revealed the popular mechanisms of preeminent salt resistance/tolerance and osmoprotection. These information has been obtained from cultivated microbes and their genome sequence. Hence, the knowledge derived by genes/genomes is overlook on specific adaptation strategies (Wu et al. 2009; Yang et al. 2019). The recent genomic data of the genus *Halorhodospira* have shown a combined use of both strategies of salt adaptation (Deole et al. 2013). Meta-omics technologies are also playing a forward role in the adaptation strategies of microorganisms (Subhashini et al. 2017). The first adaptation response involves accumulation of molar concentrations of KCl. The different proteins in halophilic microorganisms are extremely acidic in nature (Fukuchi et al. 2003; Coquelle et al. 2010). These microorganisms are generally difficult to grow in media containing low salt. Hence, halophiles have adapted the other strategies to remove the different salts from the cell cytoplasm and to generate or store different organic solutes (sugars, amino acids, polyols) which have no cross-reactivity with other enzymatic reactions in eukaryotic microorganisms such as algae and fungi (Oren 1999b). Interestingly, the finding of mannitol in microorganisms (e.g., *Pseudomonas putida*) is an exception (Kets et al. 1996).

In prokaryotes, the major variety of different organic solutes/compounds is detected in bacteria such as glycine betaine commonly used as osmotic solute (Wargo 2013; Cánovas et al. 1998). In recent years, the number of non-phototrophic prokaryote microorganism's de novo synthesis of glycine betaine as controlled by the external ionic salt concentration has gone up. Further the *Methanohalophilus portucalensis* and *Methanosarcina mazei* generate the glycine betaines in the presence of salt stress condition (Roeßler et al. 2002). Adaptation to salinity is also reported in rhizobia. For example, chickpea rhizobia shows different levels of tolerance to salt stresses/concentrations, and few microorganisms have the capacity to survive at salt concentration of > 500 mM (Kucuk and Kivanc 2008). Singh et al. (2016) have also revealed the adaptability of *Mesorhizobium* at 4.5% w/v NaCl concentration and in other studies, *Streptomyces* at 6.0% isolated from rotten wheat straw (Singh et al. 2019). Rubiano-Labrador et al. (2015) have found the alteration in the exoproteome of the halotolerant bacterium *Tistlia consotensis* after inducing the low and high salt stress and has been deciphered it by proteogenomics study. Large amounts of the HpnM protein were detected at low salinity which is for the biosynthesis of hopanoids, and at high salt concentration, the different osmosensing proteins remove the Na + ions and facilitate the different organic compatible solutes (i.e., glycine betaine, proline, etc.). At high salinity, *T. consotensis* adapted the strategy of activation of the synthesis of flagella and triggered a chemotactic response.

3.2.1 Accumulation of Potassium and Its Transport

There are different ion channels (e.g., K^+) which are reported previously in prokaryotes. Structures of various K^+ channels, initially a closed small bacterial channel and more recently a gated K^+ − channel (MthK) from *Methanobacter thermoautotrophicus*, have participated in the understanding of how these proteins are arranged in membranes (Zadek and Nimigean 2006). However, these K-channels do not respond to altered osmotic pressure. Rather, different protein complexes appear to regulate intracellular K in response to osmotic stress. Recent work has identified three genes required for K uptake: *trkA*, *trkH*, and *trkI* (Kraegeloh et al. 2005). The protein expressed by *trkA* would be analogous to the cytoplasmic NAD/NADH binding protein TrkA in *E. coli* that is required for K uptake by the Trk system, while the *Halomonas elongata* TrkH and TrkI are likely to be transmembrane proteins (Kindzierski et al. 2017).

In extreme salt conditions, bacterial cell wall water rushes intracellularly via hypoosmotic shock. Later, to retain the regular structure and original cell volume, cells required a rapid solute efflux cytoplasmic system. Martinac et al. (2008) have stated that mostly all the bacterial genera have specialized transmembrane channels that open for solute efflux during the cell membranes lateral pressure that dropped the access of solute below a serious value. Li et al. (2002) have suggested that mechanosensitive channels (Msc) are primary biosensors for osmoregulation in bacteria which are thought to be gated by membrane tension. They might be majorly played by hypoosmotic shock and resultantly released the solutes in speedy and non-discriminating manners.

Msc falls into three classes: (1) Mechanosensitive channel of large conductance (MscL) is a first MS ion channel to be identified by painstaking physical and biochemical means at molecular level. It is pentamer in structure and is not a preferential solute of large conducting activity (Ajouz et al. 2000; Bilston and Mylvaganam 2002). (2) The MscS, a widely studied channel, is driven by closed and open states. In closed stage, MscS doesn't allow solutes to exit from the cytoplasm, while in the open stage, it allowed the rapid efflux of solvent as well as solutes. The performance and endurance of proper open or closed channel system are very important for cell survival. MscS is a heptamer unit and is small in conductance and membrane tension sensitive (Bass et al. 2002). MscS is discriminative in nature to anions or cations. (3) MscK is a heptamer unit and is alike to MscS and is triggered by cytoplasmic K^+. Very less is known about this third Msc protein. It also symbolized KefA instead of MscK. KefA (MscK) has been revealed in mutant *E. coli* strain RQ2 that failed to grow in a high concentration of K+ supplemented medium as well as with the existence of betaine (McLaggan et al. 2002). Roberts (2005) has suggested one more osmosensor, volume-activated channels (VAC), and has suggested that it could respond as anion channels to hypoosmotic response. VAC sensors are very useful in the removal of a variable osmolyte, remarkably amino acids and polyols.

3.2.2 Accumulation of Organic Solutes and Its Transport

In high salt environments, halotolerant and halophilic organisms use a variable of solutes to counter external osmotic pressure. The solutes can be anionic, or non-charged, or zwitterionic or other same net charge solutes for osmotic balance (Roberts 2005). Koga et al. (1998) revealed that *Methanothermococcus thermolithotrophicus* grown in medium enriched with <1 M NaCl can accumulates the anionic α- and β-glutamate. At the higher external NaCl concentrations, bacterial cells switch to accumulating a zwitterion, Nε-acetyl-β-lysine. It is due to the partial equal concentration of glutamate and intracellular K^+. Hence, switching to zwitterions is to be the consequence of an impaired K^+ pump (Whitman 2001). Lai et al. (2000) reported halophilic methanogen *Methanohalophilus portucalensis* that has accumulated three zwitterions (betaine, Nε-acetyl-β-lysine, and β-glutamine) over its growth range. Though, β-glutamine is accumulated largely in comparison to the other three at the high NaCl concentration (L'Haridon et al. 2018). In *Halobacterium salinarum*, two ORFs upstream of transducer genes were found that are homologous to binding proteins for solutes and amino acids (Gonzalez et al. 2009).

Development of omics technologies opens the ways to identify the responsible proteome for the uptake of osmolytes. Several researches have been carried out on the mechanism of betaine and ectoine uptake from medium by halophilic methanogens (Rudolph et al. 1986).

Moreover, betaine transport has been characterized in various halotolerant and halophilic bacteria and archaea (Imhoff and Rodriguez-Valera 1984; Lai and Gunsalus 1992). Basically, betaine is transported by two superfamily transporters: (a) proton motive force or sodium motive force that mediated betaine accumulation by secondary transporters and (b) ATP binding cassette or ABC transporters that couple ATP hydrolysis to uptake the betaine. Some other secondary transport systems such as high affinity BetP uptake system in *Corynebacterium glutamicum* have also been reported (Fränzel et al. 2010). However, internalization of betaine is continued even if the genes have been deleted, though the uptake of betaine is reduced significantly (Krämer and Morbach 2004). Steger et al. (2004) have identified the gene lcoP which is for the low capacity osmoregulated permease codes for a LcoP protein of BCCT family. Generally, concentrations of betaine vary from external concentration of NaCl. Imhoff and Rodriguez-Valera (1984) examined the eight halophiles grown well with the betaine concentrations in 3% (0.51 M), 10% (1.7 M) and 20% NaCl (3.4 M). In contrast, few bacteria such as *Actinopolyspora halophila* and *Halomonas elongata* as well as *Methanohalophilus portucalensis* FDF1 transport betaine into the cell for use as an osmolyte. These bacteria are potential to synthesize betaine by the process of choline oxidation glycine methylation (Nyyssola et al. 2000).

Ectoine (1,4,5,6-tetrahydro-2-methyl-4-pyrimidinecarboxylic acid) is a cyclic tetrahydropyrimidine compound and is widely used as a marker for halophile bacteria. Ectoine was first detected in a halophilic and phototrophic bacterium *Halorhodospira halophila* (Galinski et al. 1985). It internalizes proportionally to extracellular NaCl concentration of medium. Nagata and Wang (2001) showed that *Brevibacterium* sp. JCM6894 growth was stimulated by exogenous ectoine or hydroxyectoine. In *Chromohalobacter israelensis*, ectoine accumulated only in the exponential phase of growth and with >0.6 M NaCl, and it is indicated that ectoine accumulation is also growth stage dependent (Regev et al. 1990). Bacterial cells have some specialized transporter system that can recover the solute if it leaked from the cell. For example, the ectoine and hydroxyectoine (*TeaA, TeaB, TeaC*) in *H. elongata* have homology to ATP-independent periplasmic transporter family (TRAP-T). Moreover, these transporters appear to recover the leaked ectoine from the cell (Kindzierski et al. 2017). Similarly, external ectoine was transported in the *Marinococcus halophilus*, and the participated gene product EctM belongs to the BCCT family (Onraedt et al. 2003).

Poly-β-hydroxybutyrates are also used as carbon reservoirs in cells in soluble form. Poly-β-hydroxybutyrates have been detected in *Methylarcula marina*, *M. terricola*, and *Photobacterium profundum* SS9, a deep-sea bacterium, in moderate concentrations (Doronina et al. 2000; Martin et al. 2002). The accumulation of β-glutamate as well as α-glutamate in methanogens is also detected for osmotic balance. For example, the α- and β-glutamate levels increase simultaneously in *Methanothermococcus thermolithotrophicus* when the external NaCl increases (Martin et al. 2001). However, the carbohydrate-derivatized sugars (neutral) glucosylglycerol and α-mannosylglyceramide are detected only in some bacteria (Silva et al. 1999).

Roder et al. (2005) have showed the accumulation of α-glucosylglycerol by *Stenotrophomonas*, a member of the *Proteobacteria* and revealed it as a potential asset for biotechnological uses because of its ability to utilize rare carbon sources.

Though, sucrose-synthesizing enzymes are not identified in other archaea or bacteria (Curatti et al. 2000).

3.2.3 Chaperone Genes

During salinity stresses, various classes of chaperones are induced, and the co-chaperone DnaJ and the nucleotide exchange factor GrpE were comprised of DnaK machinery, whereas the co-chaperone GroES was included in the GroEL system (Brígido et al. 2012; Subhashini et al. 2017). Ubiquitous presence of ClpB belonging to the Clp family in prokaryotes and eukaryotes was previously stated by Gottesman et al. (1997), and it was revealed that they were acting as both protease and chaperone. Several osmolytes have also induced protein stability. In the cells, they appear and act as chemical chaperones and provide the mechanistic insight to protein folding for stabilization. Moreover, the catalysis of protein disaggregation and reactivation at higher salt conditions was done by interaction of ClpB chaperone system with DnaK chaperone system (Zolkiewski 1999). Domínguez-Ferreras et al. (2006) have revealed that various genes including clpB have induced and some groESL operon copies were found repressed in *Ensifer meliloti* at saline condition. Tolerance to salt in *R. tropici* was also revealed by Nogales et al. (2002), and it has been described that the dnaJ co-chaperone is involved in that. After being submitted to salt shock, induced activity in groESL operon was reported by Susin et al. (2006) in the *E. faecalis* and *C. crescentus*. Moreover, *Saccharomyces cerevisiae* responded to salt effects by ribosome-binding at Hsp70 homolog Ssb for the membrane protein biogenesis (Peisker et al. 2010). However, in chickpea *Mesorhizobium*, chaperone genes were not significantly induced in salt shock. A similar result was shown by Brígido et al. (2012), and it has been described that chaperon gene induction is greatly based on salt type, concentration, and time of exposure.

3.3 Genomics and Proteomics of Halophiles

The first haloarchaeal genome of *Halobacterium* sp. NRC-1 had opened the opportunity to researchers to probe the mechanisms of adaptation to hypersaline brine (Ng et al. 2000). Currently ≈1179 halophilic species have been reported throughout the scientific literature and search engines under the keywords such as halophiles, salt, saline, hypersaline, extremophile, or combinations of them. Among them are Archaea (21.9%), *Bacteria* (50.1%), and eukaryotes (27.9%) (Loukas et al. 2018). Due to extreme condition, only the microorganisms equipped with required molecular, biochemical, and ecological adaptive features are able to inhabit in saline and hypersaline environment. The taxonomic diversity of such microbes is therefore scarce (Oren 2001). *Gammaproteobacteria* and *Betaproteobacteria* as well as members of the *Bacteroidetes* are predominant in freshwater samples; on the other hand,

Table 3.1 Salt adaptation-related genes from bacteria

S. No.	Gene	Bacteria	References
1.	*ablA*	*Methanosarcina mazei*	Pfluger et al. (2003)
2.	*ectABC*	*Chromohalobacter salexigens*	Schubert et al. (2007)
		Bacillus halodurans	Rajan et al. (2008)
3.	*BetT*	*Pseudomonas syringae*	Chen and Beattie (2008)
4.	*proH*, *proJ*, and *proA*	*Halobacillus halophilus*	Saum and Muller (2007)
5.	*acdS*	*Halobacillus* sp. *SL3 and Bacillus halodenitrificans PU62*	Ramadoss et al. (2013)
6.	*Mpgsmt-sdmt*	*Methanohalophilus portucalensis*	Lai and Lai (2011)
7.	*KatE*, HPT, and NPTII	*E. coli*	Prodhan et al. (2008)
8.	*BetS*, betC genes	*Sinorhizobium meliloti*	Østerås et al. (1998), Boscari et al. (2002)
9.	*katE*	*E. coli* K12	Islam et al. (2013)
10.	*ectABC*	*Chromohalobacter salexigens*	Schubert et al. (2007)
11.	*ectABC*	*Bacillus halodurans*	Rajan et al. (2008)
12.	*codA*	*Arthrobacter globiformis*	Fan et al. (2004)
13.		ATCC 8010	
14.	*ectABC*	*Chromohalobacter salexigens*	Calderon et al. (2004)
15.	*a-Aminoisobutyric acid (AIB)*	*Vibrio costicola*	Kushner et al.
16.			(1983)
17.	*OpuC* and *OpuB*	*Listeria monocytogenes*	Fraser et al. (2000)
18.	*ablA*	*Methanosarcina mazei Go¨1*	Pfluger et al. (2003)
19.	*ectABC*	*Bacillus species*	Kuhlmann and Bremer (2002)
20.	*otsA* and *otsB*	*E. coli*	Joseph et al. (2010)
21.	*PDH45*	*Escherichia coli* BL21cells	Tajrishi et al. (2011)
22.	*otsA/otsB*, *mpgS/mpgP*	*Thermus thermophilus*	Alarico et al. (2005)
23.	*BetT*	*Pseudomonas syringae*	Chen and Beattie (2008)
24.	*OtsBA* and *TreYZ*	*E. coli*	Padilla et al. (2004)
25.	*proH*, *proJ*, and *proA*	*Halobacillus halophilus*	Saum and Muller (2007)
26.	*acdS*	*Halobacillus* sp. *SL3 and Bacillus halodenitrificans PU62*	Ramadoss et al. (2013)
27.	Bacterial mannitol 1-phosphate dehydrogenase (*mtl*D) gene	*mtlD*-producing bacteria	Rahnama et al. (2011)

(continued)

Table 3.1 (continued)

S. No.	Gene	Bacteria	References
28.	codA	*Arthrobacter globiformis*	Goel et al. (2011)
29.	Orthologs of *nqrA*, *trkA2*, *nadA*, and *gdhB*	*Halomonas beimenensis*	Chen et al. (2017)
30.	groEL, dnaKJ, and clpB	*Mesorhizobium* sp.	Brígido et al. (2012)

the *Alphaproteobacteria*, *Cyanobacteria*, and chloroplast genes dominate in marine samples (Langenheder et al. 2003; Bernhard et al. 2005). Among the important halophilic adaptation at the genetic level, the abundant nucleotides are partially attached to specific amino acids at high saline or salt concentration. The identical codon usage in halophiles also seems to have congregated to a single pattern regardless of their long-term evolutionary lineages and history (Paul et al. 2008). A list of genes characterized from agriculturally important halotolerant/halophilic bacteria has been given in Table 3.1. Classical methods of protein purification involved low salt condition so initially low number of proteins could be identified (Eisenberg and Wachtel 1987).

At the proteome level, it includes convergent evolution toward a specific proteome composition, characterized by low hydrophobicity; overrepresentation of acidic residues, especially Asp; higher usage of Val and Thr; lower usage of Cys; and a lower propensity for helix formation as well as a higher tendency for coil structure (Paul et al. 2008). Pade and Hagemann (2015) have reviewed the effect of acclimation of salt (high and low) on the metabolome of model cyanobacteria *Synechocystis* sp. PCC 6803. He proposed that at low salinity level, enzymes are present intracellularly but inactive due to their binding to the phosphate backbone of DNA and might be RNA and enzymes releases only after adding the of ions favorably KCl. It also activated the GG synthesis system. Marin et al. (2002) proposed that GgpS level is increased in the cells when salt adaptation reached at saturation level that is because of the release of ion-mediated small repressor protein GgpR through the *ggpS* promoter (Klähn et al. 2010). Nonetheless, it is still mysterious that proposed classic model of salt titration will same to other proteins/enzymes participated in salt acclimation or not.

3.4 Relevance and Future Prospects

In recent years, the biotechnology industries exploited the halophilic microorganisms by applying a combination of different molecular techniques and chemotaxonomic approach. The technical and applied uses of bacteriorhodopsin include holography, modulators, etc. Cultivation of microbes in saline media could also be beneficiary for industry, agriculture, and HRD sector. Carrieri et al. (2010) have stated that "sodium stress cycling" could be helpful in the product yields in autofermentation. For example, high salt conditions are helpful in achievement of increased carbohydrate content which have served as internal compatible solutes. Halotolerant microbes have an important role in the production of fermented food products and supplements in modern food biotech industries. Moreover, they are much accepted choice for the study of life origin and potential capacity of life on other planet in universe where similar conditions of extreme salinity are reported. The bioremediation of different pollutants including toxic wastages and generation of alternative source of energy are few other applications of the halophiles.

3.5   Conclusion

Halophiles offer a magnitude of applied usage in biotechnology industry, and their study is sure to unleash the unexplored world of microbial life under extremes of salinity. The chapter summarizes the response and adaptation phenomena of halophiles and other microbiota in salt stresses and specified many biochemical and molecular pathways which possibly show the changes in genetic expression and role of proteins transporters in salt adaptation.

Acknowledgments   Authors are thankful to Dr. Raghvendra Pratap Singh and Prof. A. K. Mishra for helpful discussions on the topic.

References

Ajouz B, Berrier C, Besnard M, Martinac B, Ghazi A (2000) Contributions of the different extra-membranous domains of the mechanosensitive ion channel MscL to its response to membrane tension. J Biol Chem 275:1015–1022

Alarico S, Empadinhas N, Simões C, Silva Z, Henne A, Mingote A, Santos H, da Costa MS (2005) Distribution of genes for synthesis of trehalose and mannosylglycerate in *Thermus* spp. and direct correlation of these genes with halotolerance. Appl Environ Microbiol 71:2460–2466

Banciu HL, Sorokin DY (2013) Adaptation in haloalkaliphiles and natronophilic bacteria. Polyextremophiles. Springer, Dordrecht, pp 121–178

Banciu H, Sorokin DY, Galinski EA, Muyzer G, Kleerebezem R, Kuenen JG (2004) *Thialkalivibrio halophilus* sp. nov., a novel obligately chemolithoautotrophic, facultatively alkaliphilic, and extremely salt-tolerant, sulfur-oxidizing bacterium from a hypersaline alkaline lake. Extremophiles 8:325–334

Bass RB, Strop P, Barclay M, Rees D (2002) Crystal structure of *Escherichia coli* MscS, a voltage-modulated and mechanosensitive channel. Science 298:1582–1587

Benlloch SA, López-López EO, Casamajor L, Øvreas V, Goddard FL et al (2002) Prokaryotic genetic diversity throughout the salinity gradient of a coastal solar saltern. Environ Microbiol 4:349–360

Bernhard AE, Colbert D, McManus J, Field KG (2005) Microbial community dynamics based on 16S rRNA gene profiles in a Pacific Northwest estuary and its tributaries. FEMS Microbiol Ecol 52:115–128

Bilston LE, Mylvaganam K (2002) Molecular simulations of the large conductance mechanosensitive (MscL) channel under mechanical loading. FEBS Lett 512:185–190

Boscari A, Mandon K, Dupont L, Poggi MC, Rudulier DL (2002) BetS is a major glycine betaine/proline betaine transporter required for early osmotic adjustment in *Sinorhizobium meliloti*. J Bacteriol 184:2654–2663

Brígido C, Robledo M, Menéndez E, Mateos PF, Oliveira S (2012) A ClpB chaperone knockout mutant of shows a delay in the root nodulation of chickpea plants. Mol Plant-Microbe Interact 25(12):1594–1604

Calderon MI, Vargas C, Rojo F, Iglesias-Guerra F, Csonka LN et al (2004) Complex regulation of the synthesis of the compatible solute ectoine in the halophilic bacterium *Chromohalobacter salexigens* DSM3043. Microbiology 150:3051–3063

Cánovas D, Vargas C, Csonka LN, Ventosa A, Nieto JJ (1998) Synthesis of Glycine Betaine from exogenous choline in the moderately halophilic bacterium *Halomonas elongate*. Appl Environ Microbiol 64:4095–4097

Carrieri D, Momot D, Brasg IA, Ananyev G, Lenz O, Bryant DA, Dismukes GC (2010) Boosting autofermentation rates and product yields with sodium stress cycling: application to production of renewable fuels by cyanobacteria. Appl Environ Microbiol 76:6455–6462

Chen C, Beattie GA (2008) *Pseudomonas syringae* BetT is a low-affinity choline transporter that is responsible for superior osmoprotection by choline over glycine betaine. J Bacteriol 190(8):2717–2725

Chen YH, Lu CW, Shyu YT, Lin SS (2017) Revealing the saline adaptation strategies of the halophilic bacterium *Halomonas beimenensis* through high-throughput omics and transposon mutagenesis approaches. Sci Rep 7(1):13037

Coquelle N, Talon R, Juers DH, Girard E, Kahn R et al (2010) Gradual adaptive changes of a protein facing high salt concentrations. J Mol Biol 404:493–505

Curatti L, Porchia AC, Herrera-Estrella L, Salerno GL (2000) A prokaryotic sucrose synthase gene (susA) isolated from a filamentous nitrogen-fixing cyanobacterium encodes a protein similar to those of plants. Planta 211:729–735

DasSarma S, Arora P (2001) Halophiles. Encyclopedia of life sciences. Nature Publishing Group, Dordrecht, pp 1–9

Deole R, Challacombe J, Raiford DW, Hoff WD (2013) An extremely halophilic proteobacterium combines a highly acidic proteome with a low cytoplasmic potassium content. J Biol Chem 288(1):581–588

Domínguez-Ferreras A, Perez-Arnedo R, Becker A, Olivares J, Soto MJ et al (2006) Transcriptome profiling reveals the importance of plasmid pSymB for osmoadaptation of *Sinorhizobium meliloti*. J Bacteriol 188:7617–7625

Dong H (2008) Links between geological processes, microbial activities & evolution of life. In: Dilek Y et al (eds.). Springer, Berlin

Doronina NV, Trotsenko YA, Tourova TP (2000) *Methylarcula marina* gen. nov., sp. nov. and *Methylarcula terricola* sp. nov.: novel aerobic, moderately halophilic, facultatively methylotrophic bacteria from coastal saline environments. Int J Syst Evol Microbiol 50:1849–1859

Eisenberg H, Wachtel EJ (1987) Structural studies of halophilic proteins, ribosomes, and organelles of bacteria adapted to extreme salt concentrations. Biophys Biophys Chem 16:69–92

Fan F, Mahmoud G, Giovanni G (2004) Cloning, sequence analysis, and purification of choline oxidase from *Arthrobacter globiformis*: a bacterial enzyme involved in osmotic stress tolerance. Arch Biochem Biophys 421:149–148

Fränzel B, Trötschel C, Rückert C, Kalinowski J, Poetsch A, Wolters DA (2010) Adaptation of *Corynebacterium glutamicum* to salt-stress conditions. Proteomics 10:445–457

Fraser KR, Harvie D, Coote PJ, O Byrne CP (2000) Identification and characterization of an ATP binding cassette l-Carnitine transporter in *Listeria monocytogenes*. Appl Environ Microbiol 66(11):4696–4704

Fukuchi S, Yoshimune K, Wakayama M, Moriguchi M, Nishikawa K (2003) Unique amino acid composition of proteins in halophilic bacteria. J Mol Biol 327:347–357

Galinski EA, Pfeiffer HP, Trüper HG (1985) 1,4,5,6-Tetrahydro-2-methyl-4-pyrimidinecarboxylic acid. A novel cyclic amino acid from halophilic phototrophic bacteria of the genus *Ectothiorhodospira*. Eur J Biochem 49(1):135–139

Goel D, Singh AK, Yadav V, Babbar SB, Murata N, Bansal KC (2011) Transformation of tomato with a bacterial *cod*A gene enhances tolerance to salt and water stresses. J Plant Physiol 168(11):1286–1294. https://doi.org/10.1016/j.jplph.2011.01.010

Gonzalez O, Gronau S, Pfeiffer F, Mendoza E, Zimmer R, Oesterhelt D (2009) Systems analysis of bioenergetics and growth of the extreme halophile *Halobacterium salinarum*. PLoS Comput Biol 5(4):e1000332

Gottesman S, Wickner S, Maurizi MR (1997) Protein quality control: triage by chaperones and proteases. Genes Dev 11:815–823

Imhoff JF, Rodriguez-Valera F (1984) Betaine is the main compatible solute of halophilic eubacteria. J Bacteriol 160:478–479

Islam MS, Azam MS, Sharmin S, Sajib AA, Alam MM et al (2013) Improved salt tolerance of jute plants expressing the *kat*E gene from *Escherichia coli*. Turk J Biol 37:206–211

Joseph TC, Rajan LA, Thampuran N, James R (2010) Functional characterization of Trehalose biosynthesis genes from *E. coli*: an Osmolyte involved in stress tolerance. Mol Biotechnol 46:20–25

Kets EP, Galinski EA, de Wit M, de Bont JA, Heipieper HJ (1996) Mannitol, a novel bacterial compatible solute in *Pseudomonas putida* S12. J Bacteriol 178(23):6665–6670

Kindzierski V, Raschke S, Knabe N, Siedler F, Scheffer B et al (2017) Osmoregulation in the halophilic bacterium *Halomonas elongata*: a case study for integrative systems biology. PLoS One 12(1):e0168818

Klähn S, Höhne A, Simon E, Hagemann M (2010) The gene *ssl3076* encodes a protein mediating the salt-induced expression of *ggpS* for the biosynthesis of the compatible solute glucosylglycerol in *Synechocystis* sp. strain PCC 6803. J Bacteriol 192:4403–4412

Koga Y, Morii H, Akagawa-Matsushita M, Ohga I (1998) Correlation of polar lipid composition with 16S rRNA phylogeny in methanogens. Further analysis of lipid component part. Biosci Biotechnol Biochem 62:230–236. https://doi.org/10.1271/bbb.62.230

Konings WN, Albers S-V, Koning S, Driessen AJM (2002) The cell membrane plays a crucial role in survival of bacteria and archaea in extreme environments. Antonie Van Leeuwenhoek 81:61–72

Kraegeloh A, Amendt B, Kunte HJ (2005) Potassium transport in a halophilic member of the Bacteria domain: identification and characterization of the K⁺ uptake systems TrkH and TrkI from *Halomonas elongata* DSM 2581T. J Bacteriol 187(3):1036–1043. https://doi.org/10.1128/JB.187.3.1036-1043.2005

Krämer R, Morbach S (2004) BetP of *Corynebacterium glutamicum*, a transporter with three different functions: betaine transport, osmosensing, and osmoregulation. Biochem Biophys Acta 1658:31–36

Kucuk C, Kivanc M (2008) Survival and symbiotic effectivity of lyophilized root nodule bacteria. Pak J Biol Sci 11(16):2032–2035

Kuhlmann AU, Bremer E (2002) Osmotically regulated synthesis of the compatible solute Ectoine in *Bacillus pasteurii* and related *Bacillus* spp. Appl Environ Microbiol 68(2):772–783. https://doi.org/10.1128/AEM.68.2.772-783.2002

Kushner DJ, Hamaide F, Macleodr A (1983) Development of salt-resistant active trans- port in a moderately halophilic bacterium. J Bacteriol 153:1163–1171

L'Haridon S, Corre E, Guan Y, Vinu M, La Cono V, Yakimov M, Stingl U, Toffin L, Jebbar M (2018) Complete genome sequence of the halophilic methylotrophic methanogen archaeon *Methanohalophilus portucalensis* strain FDF-1T. Genome Announc 6:e01482–e01417. https://doi.org/10.1128/genomeA.01482-17

Lai MC, Gunsalus RP (1992) Glycine betaine and potassium ion are the major compatible solutes in the extremely halophilic methanogen *Methanohalophilus* strain Z7302. J Bacteriol 174:7474–7477

Lai MC, Hong TY, Robert PG (2000) Glycine betaine transport in the obligate halophilic archaeon methanohalophilus portucalensis. J Bac 182(17):5020–5024

Lai SJ, Lai MC (2011) Characterization and regulation of the Osmolyte Betaine synthesizing enzymes GSMT and SDMT from halophilic methanogen *Methanohalophilus portucalensis*. PLoS One 6(9):e25090

Langenheder S, Kisand V, Wikner J, Tranvik LJ (2003) Salinity as a structuring factor for the composition and performance of bacterioplankton degrading riverine DOC. FEMS Microbiol Ecol 45:189–202

Li Y, Moe PC, Chandrasekaran S, Booth IR, Blount P (2002) Ionic regulation of MscK, a mechanosensitive channel from *Escherichia coli*. EMBO J 21(20):5323–5330

Loukas A, Kappas I, Abatzopoulos TA (2018) HaloDom: a new database of halophiles across all life domains. J Biol Res (Thessaloniki) 25(2):1–8. https://doi.org/10.1186/s40709-017-0072-0

Marin K, Huckauf J, Fulda S, Hagemann M (2002) Salt-dependent expression of glucosylglycerol-phosphate synthase, involved in osmolyte synthesis in the cyanobacterium *Synechocystis* sp. strain PCC 6803. J Bacteriol 184:2870–2877

Martin DD, Ciulla RA, Robinson PM, Roberts MF (2001) Switching osmolyte strategies: response of *Methanococcus thermolithotrophicus* to changes in external NaCl. Biochim Biophys Acta 1524:1–10

Martin DD, Bartlett DH, Roberts MF (2002) Solute accumulation in the deep-sea bacterium *Photobacterium profundum*. Extremophiles 6:507–514

Martinac B, Saimi Y, Kung C (2008) Ion channels in microbes. Physiol Rev 88(4):1449–1490. https://doi.org/10.1152/physrev.00005.2008

McLaggan D, Jones MA, Gouesbet G, Levina N, Lindey S et al (2002) Analysis of the kefA2 mutation suggests that KefA is a cation-specific channel involved in osmotic adaptation in *Escherichia coli*. Mol Microbiol 43(2):521–536

Nagata S, Wang YB (2001) Accumulation of ectoine in the halotolerant *Brevibacterium* sp. JCM 6894. J Biosci Bioeng 91(3):288–293

Ng WV, Kennedy SP, Mahairas GG, Berquist B, Pan M et al (2000) Genome sequence of *Halobacterium* sp. NRC-1. PNAS USA 97(22):12176–12181

Nogales J, Campos R, BenAbdelkhalek H, Olivares J, Lluch C, Sanjuan J (2002) *Rhizobium tropici* genes involved in free-living salt tolerance are required for the establishment of efficient nitrogen-fixing symbiosis with *Phaseolus vulgaris*. Mol Plant-Microbe Interact 15:225–232

Nyyssola A, Kerovuo J, Kaukinen P, von Weymarn N, Reinikainen T (2000) Extreme halophiles synthesize betaine from glycine by methylation. J Biol Chem 275(29):22196–22201

Onraedt A, Walcarius B, Soetaert W, Vandamme EJ (2003) Dynamics and optimal conditions of intracellular ectoine accumulation in *Brevibacterium* Sp. Commun Agric Appl Biol Sci 68:241–246

Oren A (1999a) Microbiology and biogeochemistry of hypersaline environments. CRC Press, New York

Oren A (1999b) Bioenergetic aspects of halophilism. Microbiol Mol Biol Rev 63(2):334–348

Oren A (2001) In: Dworkin M, Falkow S, Rosenberg E, Schleifer K-H, Stackebrandt E (eds) The Prokaryotes. A handbook on the biology of bacteria: ecophysiology, isolation, identification, applications. Springer-Verlag, New York

Oren A (2002) Diversity of halophilic microorganisms: environments, phylogeny, physiology, and applications. J Ind Microbiol Biotech 28:56–63

Østerås M, Boncompagni E, Vincent N, Poggi MC, Le Rudulier D (1998) Presence of a gene encoding choline sulfatase in *Sinorhizobium meliloti* bet operon: choline-*O*-sulfate is metabolized into glycine betaine. PNAS USA 95:11394–11399. https://doi.org/10.1073/pnas.95.19.11394

Pade N, Hagemann M (2015) Salt acclimation of cyanobacteria and their application in biotechnology. Life (Basel, Switzerland) 5(1):25–49

Padilla L, Morbach S, Kramer R, Agosin E (2004) Impact of heterologous expression of *Escherichia coli* UDP-GlucosePyrophosphorylase on Trehalose and glycogen synthesis in *Corynebacterium glutamicum*. App Environ Microbiol 70(7):3845–3854

Paul S, Bag SK, Das S, Harvill ET, Dutta C (2008) Molecular signature of hypersaline adaptation: insights from genome and proteome composition of halophilic prokaryotes. Genome Biol 9:R70. https://doi.org/10.1186/gb-2008-9-4-r70

Peisker K, Chiabudini M, Rospert S (2010) The ribosome-bound Hsp70 homolog Ssb of *Saccharomyces cerevisiae*. Biochim Biophys Acta 1803:662–672

Pfluger K, Baumann S, Gottschalk G, Lin W, Santos H, Muller V (2003) Lysine-2, 3-Aminomutase and -lysine acetyltransferase genes of Methanogenic archaea are salt induced and are essential for the biosynthesis of N -acetyl- -lysine and growth at high salinity. App Environ Microbiol 69(10):6047–6055

Prajakta BM, Suvarna PP, Singh RP, Rai AR (2019) Potential biocontrol and superlative plant growth promoting activity of indigenous *Bacillus mojavensis* PB-35(R11) of soybean (*Glycine max*) rhizosphere. SN Appl Sci 1:1143. https://doi.org/10.1007/s42452-019-1149-1

Prodhan SH, Hossain A, Kenji N, Atsushi K, Hiroko M (2008) Improved salt tolerance and morphological variation in indica rice (*Oryza sativa* L.) transformed with a catalase gene from *E. coli*. Plant Tissue Cult Biotech 18(1):57–63

Rahnama H, Vakilian H, Fahimi H, Ghareyazie B (2011) Enhanced salt stress tolerance in transgenic potato plants (*Solanum tuberosum* L.) expressing a bacterial *mtlD* gene. Acta Physiol Plant 33:1521–1532

Rajan LA, Joseph TC, Thampuran N, James R, Ashok Kumar K et al (2008) Cloning and heterologous expression of ectoine biosynthesis genes from *Bacillus halodurans* in *Escherichia coli*. Biotechnol Lett 30:1403–1407

Ramadoss D, Lakkineni VK, Bose P, Ali S, Annapurna K (2013) Mitigation of salt stress in wheat seedlings by halotolerant bacteria isolated from saline habitats. SpringerPlus 2:6

Regev R, Peri I, Gilboa H, Avi-Dor Y (1990) 3C NMR study of the interrelation between synthesis and uptake of compatible solutes in two moderately halophilic eubacteria bacterium Ba1 and *Vibro costicola*. Arch Biochem Biophys 278(1):106–112

Rhodes ME, Fitz-Gibbon ST, Oren A, House CH (2010) Amino acid signatures of salinity on an environmental scale with a focus on the Dead Sea. Environ Microbiol 12:2613–2623. https://doi.org/10.1111/j.1462-2920.2010.02232.x

Roberts MF (2005) Organic compatible solutes of halotolerant and halophilic microorganisms Aquat. Saline Syst 1:5. https://doi.org/10.1186/1746-1448-1-5

Roder A, Hoffmann E, Hagemann M, Berg G (2005) Synthesis of the compatible solutes glucosylglycerol and trehalose by salt-stressed cells of *Stenotrophomonas* strains. FEMS Microbiol Lett 243:219–226

Roeßler M, Pflüger K, Flach H, Lienard T, Gottschalk G, Müller V (2002) Identification of a salt-induced primary transporter for Glycine Betaine in the Methanogen *Methanosarcina mazei* Gö1. Appl Environ Microbiol 68:2133–2139

Rubiano-Labrador C, Bland C, Miotello G, Armengaud J, Baena S (2015) Salt stress induced changes in the exoproteome of the halotolerant bacterium *Tistlia consotensis* deciphered by Proteogenomics. PLoS One 10(8):e0135065. https://doi.org/10.1371/journal.pone.0135065

Rudolph AS, Crowe JH, Crowe LM (1986) Effects of three stabilising agents – proline, betaine and trehalose – on membrane phospholipids. Arch Biochem Biophys 245:134–143

Saum SH, Muller V (2007) Salinity- dependent switching of osmolyte strategies in a moderately halophilic bacterium: glutamate induces proline biosynthesis in *Halobacillus halophilus*. J Bacteriol 189(19):6968–6975

Schubert T, Maskow T, Benndorf D, Harms H, Breuer U (2007) Continuous synthesis and excretion of the compatible solute ectoine by a transgenic, nonhalophilic bacterium. Appl Environ Microbiol 73(10):3343–3347

Silva Z, Borges N, Martins LO, Wait R, da Costa MS et al (1999) Combined effect of the growth temperature and salinity of the medium on the accumulation of compatible solutes by *Rhodothermus marinus* and *Rhodothermus obamensis*. Extremophiles 3:163–172

Singh RP, Manchanda G, Singh RN, Srivastava AK, Dubey RC (2016) Selection of alkalotolerant and symbiotically efficient chickpea nodulating rhizobia from North-West Indo Gangetic Plains. J Basic Microbiol 56:4–25

Singh RP, Manchanda G, Maurya IK, Maheshwari NK, Tiwari PK, Rai AR (2019) *Streptomyces* from rotten wheat straw endowed the high plant growth potential traits and agro-active compounds. Biocatal Agric Biotechnol 17:507–513. https://doi.org/10.1016/j.bcab.2019.01.014

Sorokin DY et al (2014) Microbial diversity and biogeochemical cycling in soda lakes. Extremophiles 18:791–809

Steger R, Weinand M, Kr€amer R, Morbach S (2004) LcoP, an osmoregulated betaine/ectoine uptake system from *Corynebacterium glutamicum*. FEBS Lett 573:155–160

Subhashini DV, Singh RP, Manchanda G (2017) OMICS approaches: tools to unravel microbial systems. Directorate of Knowledge Management in Agriculture, Indian Council of Agricultural Research. ISBN: 9788171641703. https://books.google.co.in/books?id=vSaLtAEACAAJ

Susin MF, Baldini RL, Gueiros F, Gomes SL (2006) GroES/GroEL and DnaK/DnaJ have distinct roles in stress responses and during cell cycle progression in *Caulobacter crescentus*. J Bacteriol 188:8044–8053

Tajrishi MM, Vaid N, Tuteja R, Tuteja N (2011) Overexpression of a pea DNA helicase 45 in bacteria confers salinity stress tolerance. Plant Signal Behav 6(9):1271–1275

Van de Vossenberg JLCM, Driessen AJM, Grant WD, Konings WN (1999) Lipid membranes from halophilic and alkali-halophilic archaea have a low H^+ and Na^+ permeability at high salt concentration. Extremophiles 3:253–257

Vorgias C, Antranikian G (2004) Extremophiles: pH, temperature, and salinity. In: Bull A (ed) Microbial diversity and bioprospecting. ASM Press, Washington, DC, pp 146–153. https://doi.org/10.1128/9781555817770.ch14

Wargo MJ (2013) Homeostasis and catabolism of choline and Glycine Betaine: lessons from *Pseudomonas aeruginosa*. Appl Environ Microbiol 79(7):2112–2120

Whitman WB (2001) Genus II. *Methanothermococcus* gen. Nov. In: Boone DR, Castenholz RW (eds) Bergey's manual of systematic bacteriology volume 1: the archaea and the deeply branching and phototrophic Bacteria, 2nd edn. Springer, New York. ISBN 978-0-387-98771-2

Wu D, Hugenholtz P, Mavromatis K, Pukall RD, Dalin E et al (2009) A phylogeny-driven genomic encyclopaedia of Bacteria and archaea. Nature 462:1056–1060. https://doi.org/10.1038/nature08656

Yakimov MM, La CV, Slepak VZ, La SG, Arcadi E et al (2013) Microbial life in the Lake Medee, the largest deep-sea salt-saturated formation. Sci Rep 3:3554. https://doi.org/10.1016/0301-0104(89)87026-0

Yang J, Ma L, Jiang H, Wu G, Dong H (2016) Salinity shapes microbial diversity and community structure in surface sediments of the Qinghai-Tibetan Lakes. Sci Rep 6:25078. https://doi.org/10.1038/srep25078

Yang Y, Singh RP, Lan X, Zhang CS, Sheng DH et al (2019) Whole transcriptome analysis and gene deletion to understand the chloramphenicol resistance mechanism and develop a screening method for homologous recombination in *Myxococcus xanthus*. Microb Cell Factories 18:1475–2859. https://doi.org/10.1186/s12934-019-1172-3

Yanhe M, Erwin AG, William DG, Oren A, Ventosa A (2010) Halophiles 2010: life in saline environments. Appl Environ Microbiol 76(21):6971–6981. https://doi.org/10.1128/AEM.01868-10

Zadek B, Nimigean CM (2006) Calcium-dependent gating of MthK, a prokaryotic potassium channel. J Gen Physiol 127(6):673–685. https://doi.org/10.1085/jgp.200609534

Zolkiewski M (1999) *ClpB* cooperates with *DnaK*, *DnaJ*, and *GrpE* in suppressing protein aggregation. A novel multi-chaperone system from *Escherichia coli*. J Biol Chem 274:28083–28086

Rhizobia at Extremes of Acidity, Alkalinity, Salinity, and Temperature

4

Junjie Zhang, Dipti Singh, Chen Guo, Yimin Shang, and Shanshan Peng

Abstract

Symbiosis and nitrogen fixation are an utmost requirement in agricultural system and global nitrogen (N) cycling. However, the soil's harsh conditions such as acidity, alkalinity, salinity, and temperature are the primary challenges to plant-microbe interaction. Because conditions altered rapidly in soil and thus associated bacteria is failed in selection and associate with compatible host. Among the soil microbiota, rhizobia are a well-known group of bacteria for their ability to fix atmospheric nitrogen via the mechanism of symbioses with leguminous plants. These are opportunistic endosymbionts as well as saprophytic bacteria. Though soil abiotic factors have altered the activity of rhizobial populations negatively, research have proven that soil pH, salinity, and temperature are the major abiotic factors in determining the rhizobial population and their symbiotic performance. Hence, the following chapter described the integrative view of adaptation mechanism and physiological responses of rhizobia in acid, alkalinity, salinity, and temperature stress.

Keywords

Rhizobia · Adaptation · Soil · Abiotic stress · Symbiosis

J. Zhang (✉)
College of Food and Bioengineering, Zhengzhou University of Light Industry, Zhengzhou, Henan Province, People's Republic of China

Collaborative Innovation Center for Food Production and Safety of Henan Province, Zhengzhou, Henan Province, People's Republic of China

D. Singh
Department of Microbiology, V.B.S. Purvanchal University, Jaunpur, Uttar Pradesh, India

C. Guo · Y. Shang · S. Peng
College of Food and Bioengineering, Zhengzhou University of Light Industry, Zhengzhou, Henan Province, People's Republic of China

4.1   Introduction

Rhizobia are rod-shaped, motile, chemoorganotrophic, gram-negative bacteria, belonging to *Rhizobiaceae* family, discovered and described in 1889 by Frank. Rhizobia have the capability to invade leguminous plant (Leguminosae) root hairs in temperate and subtropical zone of land. They provoke the root nodulation wherein the rhizobia live as intracellular symbionts. In root nodules, rhizobia live as bacteroids (pleomorphic forms) and preferentially involved in fixing atmospheric nitrogen (N_2) into a combined form (ammonia) for the utilization of host plant (Singh et al. 2020). Rhizobia are primarily recognized for their symbiosis with legumes and nitrogen fixation, and hence they are an important biological tool of sustainable cropping system and a conditioner of soil. When environmental N_2 is limited, the rhizobia interact with the roots of legumes and produce symbiotic nodules. Inside the nodules, N_2 is reduced to ammonium (NH_4^+), and reciprocally rhizobia obtain nutrients from the plant (Masciarelli et al. 2014). Two types of nodule formed by rhizobia-plant symbiosis: determinate and indeterminate. The determinate nodules are formed by "slow-growing" rhizobia and are spherical in shape because meristematic activity ceases early, while the indeterminate nodule is caused by fast-growing rhizobia and are cylindrical in shape (Hirsch 1992). Moreover, environment stresses such as acidity, alkalinity, salinity, temperature, etc. are very challenging to rhizobia (Sanja et al. 2016; Zhang et al. 2016a, b, 2017). In stress conditions, rhizobial population determined by the above described abiotic factors and affected negatively to their growth, survival, and performance.

A symbiotically efficient rhizobial strain is not predictable to express its full capacity for N_2 fixation in the negative factor of salinity, soil pH, temperature, and nutrient deficiency that might execute limitations on the potency of the microbe-host interaction as well as on legumes (Brockwell et al. 1995). Inoculation of stress-tolerant rhizobia is an option for the symbiosis and N_2 fixation ability under stress conditions. Several studies revealed that after inoculation with salt-tolerant strains of rhizobia with legume hosts in saline conditions, the growth and symbiosis were found to be increased (Zou et al. 1995a, b; Shamseldin et al. 2006), though rhizobia have varied tolerance to abiotic factors (Biswas et al. 2008).

Rhizobium spp. are very sensitive to soil stresses such as extreme of pH, salts, and temperature. They can distress directly on cell growth, and hence the symbiosis as well as legume productivity is also affected (Idrissi et al. 1996; Athar and Johnson 1997; Abdelmoumen et al. 1999; Rehman and Nautiyal 2002). Surange et al. (1997) have speculated that *Rhizobium* spp. of Indian alkaline soils, sampled in summer, have the ability to tolerate the extreme of pH, salt, and temperature stress, though most of the endosymbionts (rhizobia) of significant crops are found to be sensitive to desiccation in soil as well as on seed (Vincent et al. 1962; OsangAlfiana and Alexander 1982). Therefore, it is a global demand of the world where fertilizers are scarce or expensive and soils are in harsh condition, it is compulsory to add such types of strains of rhizobia for the betterment of ecology and agriculture. Moreover, the study of such types of *Rhizobium* sp. strains with the genetic potential to abiotic stress tolerance could have enhanced the production of pulses, food, and forage legumes around the world (Brockwell et al. 1995; Peoples 1995).

In the soil, rhizobia need to manage the environmental stress as well as nutritional competition (Hinsinger et al. 2009). First, rhizobia compete with other microbial communities to stablish their own population (Kong et al. 2015; Li et al. 2016), then specified itself for targeted plants. Moreover, it needs to establish the synergism with other microbiota to fight against stress of acid, alkalinity, salinity, temperature, etc. When rhizobia infect the compatible host, the growth conditions varied according to symbiotic stages and developments. It included the effect of stress conditions (acidic and oxidative) inside the curled hair pocket (Hawkins et al. 2017), oxidative penetration and infection thread formation along with linear growth (Santos et al. 2000), and symbiotic differentiation of rhizobia after release into host-infected cells.

After several studies and reviews, biological N_2 fixation for agriculture is still in fussy stage. Howieson and Ballard (2004) have raised the issue of biological N_2 fixation by rhizobia for agriculture and its impact on soil management of arid regions. Several studies have been done for the isolation and characterization of abiotic stress-resistant rhizobia (Rhijn and Vanderleyden 1995; Chen et al. 2000; Jenkins 2003; Ji et al. 2014; Singh et al. 2016). Additionally, several techniques such as direct inoculation, seed priming, seed coating, bioformulation, etc. are applied for the enhancement of population of *Rhizobium* sp. in the soil for effective nodulation and N_2 fixation, but the abiotic stresses caused hindrance in their survival (Salema et al. 1982; Smith 1992; Trotman and Weaver 1995; Kosanke et al. 1999; Singh et al. 2011). Because the sustainability of agriculture and soil health are the basis of productivity of the crop, climate change, human arbitrary uses of land, amount of rainfall, mineral weathering, acid rain, uses of fertilizers, mine spoil, leaching of nitrogen, and plant root activity caused soil degradation as well as enhanced stress such as salinization, acidification, temperature, etc.

Nonetheless, the rhizobial-mediated N_2 fixation and their symbiosis with legumes can be enhanced by inoculation of stress-tolerant strains. Several studies revealed that legumes were grown well in saline condition after inoculating with salt-tolerant strains of rhizobia (Zou et al. 1995a, b; Shamseldin and Werner 2005). Several studies on soybean rhizobia were conducted under temperature, salinity, alkalinity, and acid stress. Chen et al. (2002) isolate the rhizobia from soybean (*Glycine max*) that has grown at 40 °C. Similarly, *Phaseolus vulgaris* nodule bacterium survived to 47 °C, but it lost its ability to induce nodulation at this temperature (Karanja and Wood 1988). *Sesbania aculeata* resided Rhizobium isolates were able to grow at 50 °C (Kulkarni and Nautiyal 2000). The above-described factors also indirectly lead to desiccation stress in rhizobia. Survival of rhizobia in desiccated conditions leads to a reduction in water activity and increases in salt stress. This might lead to the activation and expression of salt-responsive loci which are osmoprotectant accumulators, desiccation protectants, ROS scavenger, heat shock proteins (Hsps), chaperones, etc. (Vriezen et al. 2007). Although enumeration of a persistent and symbiotically efficient rhizobial strain with full ability to survive in limiting factors is tedious to find, the insight clarification of rhizobia strategies against these responses can highlight the way of search. Hence the proposed chapter is aiming to describe the insight of the sensitivity of rhizobia to extreme of acid, alkali, salinity, and temperature to find ways to save and enhance it in soil.

4.2 Response to Acid Stress in Rhizobia

Low pH or acidity in soil is a main threat to soil ecology and environment. Acid stress directly corelated to declination of nitrogen fixation, crop production, and plant growth promotion because of their sensitiveness to rhizobia (Graham 1992). Rhizobia are mostly sensitive to low pH and grow very slow or not at pH below 6.0 (Howieson et al. 1992). So, responses and mechanism of survival of acid-tolerant rhizobial strains are markedly important for legume productivity and nitrogen fixation (Howieson et al. 1988). Moreover, acid stress for rhizobia must be considered as an important trait for study and isolation of acid-tolerant rhizobia for effective symbiosis and nitrogen fixation as well as plant growth promotion (Howieson et al. 1988; Singh et al. 2017a, b). Day-by-day enhancement of hydrogen-ion concentrations in the nature has sturdily affected the viability, saprophytism, and symbiosis of rhizobia. For example, *S. meliloti* (Lowendorf et al. 1981), *S. medicae* (Glenn et al. 1999), and several other rhizobia (Zahran 1999) were affected in growth and symbiosis by acid stress. After several studies, it has been marked that acid stress is maleficent to soil ecology and plant growth and symbiosis as well as for rhizobia.

Every species and strain have their own optimum pH at which they grow normally. Hence rhizobial strains diverge broadly in their acidity response. For example, mutants of *R. leguminosarum* have the ability to grow at pH 4.5 (Chen et al. 1993), *S. meliloti* have viability only down to pH 5.5 (Foster 2000), *S. fredii* grows with pH range 4.0 to 9.5, but *B. japonicum* inversely never grow at the extreme range of pH (Fujihara and Yoneyama 1993). Though neutral conditions (7.0) are optimum for maximum bacteria as well as for several *Rhizobium* sp. (Glenn and Dilworth 1994), several studies have revealed the genetic mechanism of adaption to acid in rhizobia. For example, Kurchak et al. (2001) reported the ~20 genes act genes (acid tolerance) in *R. leguminosarum* for acid stress response. Alfalfa symbionts *S. medicae* have been explored widely for their performance at low pH. Resultantly, genes for acid tolerance *actA*, *actR/S*, *actP exoH*, and *exoR* (Tiwari et al. 1996a, b; Glenn et al. 1999; Vivas-Marfisi et al. 2002; Fenner et al. 2004) and some other genes, such as *phrR* and *lpiA*, were also induced at low pH (Vinuesa et al. 2003; Reeve et al. 2006).

Glenn and Dilworth (1994) have suggested a sensing mechanism for acid shock response to root nodule bacteria. For the sensing of such types of systems for environmental stresses as well as for the harsh condition, two components sensor and regulator were investigated. Tiwari et al. (1996b) have described one of it in *S. meliloti* and suggested that the genes actR and actS encode for the regulator and sensor, respectively. During the detection of external acidity, ActS derived the membrane-bound product of actS and activates ActR via phosphorylation. Further, ActR has activated the transcription of other genes within the rhizobia for acid response (Tiwari et al. 1996b). Moreover, Tiwari et al. (1996a) have speculated the role of calcium in *S. meliloti* acid tolerance. Though the facilitation process via calcium is still unknown. It has opened a way for acid stress response research that is either calcium suitable or not (Tiwari et al. 1996a). Similarly, glutathione of *R. tropici* showed involvement in acid tolerance, but the mechanism is unknown (Riccillo et al. 2000). Kiss et al. (2004) have reported the requirement of TypA for the growth of rhizobia at low pH. It is hypothesized that TypA acts as a regulator by governing the phosphorylation of proteins. Reeve et al. (1998) have proposed that acid shock induces the pH-regulated repressor (PhrR) protein. Reports revealed that exopolysaccharide (EPS) rhizobia also have a main protective role against acid response (Cunningham and Munns 1984). Under acidity, increased EPS biosynthesis has been observed in *S. meliloti*, but the glucose-6-P and phosphoglucomutase are consistent at this stage. Hellweg et al. (2009) have also reported the induction of EPS-biosynthetic genes followed by short exposure of low pH to rhizobia.

At the nutrient depletion, rhizobia undergo the stationary phase that might also protect the cells from acid stress (Thorne and Williams 1997). Moreover, acidic stress (pH < 5.0) can also lead to metal stress, and it is estimated to be approx. 30–40% of world agriculture lands (Lilienfein et al. 2003). Some metals like aluminum exist in soil in different forms. In acidic environment, aluminum is found as Al^{3+} by solubilizing into $[Al (H2O)]^{3+}$, and this stage is very toxic to several plant species (Eva et al. 2004).

However, some compounds such as phosphate (P) can aid in the survival of bacteria at low pH by acting as an internal buffer and maintained the pH by increasing the proton concentration. Booth IR (1985) has speculated that potassium (K) played an important role in pH homeostasis in *E. coli* and *Vibrio alginolyticus* via potassium-proton antiport. Higher concentration of K at internal levels might help in the survival of nodule bacteria by maintenance of pH over anion exchange. For example, acid-tolerant *S. meliloti* is able to maintain the pH at acidic conditions (O'Hara et al. 1989), whereas salt-sensitive strain is not. In wild-type strains, acid tolerance capability has been restored by adding glutathione to the nutrient medium. It was proven in *R. tropici* strain CIAT899 acid-sensitive mutant as it was unable to synthesize glutathione (Riccillo et al. 2000). Watkin et al. (2003) have reported the ability of acid tolerance in potassium- and phosphorus-accumulating *R. leguminosarum* bv. *trifolii*. Whole-genome microarray analysis of acid-treated *S. meliloti* 1021 was done and revealed that it started to respond by either up- or downregulating the specific genes or hypothetical gene (Hellweg et al. 2009).

4.3 Response to Alkali Stress in Rhizobia

Efficient rhizobial cell growth and symbiotic performance (between rhizobia and plants) are very vital factors which are directly affected by soil pH. High soil pH (>8.0) disrupts the rhizosphere community structure and root physiology which can even damage the root membrane structure. High pH is known as alkaline stress for rhizobia and disturbs the mineral ion homeostasis of rhizosphere as well as their microbial natives. Hence, alkaline stress effects tend to be far pronounced than ion poisoning or osmotic stresses. Several reports indicated that plants have markedly lower tolerance to alkali stress than salinity (Girvin 2010; Wang et al. 2012). Alkali stress can also avert the rhizobial growth and nodulation. However, *R. leguminosarum* bv. *trifolii* is reported to colonize the legume root at higher pH rate (pH 11.5) and nodulate at higher frequency. Singh et al. (2016) have reported the efficient nodulation in chickpea by *Mesorhizobium ciceri* in high alkaline condition in North-West Indo-Gangetic Plain of India. In alkaline condition, a polyamine named "Homospermidine" presents in high concentrations in nodule bacteria. Fujihara and Yoneyama (1993) reported the higher accumulation of Homospermidine in *B. japonicum* in alkaline stress, but its mechanism and function are unclear.

4.4   Response to Salt and Alkali Stress in Rhizobia

Salinity is a potential worldwide problem which is estimated to affect about 40% of the world's land of Mediterranean tropics and regions (Zahran 1999; Bouhmouch et al. 2005). Salinity in soil might be occurred by neutral or alkaline salts such as $NaCl$ and Na_2SO_4 or $NaHCO_3$ and Na_2CO_3, respectively. Excess presence of these salts forms saline and alkali stress, and resultantly it harms the crops (Yang et al. 2008). Salinity effects the overall plants as reduced growth and weak plant and fruiting bodies. Ogutcu et al. (2008) reported that chickpea is very sensitive to salinity and evident in results as declined root and shoot dry mass. Several researchers also investigated that roots are more sensitive to salinity stress than shoots in legumes such as chickpea and faba bean (Rao and Sharma 1995; Cordovilla et al. 1999; Soussi et al. 2001). Salinity also effects root exudate secretion that has altered the chemotaxis ratios of symbiotic rhizobia and is resulting in poor nodulation (Islam and Ghoulam 1981). Rhizobia are the most important group soil bacteria and fix the atmospheric nitrogen N_2 in legume plant root nodules via nodulation process (Singh et al. 2016; Subhashini et al. 2017). But, rhizobial species are sensitive to various environmental factors in which alkalinity and salinity are also accountable factor for *Rhizobium* growth and survival (Brígido et al. 2012). Soil salinity might be affecting the survival and proliferation of *Rhizobium* sp. in the soil as well as in the rhizosphere by decreasing the rhizobial cell number, infection mechanism hindrances, disturbing the nodule function, plant growth reduction, photosynthesis breakdown, and reducing the nitrogen (N) demand (Singleton et al. 1982). Resultantly, salinity limits the symbiosis mechanism of rhizobia. Poor or failure of nodulation also attributed to root hair damages (Tu 1981). Soussi et al. (1998) have found that nodulation of chickpea was declined even at very low salt concentration (NaCl 50 mM); however, dry weight of plant was affected only by 100 mM of NaCl. Salt stress can negatively influence the rhizobial cell growth, and it is varying within the strains and species. In general, fast-growing rhizobia are more salt-tolerant than slow-grower (Zou et al. 1995a, b). For example, *Mesorhizobium* strains are more sensitive to salt stress than *Rhizobium* and *Sinorhizobium* strains but have more potentiality in tolerance to salt than *Bradyrhizobium* strains (Brígido et al. 2012). Hence, it is a vital point to focus for the researcher during the isolation of salt-tolerant rhizobia. Tolerance of *Mesorhizobium* strains to NaCl was tested by Laranjo and Oliveira (2011) and displayed the significant exhibition of cell growth. Yelton et al. (1983) have revealed that *R. japonicum* strain USDA191 is more tolerant to NaCl (0.4 M) than the strain USDA110.

Rhizobia adapted the precise approaches to cope with salinity stress. It includes the accumulation of inorganic cations such as sodium, potassium, etc. and production of low-molecular-weight organic solutes intracellularly, such as betaine, ectoines, glycine, proline, polyamines, and trehalose. These compounds stabilize the protein conformation and biological membranes to protect the cell from desiccation and osmotic stress (Santos and da Costa 2002, Wdowiak-Wróbe et al. 2013). Moreover, homologous compounds of betaines, such as glycine betaine, proline betaine, and carnitine, also majorly played the role in osmoregulation in rhizobial cell under salt stress, for example, *Rhizobium* sp., *R. meliloti*, *Mesorhizobium* sp., and *S. meliloti* (Bernard et al. 1986; Lunn et al. 2014). In the case of *S. meliloti*, trehalose played as an imperative osmoprotectant (Domínguez-Ferreras et al. 2009), and in *Mesorhizobium* sp. (USDA3350 and ACMP18), some exogenous osmoprotective compounds are involved in the regulation of osmotic balance (Wdowiak-Wróbe et al. 2013). It has been observed in a previous report that plant growth-promoting trait (PGP) IAA also helps in the protection of bacterial (rhizobial) cells against salinity and oxidative stresses (Bianco et al. 2006; Singh et al. 2019; Prajakta et al. 2019). He demonstrated that salt-tolerant strain RJS9-2 produced higher concentration of IAA, while in salt-sensitive strains, it was low. It might be indicating that salt tolerance in strain RJS9-2 might be due to increased IAA production.

A common model for rhizobial response upon salt stress has been derived by several studies (Bernard et al. 1986; Santos et al. 2000; Wdowiak-Wróbel et al. 2013). It is illustrated that in the subsequent upshift of osmosis, the metabolism of rhizobial cell slows down. The genes participated in to tricarboxylic acid cycle (TCA), carbon source uptake, respiratory chains and ribosomal genes were found repressed. Though, NaCl encoded ribosomal proteins downregulates all the genes (25%). During the stress conditions, rhizobia accumulate potassium ions, a biochemically regulated mechanism. For the restoration of cell growth after stress shock, rhizobia can accumulate the carbon sources in glycogen forms. This statement is endorsed by the study of glycogen metabolism regulated genes (glgA2, glgB2, and glgX) of *S. meliloti* SMb20704, SMb21447, and SMb21446. These genes were expressed highly during salt stress and might indicate that glycogen accumulates during high salinity. Stressed rhizobial cells accumulate compatible solutes, and uptake is favored over synthesis. Finally, the stressed cells alter the long-chain macromolecules like exopolysaccharides (EPS) (Cheng and Walker 1998; Lloret et al. 1998), lipopolysaccharides (LPS), etc. (Lloret et al. 1995; Bhattacharya and Das 2003).

Ruberg et al. (2003) have done the *S. meliloti* gene expression analysis under NaCl-mediated osmotic upshift, and it was revealed the declined expression of flagellum genes (*flaA*, *flaB*, *flaC*, *flaD*) and chemotaxis genes (*mcpZ*, *mcpX*, *cheY1*, *cheW3*) that can shut down the flagella synthesis under adverse conditions to save energy for endurance. Moreover, genes involved in the biosynthesis of cysteine (cysK2), proline (smc03253), serine (serA, serC), thiamine (thiC, thiE, thiG), and others related to iron uptake were repressed. Genes involved in the transportation of small molecules (amines, amino acids, and peptides), anions, and alcohols as well as genes for surface polysaccharide biosynthesis and regulation were induced under the osmotic upshift. Furthermore, the above discussion elucidated the fundamental mechanisms of rhizobia in response to soil salinity and is vital to improving the *Rhizobium* strains with a highly efficient symbiotic ability.

4.5   Response to Temperature Stress in Rhizobia

Alteration of temperature or osmolarity variations can directly affect the rhizosphere and their best saprophytic competent, rhizobia. Rhizobia are influenced for not only their own growth but also their competitiveness and persistence (Dowling and Broughton 1986). Temperature is a predominant factor for rhizobia because they are often exposed to temperatures in both saprophytic and symbiotic life. The optimum temperature for most of rhizobia is 25–30 °C, and many are unable to grow beyond 37 °C (Zhang et al. 1995; Zahran 1999). Temperature stress affects their abundance in the rhizosphere, root hair infection, bacteroid differentiation, nodule shape, and efficiency of N_2 fixation. A review of Hungria and Vargas (2000) has exemplified that among grain legumes such as peanut (*Arachis hypogaea*), soybean of tropical conditions has no nodulation observed at 40 °C.

Oppositely, in case of *Leucaena* trees nodules were observed, the isolated *Rhizobium* strains were able to grow well at 42 °C (Hashem et al. 1998). Though the above processes typically work well over a range of ~5 °C (Zahran 1999), these temperature fluctuations make differences between crops and visibly rely on the rhizobial occupancy. Generally, temperature stress is separated into two classes: 1) heat shock and 2) cold shock.

Heat shock in rhizobia is governed by proteins encoded for heat shock proteins (HSPs). HSPs contribute to heat tolerance by protecting the bacteria cell from heat protection but didn't disturb the internal cells' temperature (Yura et al. 2000). HSPs are abundant in the cells as chaperones and proteases. HSPs' gene structure, function, and regulation are studied greatly and appeared highly conserved among prokaryotes and eukaryotes (Netzer and Hartl 1998). Münchbach et al. (1999) have highlighted that HSPs are also vital under normal growth (non-heat shock) conditions. Generally, bacteria have a low number of HSPs, but in the case of *Rhizobium*, it is an exception (Michiels et al. 1994). For example, *R. leguminosarum* contains at least three copies of the HSP gene (cpn60) that encodes for GroEL or cpn60 protein (Wallington and Lund 1994). The GroEL protein interacts with GroES (encoded by cpn10 gene), and a copy of a GroES-encoded gene is upstream of at least two of the cpn60 genes (HSP gene). Studies reported that a superfamily of six small HSPs has been found to be abundant in *Rhizobium* in which one is essential for symbiosis (Münchbach et al. 1999; Natera et al. 2000). However, it is still unclear why rhizobia comprise so many HSPs. It might be due to the need of immediate response in times of heat stress as well as minimizing the damage caused by it. The genome structure of *Rhizobium* also revealed the several copies of genes that can be homologues or paralogue that revealed the higher redundancy in their genetic system.

Cold stress response in rhizobia is led by the several cold shock proteins, controlled by CSP gene. Same as HSPs, CSPs also are mainly chaperones and proteases. Alexandre and Oliveira (2013) have no change observed in the expression dnaK and groESL (major chaperone genes) after a cold shock in mesorhizobia. Similarly, multiple groEL mutants of *S. meliloti* were unaffected under low temperatures (Bittner et al. 2007). During cold shock, bacteria have to resist itself by loss of membrane and fluidity of cytosol as well as stabilizing the secondary structures of nucleic acid (DAN/RNA) (Phadtare et al. 2000). Stabilization of DNA/RNA decreases the efficiency of replication, transcription, and translation. Response to cold shock is a rapid transient response of bacterial cell to the temperature downshift that ensures the adaptation strategies of bacterial cell to grow at low temperatures (Panoff et al. 1997). Though fluidity loss in bacteria is due to the increasing unsaturated fatty acid amount in the membrane phospholipids (Phadtare et al. 2000), majorly, an RNA chaperone CspA is found in many bacteria (Jiang et al. 1997). At downshift of 30 to 15 °C, CspA homologue, along with the three rRNA operons, is induced in *S. meliloti*. However, the function of CspA genes and products of the rRNA operons is unknown because after cold shock response, mutations made in the above-described genes displayed no alteration in cell phenotype at 15 °C when compared to wild-type strain (Gustafson et al. 2002).

Graham (1992) has reviewed that low temperatures have negatively affected nodulation. Zhang et al. (1995) have revealed that low temperature shift (25 to 15 °C) influenced the root zone infection, early nodulation, and N2 fixation by delaying or shutting down the temperature shift in soybean rhizobia. Duzan et al. (2006) reported the Nod factor production declination at 17 °C or 15 °C in *B. japonicum*. However, sustainability of legume-rhizobia nodulation and symbiosis varied from legume to legume. The symbiotic ability of *B. japonicum* isolates of the Northern USA is to be able to survive and nodulate the legumes at low temperatures, and hence it is used as a model strain in cold conditions (Zhang et al. 2003).

4.6 Conclusion

Harsh conditions essentially affect the rhizobial ecology of their interaction with host plants. Stresses can seize the growth, survival, and performance of *Rhizobium* in the soil. Moreover, they hinder the molecular signaling, attachment, infection, nodulation, and N_2 function which results in poor legume-*Rhizobium* symbiosis and which directly affects the crop yield. Selection of N_2-fixing endosymbionts (rhizobia) with capability of tolerance to range of harsh environment is important for global agriculture system. However, these stresses severely affect legumes for symbiotic gene expression and protein synthesis with rhizobia. But, the era of omics and technologies of microbial biotechnologies is now enabling the researchers to isolate the stress-tolerant potential rhizobia for efficient symbiosis and interaction with leguminous plant hosts. Moreover, the genetic intervention for adaptation mechanisms should be an important component for the future studies. Additional investigation of the described stress is urgently needed to attend deeply to recognize the rhizobia-legume interactions for the protection of *Rhizobium* ecology under harsh soil conditions.

References

Abdelmoumen H, Filaling MA, Neyra M, Belabed A, El Idrissi MM (1999) Effects of high salts concentrations on the growth of rhizobia and responses to added osmotica. J Appl Microbiol 86:889–898

Alexandre A, Oliveira S (2013) Response to temperature stress in rhizobia. Crit Rev Microbiol 39:219–228

Athar M, Johnson DA (1997) Effects of drought on the growth and survival of *Rhizobium meliloti* strains from Pakistan and Nepal. J Arid Environ 35:335–340

Bernard T, Pocard JA, Perround B, Le Rudulier D (1986) Variations in the response of salt-stressed *Rhizobium* strains to betaines. Arch Microbiol 143:359–364

Bhattacharya I, Das HR (2003) Cell surface characteristics of two halotolerant strains of *Sinorhizobium meliloti*. Microbiol Res 158:187–194

Bianco C, Imperlini E, Calogero R, Senatore B, Amoresano A et al (2006) Indole-3-acetic acid improves *Escherichia coli's* defences to stress. Arch Microbiol 185:373–382

Biswas S, Das RH, Sharma GL (2008) Isolation and characterization of a novel cross-infective rhizobial from *Sesbania aculeata* (Dhaincha). Curr Microbiol 56:48–54

Bittner AN, Foltz A, Oke V (2007) Only one of five groEL genes is required for viability and successful symbiosis in *Sinorhizobium meliloti*. J Bacteriol 189:1884–1889

Booth IR (1985) Regulation of cytoplasmic pH in bacteria. Microbiol Rev 49:359–378

Bouhmouch I, Souad-Mouhsine B, Brhada F (2005) Influence of host cultivars and Rhizobium species on the growth and symbiotic performance of *Phaseolus vulgaris* under salt stress. J Plant Physiol 162:1103–1113

Brígido C, Alexandre A, Oliveira S (2012) Transcriptional analysis of major chaperone genes in salt-tolerant and salt-sensitive mesorhizobia. Microbiol Res 167:623–629

Brockwell J, Bottomly PJ, Thies JA (1995) Manipulation of rhizobia microflora for improving legume productivity and soil fertility. A critical assessment. Plant Soil 174:143–180

Chen H, Richardson AE, Rolfe BG (1993) Studies of the physiology and genetic basis of acid tolerance in *Rhizobium leguminosarum* biovar *trifolii*. Appl Environ Microbiol 59:1798–1804

Chen WM, Lee TM, Lan CC, Cheng CP (2000) Characterization of halotolerant rhizobia isolated from root nodules of *Canavalia rosea* from seaside areas. FEMS Microbiol Ecol 34:9–16

Chen LS, Figueredo A, Pedrosa FO, Hungria M (2002) Genetic characterization of soybean rhizobia in Paraguay. Appl Environ Microbiol 66:5099–5103

Cheng HP, Walker GC (1998) Succinoglycan production by *Rhizobium meliloti* is regulated through the ExoS-ChvI two-component regulatory system. J Bacteriol 180:20–26

Cordovilla MP, Ligero F, Lluch C (1999) Effect of salinity on growth, nodulation and nitrogen assimilation in nodules of faba bean (*Vicia faba* L.). Appl. Soil Ecol 11:1–7

Cunningham SD, Munns DN (1984) The correlation of the exopolysaccharide production and acid-tolerance in *Rhizobium*. Soil Sci Soc Am J 48:1273–1276

Domínguez-Ferreras A, Soto MJ, Pérez-Arnedo R, Olivares J, Sanjuán J (2009) Importance of tre-halose biosynthesis for *Sinorhizobium meliloti* osmotolerance and nodulation of alfalfa roots. J Bacteriol 191:7490–7499

Dowling DN, Broughton WJ (1986) Competition for nodulation of legumes. Annu Rev Microbiol 40:131–157

Duzan HM, Mabood F, Souleimanov A, Smith DL (2006) Nod Bj-V (C18:1, MeFuc) production by *Bradyrhizobium japonicum* (USDA110, 532C) at suboptimal growth temperatures. J Plant Physiol 163:107–111

Eva D, Helga A, Eva SB, Jozesf F, Fodor FB et al (2004) Aluminum toxicity, Al tolerance and oxi-dative stress in an Al- sensitive wheat genotype and in Al-tolerant lines developed by in-vitro microspore selection. Plant Sci 166:583–591

Fenner BJ, Tiwari RP, Reeve WG, Dilworth MJ, Glenn AR (2004) *Sinorhizobium medicae* genes whose regulation involves the ActS and/or ActR signal transduction proteins. FEMS Microbiol Lett 236:21–31. https://doi.org/10.1016/j.femsle.2004.05.016

Foster JW (2000) Microbial responses to acid stress. In: Storz G, Hengge-Aronis R (eds) Bacterial stress response. ASM Press, Washington, DC, pp 9–115

Frank B (1889) Ueber die Pilzsymbiose der Leguminosen. Ber Dtsch Bot Ges 7:332–346

Fujihara S, Yoneyama T (1993) Effects of pH and osmotic stress on cellular polyamine contents in the soybean *Rhizobia fredii* P220 and *Bradyrhizobium japonicum* A1017. Appl Environ Microbiol 59:1104–1109

Girvin R (2010) Effects of saline and alkaline stress on germination, seedling growth, and ion bal-ance in wheat. Agron J 102:1252–1260. https://doi.org/10.2134/agronj2010.0022

Glenn AR, Dilworth MJ (1994) The life of root nodule Bacteria in the acidic underground. FEMS Microbiol Lett 123:1–10

Glenn AR, Reeve WG, Tiwari RP, Dilworth MJ (1999) Acid tolerance in root nodule bacteria. Novartis Found Symp 221:112–126

Graham PH (1992) Stress tolerance in *Rhizobium* and *Bradyrhizobium*, and nodulation under adverse soil condition. Can J Microbiol 38:475–484

Gustafson AM, O'Connell KP, Thomashow MF (2002) Regulation of *Sinorhizobium meliloti* 1021 rrnA-reporter gene fusions in response to cold shock. Can J Microbiol 48:821–830

Hashem FM, Swelim DM, Kuykendall LD, Mohamed AI, Abdel-Wahab SM et al (1998) Identification and characterization of salt- and thermo-tolerant Leucaena-nodulating *Rhizobium* strains. Biol Fertil Soil 27:335–341

Hawkins JP, Geddes BA, Oresnik IJ (2017) Succinoglycan production contributes to acidic pH tolerance in *Sinorhizobium meliloti* Rm1021. Mol Plant-Microbe Interact 30(12):1009–1019. https://doi.org/10.1094/MPMI-07-17-0176-R

Hellweg C, Pühler A, Weidner S (2009) The time course of the transcriptomic response of *Sinorhizobium meliloti* 1021 following a shift to acidic pH. BMC Microbiol 9:37. https://doi.org/10.1186/1471-2180-9-37

Hinsinger P, Bengough AG, Vetterlein D, Young IM (2009) Rhizosphere: biophysics, biogeochem-istry and ecological relevance. Plant Soil 321:117–152

Hirsch AM (1992) Developmental biology of legume nodulation. New Phytol 122:211–237

Howieson J, Ballard R (2004) Optimising the legume symbiosis in stressful and competitive environments within southern Australia—some contemporary thoughts. Soil Biol Biochem 36:1261–1273

Howieson JG, Ewing MA, D'Antuono MF (1988) Selection for acid tolerance in Rhizobium meliloti. Plant Soil 105(2):179–188

Howieson JG, Robson AD, Abbott LK (1992) Acid-tolerant species of *Medicago* produce root exudates at low pH which induce the expression of nodulation genes in *Rhizobium meliloti*. Aust J Plant Physiol 19:287–296

Hungria M, Vargas MAT (2000) Environmental factors affecting N2 fixation in grain legumes in the tropics, with an emphasis on Brazil. Field Crops Res 65:151–164

Idrissi MME, Aujjar N, Dessaux Y, FilalingMaltouf A (1996) Characterization of rhizobia isolated from Carob tree (*Ceratonia siliqua*). J Appl Biotechnol 80:165–173

Islam R, Ghoulam W (1981) Screening of several strains of faba bean Rhizobium for tolerance to salinity. FABIS-Newslett (ICARDA) 3:34

Jenkins MB (2003) Rhizobial and bradyrhizobial symbionts of mesquite from the Sonoran Desert: salt tolerance, facultative halophily and nitrate respiration. Soil Biol Biochem 35:1675–1682

Ji SH, Gururani MA, Chun SC (2014) Isolation and characterization of plant growth promoting endophytic diazotrophic bacteria from Korean rice cultivars. Microbiol Res 169(1):83–98

Jiang W, Hou Y, Inouye M (1997) CspA, the major cold-shock protein of Escherichia coli, is an RNA chaperone. J Bacteriol 272:196–202

Karanja NK, Wood M (1988) Selecting Rhizobium phaseoli strains for use with beans (*Phaseolus vulgaris* L.) in Kenya: tolerance of high temperature and antibiotic resistance. Plant Soil 112:115–122

Kiss E, Huguet T, Poinsot V, Batut J (2004) The typA is required for stress adaptation as well as for Symbiosis of *Sinorhizobium meliloti* 1021 with certain *Medicago truncatula* lines. Mol Plant-Microbe Interact 17:235–244

Kong Z, Mohamad OA, Deng Z, Liu X, Glick BR et al (2015) Rhizobial symbiosis effect on the growth, metal uptake, and antioxidant responses of *Medicago lupulina* under copper stress. Environ Sci Pollut Res 22:12479–12489

Kosanke JW, Osburn RM, Smith RS, LiphaTech Inc (1999) Process for preparation of bacterial agricultural products. Canadian patent 2,073:507

Kulkarni S, Nautiyal CS (2000) Crossing the limits of *Rhizobium* existence in extreme conditions. Curr Microbiol 41:402–409

Kurchak ON, Provorov NA, Simarov BV (2001) Plasmid pSym1-32 of *Rhizobium leguminosarum* bv. *viceae* controlling nitrogen fixation activity, effectiveness of Symbiosis, competitiveness and acid tolerance. Russian J Genet 37:1025–1031

Laranjo M, Oliveira S (2011) Tolerance of Mesorhizobium type strains to different environmental stresses. Anton Leeuw 99:651–662

Li Z, Zu C, Wang C, Yang J, Yu H et al (2016) Different responses of rhizosphere and non-rhizosphere soil microbial communities to consecutive *Piper nigrum* L. monoculture. Sci Rep 6(1):35825. https://doi.org/10.1038/srep35825

Lilienfein J, Qualls RG, Uselman SM, Bridgham SD (2003) Soil formation and organic matter accretion in a young andesitic chronosequence at Mt. Shasta, California. Geoderma 116:249–264

Lloret J, Bolanos L, Mercedes LM, Peart JM, Brewin NJ et al (1995) Ionic stress and osmotic pressure induce different alterations in the lipopolysaccharide of a *Rhizobium meliloti* strain. Appl Environ Microbiol 61:3701–3704

Lloret J, Wulff B, Rubio JM, Downie JA, Bonilla I et al (1998) Exopolysaccharide II production is regulated by salt in the halotolerant strain *Rhizobium meliloti* EFB1. Appl Environ Microbiol 64:1024–1028

Lowendorf HS, Baya AM, Alexander M (1981) Survival of *Rhizobium* in acid soils. Appl Environ Microbiol 42:951–957

Lunn JE, Delorge I, Figueroa CM, Van Dijck P, Stitt M (2014) Trehalose metabolism in plants. Plant J 79:544–567

Masciarelli O, Llanes A, Luna V (2014) A new PGPR co-inoculated with *Bradyrhizobium japonicum* enhances soybean nodulation. Microbiol Res 169:609–615

Michiels J, Verreth C, Vanderleyden J (1994) Effects of temperature stress on bean-nodulating *Rhizobium* strains. Appl Environ Microbiol 60:1206–1212

Münchbach M, Nocker A, Narberhaus F (1999) Multiple small heat shock proteins in rhizobia. J Bacteriol 181:83–90

Natera SHA, Guerreiro N, Djordjevic MA (2000) Proteome analysis of differentially displayed proteins as a tool for the investigation of Symbiosis. Mol Plant Microbe Interact 13:995–1009

Netzer WJ, Hartl FU (1998) Protein folding in the cytosol: chaperonin- Dependent and independent mechanisms. Trends Biochem Sci 23:68–73

O'Hara GW, Goss TJ, Dilworth MJ, Glenn AR (1989) Maintenance of intracellular pH and acid tolerance in *Rhizobium meliloti*. Appl Environ Microbiol 55:1870–1876

Ogutcu H, Algur ÖF, Elkoca E, Kantar F (2008) The determination of symbiotic effectiveness of rhizobium strains isolated from wild chickpea collected from high altitudes in Erzurum. Turk J Agric For 32:241–248

OsangAlfiana LO, Alexander M (1982) Differences among cowpea *Rhizobium* in tolerance to high temperature and desiccation in soil. Appl Environ Microbiol 43:435–439

Panoff JM, Corroler D, Thammavongs B, Boutibonnes P (1997) Differentiation between cold shock proteins and cold acclimation proteins in a mesophilic gram-positive bacterium, *Enterococcus faecalis* JH2-2. J Bacteriol 179:4451–4454

Peoples MB (1995) Biological nitrogen fixation: an efficient source of nitrogen for sustainable agriculture production. Plant Soil 174:3–28

Phadtare S, Yamanaka K, Inouye M (2000) The cold shock response. In: Storz G, Hengge-Aronis R (eds) Bacterial stress response. ASM Press, Washington, DC, pp 33–45

Prajakta BM, Suvarna PP, Singh RP, Rai AR (2019) Potential biocontrol and superlative plant growth promoting activity of indigenous *Bacillus mojavensis* PB-35(R11) of soybean (*Glycine max*) rhizosphere. SN Appl Sci 1:1143. https://doi.org/10.1007/s42452-019-1149-1

Rao DLN, Sharma PC (1995) Alleviation of salinity stress in chickpea by *Rhizobium* inoculation or nitrate supply. Biol Plant 37:405–410

Reeve WG, Tiwari RP, Wong CM, Dilworth MJ, Glenn AR (1998) The transcriptional regulator gene phrR in Sinorhizobium meliloti WSM419 is regulated by low pH and other stresses. Microbiology 144:3335–3342

Reeve WG, Bräu L, Castelli J, Garau G, Sohlenkamp C et al (2006) The *Sinorhizobium medicae* WSM419 lpiA gene is transcriptionally activated by FsrR and required to enhance survival in lethal acid conditions. Microbiology 152:3049–3059

Rehman A, Nautiyal CS (2002) Effect of drought on the growth and survival of the stress-tolerant bacterium *Rhizobium* sp. NBRI2505 sesbania and its drought-sensitive transposon Tn5 mutant. Curr Microbiol 45(5):368–377

Rhijn PV, Vanderleyden J (1995) The rhizobium -plant Symbiosis. Microbiol Rev 59(1):124–142

Riccillo PM, Muglia CJ, de Bruijn FJ, Roe AJ, Booth IR (2000) Glutathione is involved in environmental stress responses in *Rhizobium tropici*, including acid tolerance. J Bacteriol 182:1748–1753

Ruberg S, Tian ZX, Krol E, Linke B, Meyer F (2003) Construction and validation of a *Sinorhizobium meliloti* whole genome DNA microarray: genome-wide profiling of osmoadaptive gene expression. J Biotechnol 106:255–268

Salema MP, Parker CA, Kirby DK, Chatel DL (1982) Death of rhizobia on inoculated seed. Soil Biol Biochem 14:13–14

Sanja K, Hulak N, Sikora S (2016) Environmental stress response and adaptation mechanisms in rhizobia. Agric Conspec Sci 81(1):15–19

Santos H, da Costa MS (2002) Compatible solutes of organisms that live in hot saline environments. Environ Microbiol 20024:501–509

Santos R, Herouart D, Puppo A, Touati D (2000) Critical protective role of bacterial superoxide dismutase in rhizobium-legume symbiosis. Mol Microbiol 38:750–759

Shamseldin A, Werner D (2005) High salt and high pH tolerance of new isolated Rhizobium etli strains from Egyptian soils. Curr Microbiol 50(1):11–16

Shamseldin A, Nyalwidhe J, Werner DA (2006) Proteomic approach towards the analysis of salt tolerance in *Rhizobium etli* and *Sinorhizobium meliloti* strains. Curr Microbiol 52:333–339

Singh RP, Singh RN, Srivastava AK, Kumar S, Dubey RC et al (2011) Structural analysis and 3D-modelling of fur protein from *Bradyrhizobium japonicum*. J Appl Sci Environ Sanit 6(3):357–366

Singh RP, Manchanda G, Singh RN, Srivastava AK, Dubey RC (2016) Selection of alkalotolerant and symbiotically efficient chickpea nodulating rhizobia from North-West Indo Gangetic Plains. J Basic Microbiol 56:14–25. https://doi.org/10.1002/jobm.201500267

Singh RP, Manchanda G, Anwar MN, Zhang JJ, Li YZ (2017a) Mycorrhiza – helping plants to navigate environmental stresses. In: Kashyap PL, Srivastava AK, Tiwari SP, Kumar S (eds) Microbes for climate resilient agriculture. https://doi.org/10.1002/9781119276050.ch10

Singh RP, Manchanda G, Li ZF, Rai AR (2017b) Insight of proteomics and genomics in environmental bioremediation. In: Bhakta JN (ed) Handbook of research on inventive bioremediation techniques. IGI Global, Hershey. https://doi.org/10.4018/978-1-5225-2325-3

Singh RP, Manchanda G, Maurya IK, Maheshwari NK, Tiwari PK et al (2019) *Streptomyces* from rotten wheat straw endowed the high plant growth potential traits and agro-active compounds. Biocatal Agric Biotechnol 17:507–513. https://doi.org/10.1016/j.bcab.2019.01.014

Singh RP, Manchanda G, Yang Y, Singh D, Srivastava AK et al (2020) Deciphering the factors for nodulation and Symbiosis of Mesorhizobium associated with *Cicer arietinum* in Northwest India. Sustainability 12:1–17

Singleton PW, Elswaify SA, Bohlool BB (1982) Effect of salinity on rhizobium growth and survival. Appl Environ Microbiol 44:884–890

Smith RS (1992) Legume inoculant formulation and application. Can J Microbiol 38:485–492

Soussi M, Ocana A, Lluch C (1998) Effects of salt stress on growth, photosynthesis and nitrogen fixation in chick-pea (Cicer arietinum L.). J Exp Bot 49:1329–1337

Soussi M, Khadri M, Lluch C, Ocana A (2001) Carbon metabolism and bacteroid respiration in nodules of chick-pea (*Cicer arietinum* L.) plants grown under saline conditions. Plant Biosyst 135:157–164

Subhashini DV, Singh RP, Manchanda G (2017) OMICS approaches: tools to unravel microbial systems. Directorate of Knowledge Management in Agriculture, Indian Council of Agricultural Research. ISBN: 9788171641703. https://books.google.co.in/books?id=vSaLtAEACAAJ

Surange S, Wollum AG, Kumar N, Nautiyal CS (1997) Characterisation of *Rhizobium* from root nodules of leguminous trees growing in alkaline soils. Can J Microbiol 43:891–894

Thorne SH, Williams HD (1997) Adaptation to nutrient starvation in *Rhizobium leguminosarum* bv. *phaseoli*: analysis of survival, stress resistance and changes in macromolecular synthesis during entry to and exit from stationary phase. J Bacteriol 179:6894–6901

Tiwari RP, Reeve WG, Dilworth MJ, Glenn AR (1996a) An essential role for actA in acid tolerance of *Rhizobium meliloti*. Microbiology 142:601–610

Tiwari RP, Reeve WG, Dilworth MJ, Glenn AR (1996b) Acid tolerance in *Rhizobium meliloti* strain WSM419 involves a two- component sensor-regulator system. Microbiology 142:1693–1704

Trotman AP, Weaver RW (1995) Tolerance of clover rhizobia to heat and desiccation stresses in soil. Soil Sci Soc Am J 59:466–470

Tu JC (1981) Effect of salinity on rhizobium-root hair interaction, nodulation and growth of soybean. Can J Plant Sci 61:231–239

Vincent JM, Thompson JA, Donovan KO (1962) Death of root- nodule bacteria on drying. Aust J Agric Res 13:258–270

Vinuesa P, Neumann-Silkow F, Pacios-Bras C, Spaink HP, Martínez-Romero E et al (2003) Genetic analysis of a pH-regulated operon from *Rhizobium tropici* CIAT899 involved in acid tolerance and nodulation competitiveness. Mol Plant-Microbe Interact 16:159–168. https://doi.org/10.1094/MPMI.2003.16.2.159

Vivas-Marfisi A, Tiwari R, Dilworth M, Glenn A (2002) In nitrogen fixation: from molecules to crop productivity. In: Pedrosa FO, Hungria MY, Geoffrey N, William E (eds) Current plant science and biotechnology in agriculture. Ch. 272, vol 38. Kluwer Academic, New York, pp 487–487

Vriezen JAC, De Bruijn FJ, Nüsslein K (2007) Responses of rhizobia to desiccation in relation to osmotic stress, oxygen, and temperature. Appl Environ Microbiol 73(11):3451–3459

Wallington EJ, Lund PA (1994) *Rhizobium leguminosarum* Contains Multiple Chaperonin (cpn60) Genes. Microbiology 140:113–122

Wang XP, Chen WC, Ying Z, Han JY, Jing Y et al (2012) Comparison of adaptive strategies of alfalfa (*Medicago sativa* L.) to salt and alkali stresses. Aust J Crop Sci 6:309–315

Watkin ELJ, O'Hara GW, Glenn AR (2003) Physiological responses to acid stress of an acid-soil tolerant and an acid-soil sensitive strain of *Rhizobium leguminosarum* biovar trifolii. Soil Biol Biochem 35:621–624

Wdowiak-Wróbel S, Leszcz A, Małek W (2013) Salt tolerance in Astragalus cicer microsymbionts: the role of glycine betaine in osmoprotection. Curr Microbiol 66:428–436

Yang CW, Wang P, Li CY, Shi DC, Wang DL (2008) Comparison of effects of salt and alkali stresses on the growth and photosynthesis of wheat. Photosynthetica 46:107–114. https://doi.org/10.1007/s11099-008-0018-8

Yelton MM, Yang SS, Edie SA, Lim ST (1983) Characterization of an effective salt-tolerant, fast-growing strain of *Rhizobium japonicum*. Microbiology 129:1537–1547

Yura T, Kanemori M, Morita MT (2000) The heat shock response: regulation and function. In: Storz G, Hengge- Aronis R (eds) Bacterial stress response. ASM Press, Washington, DC, pp 3–18

Zahran HH (1999) Rhizobium-legume symbiosis and nitrogen fixation under severe conditions and in an arid climate. Microbiol Mol Biol Rev 63(4):968–989

Zhang F, Lynch DH, Smith DL (1995) Impact of low root temperatures in soybean [*Glycine max* (L) Merr] on nodulation and nitrogen fixation. Environ Exp Bot 35:279–285

Zhang H, Prithiviraj B, Charles TC, Driscoll BT, Smith DL (2003) Low temperature tolerant *Bradyrhizobium japonicum* strains allowing improved nodulation and nitrogen fixation of soybean in a short season (cool spring) area. Eur J Agron 19:205–213

Zhang JJ, Jing XY, de Lajudie P, Ma C, He PX et al (2016a) Association of white clover (*Trifolium repens* L.) with rhizobia of sv. *trifolii* belonging to three genomic species in alkaline soils in North and East China. Plant Soil 407:417–427

Zhang JJ, Yang X, Chen G, de Lajudie P, Singh RP et al (2016b) *Mesorhizobium muleiense* and *Mesorhizobium* gsp. nov. are symbionts of Cicer arietinum L. in alkaline soils of Gansu, Northwest China. Plant Soil 410:103–112

Zhang JJ, Xu Y, Guo C, de Lajudie P, Singh RP, Wang E, Chen W (2017) *Mesorhizobium muleiense* and *Mesorhizobium* gsp. nov. are symbionts of Cicer arietinum L. in alkaline soils of Gansu, Northwest China. Plant Soil 410(1–2):103–112

Zou N, Dart PJ, Marcar NE, Bushby HVA (1995a) Interaction of salinity and rhizobial strain on growth and nitrogen fixation by *Acacia ampliceps*. Soil Biol Biochem 27:409–413

Zou N, Dart PJ, Marcar NE, Bushby HVA (1995b) Interaction of salinity and rhizobial strain on growth and nitrogen fixation by *Acacia ampliceps*. Soil Biol Biochem 27:4094–4013

Mechanism of Microbial Adaptation and Survival Within Psychrophilic Habitat

5

Xiuling Ji and Yunlin Wei

Abstract

Microorganisms which are able to grow in environments with low temperature (5 °C or below) are known as psychrophiles. They are widely distributed in the world and play significant roles in biological evolution and maintain the balance of the Earth's biosphere. Their habitats include natural frozen soil, ocean, lake, glacier, polar, sea ice, and other artificial environmental areas. The mechanisms of microbial cold adaptation and survival have attracted more attentions in the past decades. Well-known strategies include the accumulation of low molecular mass carbohydrate cryoprotectants, adjusting membrane lipid composition, use of cold-active small molecules or antifreeze proteins that inhibit ice recrystallization, and synthesis of cold-adapted enzymes. Understanding the molecular mechanism and adaptation mechanism of psychrophilic microorganisms will provide new pathway for biological evolution and benefit in applications of psychrophilic microbes in the future.

Keywords

csp gene · Chaperones · Antifreeze proteins · Proteome · Enzymes

5.1 Introduction

Cold-adapted microorganisms (psychrophiles and psychrotrophs) are defined as those with the ability to grow and survive at 15 °C and lower. These microorganisms have the ability to enduringly colonize with the cold environments successfully (Morita 1975; D'Amico et al. 2006). The upgradation in the discovery of

X. Ji (✉) · Y. Wei
Faculty of Life Science and Technology, Kunming University of Science and Technology, Kunming, People's Republic of China

© Springer Nature Singapore Pte Ltd. 2020
R. P. Singh et al. (eds.), *Microbial Versatility in Varied Environments*,
https://doi.org/10.1007/978-981-15-3028-9_5

cold-tolerant microorganisms has extended the known range of environmental circumstances which support the life of microorganisms. However, despite their wide distribution, little courtesy had been paid to these microorganisms which is mainly due to their extreme inhabited environment. But now, the important roles of these organisms and their products are recognized in the food industry, laundry business, chemical industry, agriculture, and even in astrobiology (Rampelotto 2010; Feller 2013; Subhashini et al. 2017). To adapt and survive in cold conditions, psychrophilic microorganisms developed special mechanisms, such as antifreeze proteins or cold shock proteins, cold-active enzymes, increased membrane flexibility, antibiofilm molecules, surfactants, glycoproteins, polysaccharides, production of outer membrane vesicles, secondary metabolic pathway, and cell envelope modification (Santiago et al. 2016; Tribelliand and López 2018; Anwar et al. 2019).

Low-temperature microbes have formed a series of low-temperature adaptation mechanisms in the long-term biological evolution process, mainly in the fine composition and structural changes at the level of cell membrane, protein, and enzyme molecules, thus making up for the adverse effects of low temperature on growth. Gottesman et al. have reviewed the bacterial adaptation mechanism at the declined temperature ranges from 37 °C to 10 °C, and proposed that a set of cold shock proteins increased while the synthesis and cell growth of other proteins stopped gradually.

Moreover, the adaptation of microorganisms under low temperatures should be considered for psychrophile definition (Feller and Gerday 2003, Cavicchioli 2016), but different features related to life in cold conditions have been described in a different way. Hence, the mechanism and strategies of psychrophilic bacteria and their novel features for a cold lifestyle were analyzed in the chapter.

5.2   Cold-Adapted Bacteria Strains, Plasmid, Transformation Methods, and Efficiency Promoters Used in Cold-Adapted Bacteria

5.2.1   Cold-Adapted Bacteria Strains

The surface temperature experienced in most of the world yearly was estimated to be 14 °C (57 °F) (NASA 2010). Meyer et al. (2004) have reported the dramatic temporal changes in microclimatic properties in alpine soils and have proposed that these bacteria encounter uncommon shifting selection gradients. Cold-tolerant strains of *Pseudomonas* species were generously and predominantly found in the alpine soil community (Lipson et al. 1999, 2000). Psychrophiles were rich in Arctic sediments in polar water, Antarctic and Arctic sea ice, as well as melt pools on top of Arctic ice floes (Knoblauch et al. 1999). In the northern of the Fram Strait, belonging to the Arctic Ocean, the amount and activities of bacteria were found at 1 °C exceeding those at 22 °C. Psychrophiles also were dominant in early autumn, while the prominent reduced distinctly and chlorophyll lessened in winter surface water of the Weddell Sea belonging to the Southern Ocean. True cold-adapted organisms have also been found even in the temperate sediments. In winter, there is a marked shift to psychrophilic communities. Compared with Arctic isolates, Antarctic cold-adapted isolates were more which had revealed that cold-adapted bacteria have played the role in the global marine environment (Mocali et al. 2017). Many isolated psychrophilic microbes have become a unique branch of cold adaptation, among which representatives are also found in the Antarctic and the Swedish tundra. Proteolytic bacteria (six) that belong to *Pseudoalteromonas* spp. were isolated from Aleutian margin sediments, and other three cold-adapted Antarctic marine bacteria, identified as *Pseudoalteromonas*, *Psychrobacter*, and *Vibrio*, were screened (Xiong et al. 2007). More interestingly, isolates *Rahnella aquatilis* BS1 and *Buttiauxella* HS39 were able to break down lactose at low temperatures (Park et al. 2006). The cold-adapted Gram-negative rod bacterium *Pseudoalteromonas* sp. MB-1 was isolated from mud of the bottom of the Huanghai province which has secreted the cold-adapted cellulase (You and Wang 2005). Methanogenic bacteria and their metabolic products in the Earth's permafrost offer many analogues which could serve to seek probable ecosystems and possible inhabitants of cold alien bodies without oxygen. The cold-adapted *Pseudomonas* sp. B11-1 strain was isolated and identified from Alaskan soil (Choo et al. 1988). Several psychrotrophic isolates obtained from algal-rich Antarctic and Southern Ocean samples formed three unique groups in the *Flavobacteria* class, which are phylogenetically different from other species: *Algoriphagus ratkowskyi* gen. nov., sp. nov. and *Brumimicrobium glaciale* gen. nov., sp. nov. were isolated from sea ice, while *Cryomorphaignava* gen. nov., sp. nov. was isolated from the Southern Ocean (John et al. 2003).

Cold-adapted bacteria have also been reported for the bioremediation of wasted environment. It is very difficult to degrade petroleum hydrocarbon pollutants with physical or chemical methods in cold environments, but indigenous psychrotrophic microbes could degrade contaminants. Two cold-adapted *Rhodococcus* sp., isolated from hydrocarbon-contaminated alpine soils, could degrade large amounts of phenol (Margesin et al. 2005). The feasibility of pollutant removal at low temperature has been displayed, and the biodegradation of phenol by psychrotrophic *Pseudomonas* isolates was verified at 14 °C or 4 °C with 2-undecanone, diethyl sebacate, or 2-decanone as organic phases (Guieysse et al. 2005). Soares et al. (2003) have isolated 3 strains from contaminated soil which can mineralize nonylphenol as the sole carbon source. They were identified as 2 psychrophilic Pseudomonas spp. strains and psychrotrophic Stenotrophomonas sp. strain, respectively. It was the first report of using cold-adapted microorganisms to mineralize nonylphenol.

Not only cold-adapted bacteria have been widely isolated, identified, and used, but also their derived products and enzymes have been studied extensively. *Photobacterium lipolyticum* sp. nov. M37, isolated from the intertidal bank of the Yellow Sea (Korea), showed lipolytic activity and exhibited the maximum lipase activity at 25 °C and could keep lipase activity at 5–25°C with an activation energy of 2.07 kcal/mol. Accordingly, M37 was characterized as a new cold-adapted enzyme (Jung et al. 2008). It showed that the monomer and dimer with different structure isozymes of isocitrate dehydrogenase (IDH; EC 1.1.1.42) were existed in a cold-active *Colwellia psychrerythraea* bacterium (Ewert and Deming 2011). A lysozyme from Antarctic bacterium strain 643 (VAB) in *Vibrio* is currently the only enzyme capable of growing in a family of thermal lysozymes in extremely cold habitats (Adekoya et al. 2006). *Colwellia psychrerythraea* 34H is a model for studying life in permanently low-temperature environments, revealing important functions for carbon and nutrient cycling, bioremediation, secondary metabolite production, and cold-active enzymes (Czajka et al. 2018). Antarctic psychrotroph *Pseudoalteromonas* sp. strain AS-11 was isolated, and secreted psychrophilic alkaline serine protease Apa1 was studied (Dong et al. 2005). The marine, psychrotolerant, rod-shaped, and Gram-negative *Pseudoalteromonas* sp. 22b was isolated from the alimentary tract of krill *Thysanoessa macrura* (Antarctic), which efficiently hydrolyzes lactose at 0–20 °C (Turkiewicz et al. 2003). The strain synthesizes an intracellular cold-adapted β-galactosidase, and the recombinant strain of *Pseudoalteromonas* sp. 22b-galactosidase has been applied at refrigeration.

5.2.2   Storage of Cold-Adapted Bacteria

Various techniques are adapted for the storage of bacteria like subculture, drying, freeze drying, and cryopreservation in liquid nitrogen, but all the methods are suitable for the storage of cold-adapted bacteria in the long term. Freeze drying (lyophilization) is a way of long-time preservation (10–50 years) for the viability and stability of cold-adapted bacteria. In the lyophilization process, the bacterial cells are suspended in a solution that contains protective agents that act as cryoprotectant, such as glycerol, trehalose, or skim milk. It has been reported that the use of skim milk in combination with glycerol is an excellent cryoprotectant for the preservation of bacteria (Cody et al. 2008). Then the water is removed by the sublimation of the frozen sample of bacteria. Berny and Hennebert (1991) reported that the cooling rate and protecting media are two very significant factors that affect the viability and stability of cold-adapted bacteria during lyophilization. The freeze-dried culture is further stored in vacuum-sealed ampoules in a refrigerator. It is considered to be the best way of cryopreservation for long-time preservation of cold-adapted bacteria viability and stability. Psychrotolerant bacteria in low-temperature culture showed better adaptability to cryopreservation and recover quickly.

5.2.3   Plasmid Used in Cold-Adapted Bacteria

Plasmids are used for the construction of expression vectors in cold-adapted bacteria, and they can mainly fall into three classes. First is the construction of low-temperature expression vectors based on cold-adapted bacteria's own plasmids. Plasmids are very important systems for studying biological processes. In addition, these extrachromosomal components may play a key role in the adaptation of bacteria to environmental changes. At present, more and more indigenous plasmids have been isolated and characterized from cold-adapted bacteria, such as plasmid pMtBL isolated from the Antarctic Gram-negative bacterium *P. haloplanktis* TAC125 and pTAUp and pTADw obtained from Gram-negative *Psychrobacter* sp. TA144 (Duilio et al. 2004). Plasmids pTM41, pTM121, pTM164, and pTM172 were isolated from heterotrophic bacteria originating from coastal California marine sediments (Sobecky et al. 1998).

Second, it also has been demonstrated that natural broad-host-range plasmids can be used in cold-adapted bacteria. Plasmids are thought to have wide host ranges and may have a significant effect on the structure and function of microbial community due to their ability to replicate and maintain in distant relatives. As previously described, plasmids pTM41, pTM121, pTM164, and pTM172 have been used in cold-adapted bacteria; in fact, they are broad-host-range plasmids (Sobecky et al. 1998). Plasmid RK2 belongs to the IncP-1 incompatibility group, while RSF1010, R300B, and R117 plasmids belong to the incompatibility group IncQ and have significant replication characteristics in most Gram-negative bacteria (Schmidhauser and Helinski 1985). It was reported that the above three plasmids existed in *Acetobacter xylinum*, *Acinetobacter calcoaceticus*, *Pseudomonas* sp., *Azotobacter vinelandii*, and *Providencia* sp. So, they are potential cloning vectors (Aakvik et al. 2009).

Third, the most applauded expression vectors should be chimeric plasmids which comprise regulation elements, selection marker gene, and replication elements from different plasmids, respectively. Plasmids pJRD215, pDSK509, and pDSK519 were constructed from RSF1010 (Keen et al. 1998). Plasmid pJRD215 includes 2 antibiotic resistance genes, the wide-host-range replicon, RSF1010 mobilization function, phage λ *cos* region, and at least 23 special restriction endonuclease sites. Thus, pJRD215 is smaller than RSF1010-based vectors and includes more available cloning sites and is a wide-host-range cosmid vector. Plasmid pPK415, a derivative of pPK404 with several new polylinker cloning sites, remains the *lac* promoter of pPK404, and the plasmid has been shown to be important for subcloning and maintaining small fragments of DNA in *P. syringae* pv. *glycinea* and other *P. syringae* pathogens. Plasmid pJB3 is the smallest replicon derivative of the wide-host-range plasmid RK2 (Hashimoto-Gotoh and Timmis 1981). The minimal replicon is composed of the origin of vegetative replication (*oriV*) and the *trfA* gene whose protein product plays a priming role in *oriV* iterons. Moreover, pJB3 retains a functional *oriT* leading to be fixed through RK2 mobilization and delivery functions. The plasmid harbors ampicillin resistance and possesses the *lac* promoter and *E. coli* operator. Plasmid pKT240 is derived from the wide-host-range plasmid RSF1010 which is larger (8.7 kb) and contains very few unique restriction sites for cloning, so it is not a good cloning vector. The plasmid includes the *rep* gene, *oriV* gene, and an *oriT*, permitting it to be activated by the mob and RK2 transfer functions on the plasmid. The plasmid harbors ampicillin and kanamycin resistance genes. Plasmids pJB3 and pKT240 are part of two different incompatible groups, so they could coexist in the same bacterial cell. Plasmid pBBR122 is derived from a plasmid which originally was isolated from *Bordetella bronchiseptica* (Cedric et al. 2001). It possesses wide-host-range and is not part of the known incompatible groups. Plasmid pBBR122 was designed to include chloramphenicol and kanamycin resistance genes. The cold-active cloneQ shuttle vector was sequentially inserted to modify the strong rho-independent transcription terminator and promoterless β-galactosidase gene originating from *P. haloplanktis* TAE79 (*Ph*TAE79lacZ), generating the promoter-trap plasmid pPLB. Plasmid pP# was constructed by the pPLB recombinant vectors containing the P# promoter fragments (Hoyoux et al. 2001).

5.2.4 Genetic Transformation Method for Cold-Adapted Bacteria

Several effective bacterial cell transformation methods have been used. These methods include treating cells with $CaCl_2$ or hexammine cobalt chloride. Substances, such as polyethylene glycol (PEG), could also lead to bacterial cell transformation. Electroporation method in cold-adapted bacteria has been described here. The bacterial cells were grown at 15 °C to the early log phase in optimal medium, cooled it (10 min), and then later cells have been harvested (at 4000 rpm and 4 °C). Further, an osmoprotective solution (OPS), consisting of 137 mM sucrose, 1 mM HEPES buffer, pH 7.8, and 10% glycerol, was used to resuspend and wash the cell pellet; the cells can be directly used for electroporation or can be quickly frozen on dry ice-methanol and stored at −80 °C. The cells could remain active for at least 6 months. It offers a simple method to prepare the storage of cold-adapted bacterial cells.

Then, purified plasmid DNA (0.5 μg) is mixed with 100 μl of the prepared cell thoroughly and left on ice (for 1 min). Electroporation is performed in 0.2 mm cuvette in a Gene Pulser 11 apparatus (Bio-Rad Lab., USA): capacitance 25 farad, resistance 200 Ω, field strength 2500 volts. After the electrical pulse, the cells were gently mixed in 1 ml suitable medium. An important problem was that the conjugation phase must take precedence at temperatures lower than 15 °C. The usual binding temperature for mesophilic organisms is 37 °C, which is too high for low-temperature bacteria. At 15 °C, the cells were exposed to good aeration for 2 h and then electroplated on selective agar plates. After incubation at 15 °C for 3 days, the colony number was scored.

5.2.5 Efficiency Promoters Used in Cold-Adapted Bacteria

Duilio et al. (2004) have taken many efforts on constructing a psychrophilic expression system of heterologous protein in cold-adapted bacteria with natural plasmids as cloning vectors. In order to study the likelihood of gaining recombinant proteins in the psychrophilic host cell, the psychrophilic α-amylase from the *P. haloplanktis* TAB23 of Antarctic bacterium was used as an enzyme production model. It showed that the recombinant PhTAC125 cells could not only produce but also secrete cold-active enzyme efficiently. This system provides a possibility for the isolation of constitutive cold-adapted promoters and the construction of low-temperature expression systems of the homologous/heterologous protein at low temperatures. The expression system was the first example on behalf of foreign protein production on account of true cold-adapted replicons. Plasmid pMtBL, the novel cold-adapted replication element, isolated from the Antarctic *P. haloplanktis* TAC125, had a wide host range. Plasmid pMtBL was cloned into a mesophilic construction to obtain a low-temperature expression vector that could promote the production of *P. haloplanktis* A23 α-amylase in a cold-adapted bacterium. Tutino et al. (2002) have for the first time reported on the cold-active protein's successful recombination in an Antarctic psychrophilic host. When a cloning vector is changed to an expression vector, the transcriptional and translation regulatory sequences are inserted, so the relevant aspartate aminotransferase gene signals which were isolated from *P. haloplanktis* TAC125 were inserted to generate the pFF vector. The psychrophilic bacterium *P. haloplanktis* TAC125 acted as the receptor for a biodegradative gene in mesophilic *P. stutzeri* OX1 (Siani et al. 2006).

Next, efficient promoters used in cold-adapted bacteria have been isolated artificially by the construction of genomic DNA library. The genomic DNA fragments of *P. haloplanktis* TAC125 were cloned in a shuttle vector, and recombinant clones containing promoters were chosen to express a promoterless *lacZ* gene (Duilio et al. 2004). A consistent promoter sequence of *P. haloplanktis* TAC125 was identified through the comparison of different active promoters. The pUCRP-induced cold adaptation gene expression vector was constructed through cloning the DNA fragment of the P (pSHAb0363) vector cold promoter. Two "difficult" proteins were used as the model systems to test the performance of the induction system for psychrophilic and mesophilic protein production. Data showed that both psychrophilic α-galactosidase from *Ph*TAE19 and mesophilic d-glucosidase from *Saccharomyces cerevisiae* (56–58) are produced in *Ph*TAC125 with fine yields and fully soluble catalytic activity. Therefore, an induced cold expression system has been successfully constructed in the Antarctic cold-adapted *Ph*TAC125 bacterium cell, which can effectively produce cold-active proteins and mesophilic proteins (Papa et al. 2007). The cold-inducible expression systems have obvious advantages in application, for example, cell can grow at the natural low temperature, and fine-tuning the recombinant expression via the inducible promoter is possible which is very useful in laboratory application. In this regard, for the sake of economy and the use of bioreactors, the better performances of the system in the smallest medium must be emphasized.

5.3   Mechanisms for Cold Adaptation

5.3.1   Cold Shock Proteins or Antifreeze Proteins

When the ambient temperature suddenly drops, the microorganisms will induce the expression of a variety of proteins. These proteins are called cold shock proteins (Csps) (Jones et al. 1987). Csps is a small nucleic acid-binding protein between 65 and 75 amino acids, widely distributed in prokaryotic and eukaryotic organisms.

Csps are involved in the intracellular transcription, translation, protein folding, and regulation of membrane fluidity. Cold acclimation proteins (Caps), a subgenus of cold shock proteins, are expressed only in low-temperature microorganisms. Cold-adapted proteins are overexpressed when microorganisms are exposed to low temperatures for long periods of time (Wang et al. 2014).

Antifreeze proteins (AFPs) help in the survival of psychrophiles in ice. It could change the crystal structure, prevent the recrystallization of extracellular fine ice crystals, and allow the organism to survive at low temperatures (Wilson et al. 2006). These proteins have potential applications, ranging from cryobiology to aerospace, due to their ability to avoid the structure damage in the animal or vegetal foods. Muñoz et al. (2017) have revealed that AFPs molecular dynamics simulations allowed it to establish the relation between Antarctic AFPs from isolates GU3.1.1 and AFP5.1 at the water/ice interface. Threonine residues were found to be an important factor during the interaction. The oxygen distribution function was investigated and evaluated with 150 ns molecular simulation.

5.3.2   Cold Shock Proteins of *E. coli*

It has been cleared that two major responses are commonly observed during the exposure of hasty temperature downshifts, with the membrane lipid composition changes and novel protein production. Although low-temperature bacteria are grown by enhancing the production of cold shock proteins to keep the membrane lipid fluidity, the molecular and physiologic mechanisms behind the temperature between microorganisms remain largely unknown. Two important terms must be introduced first before we discuss bacterial physiology at low temperature; they are cold shock and cold acclimation which refer to the sudden temperature decline and the steady-state growth at low temperature, respectively. When the cultures of *E. coli* at the mid-exponential phase are quickly transferred from 37 °C to lower temperature, a significant reduction was observed in the cellular protein synthesis, while most of "cold shock" proteins are synthesized and gradually the bacteria resume growth (Mujacic et al. 1999). Obviously, many Csps played key roles in gene expression, transcription regulation, and protein synthesis.

Baneyx and Mujacic (2002) had proposed that *E. coli* Csps are routinely classified into two groups based on the degree of cold induction. Class I proteins are expressed at an extremely low level at optimal growth temperature of *E. coli* (37 °C) and are rapidly induced to pretty high level when transitioned to lower temperatures. Two of them, CsdA and RbfA, are closely related to ribosome. CsdA binds to 70S subunits and shows RNA-unwinding activity. RbfA only binds to 30S subunits, served as an initiation factor or late maturation. At low temperatures, RbfA is necessary for an efficient translation of most cellular mRNAs. Additionally, Class I Csps could also include other members of NusA which belong to a transcription termination-antitermination factor, and PNPase, which is an exonuclease, is involved in mRNA turnover. Class II Csps could be detected easily at 37 °C, but their induction after cold shock is not dramatic (two to tenfold).

The most highly cold-inducible protein is the first discovered CspA (Goldstein et al. 1990; Keto-Timonen et al. 2016). It belongs to a family of nine low molecular weight ($\approx$7 kDa) paralogs, of which four (*CspA, CspB, CspG*, and *CspI*) are also the most abundant cold shock proteins in *E. coli*. It's very interesting that CspA, CspB, and CspG were synthesized in large quantities under cold shock conditions that usually protein synthesis was inhibited, such as amino acid starvation and antibiotics treatment. Among them, CspA has been well characterized and it has been defined as an RNA chaperone protein. Recent research revealed the transcription antiterminator characters of CspA and its homologs (Phadtare et al. 2002). In addition, they appear to be used as regulator to induce cold-regulated expression of four other CSPs, such as RbfA, IF2, NusA, and PNP (polynucleotide phosphorylase). Both RbfA and IF2 are employed in the translation initiation; NusA is related to RNA core polymerase and participated in transcriptional termination and transcriptional antitermination; PNP is a $3'–5'$exonuclease which associates with *Ribonuclease E* and degrades mRNA. Although CspB, CspG, and CspI showed very high degree of homology to CspA, most of their function remains unclear until now, while it is suggested by genetic studies that these Csps proteins may play analogous or complementary roles in low-temperature adaptation.

Other Csps, such as CsdA, H-NS, trigger factor and Hsc66, play unlike roles in low-temperature domestication. It has been suggested that CsdA with helix-destabilizing activity plays a crucial role in destroying mRNA secondary structures at low temperature (Charollais et al. 2004). H-NS is a rich DNA-binding protein which affects the conformation of the DNA superhelix and several unrelated gene expressions, containing the *rrn* operons. The trigger factor has the activity of peptidyl-prolyl isomerase, which can promote the folding of nascent polypeptides. Some research indicated that after incubation of 1 week at 4 °C, the overexpression of trigger factor aggrandized the survival rate of *E. coli* from 15% to 40%. Hsc66 is about 42% similar to the heat shock protein (Hsp) DnaK and prevents protein aggregation in vitro. It is suggested that the trigger factor and Hsc66 can participate in low-temperature adaptation as molecular chaperones. The properties of the trigger factor and Hsc66 as Csps suggest that low temperature negatively affected multiple steps of protein folding in *E. coli* (Phadtare 2004).

5.3.3   CspA Regulation in *E. coli*

In *E. coli*, CspA is counted as the main cold shock protein and is barely detectable at the favorable growth temperature of *E. coli* (37 °C), but after the first hour, the temperature drops to 15 °C. More than 10% of the total protein synthesis is dedicated to its production (Bae et al. 2000). After 2–4 h of temperature downshift (from 37 °C to 15 °C), CspA stability synthesis decreases and cell growth continues within a decreased doubling time. Different from heat shock genes which rigorously depend on unique promoter sequences and optional sigma factors to transcript, there is no obvious difference between *cspA* core promoter and vegetative promoters (Fig. 5.1). It can be recognized by the Eς70 holoenzyme not only at low temperatures

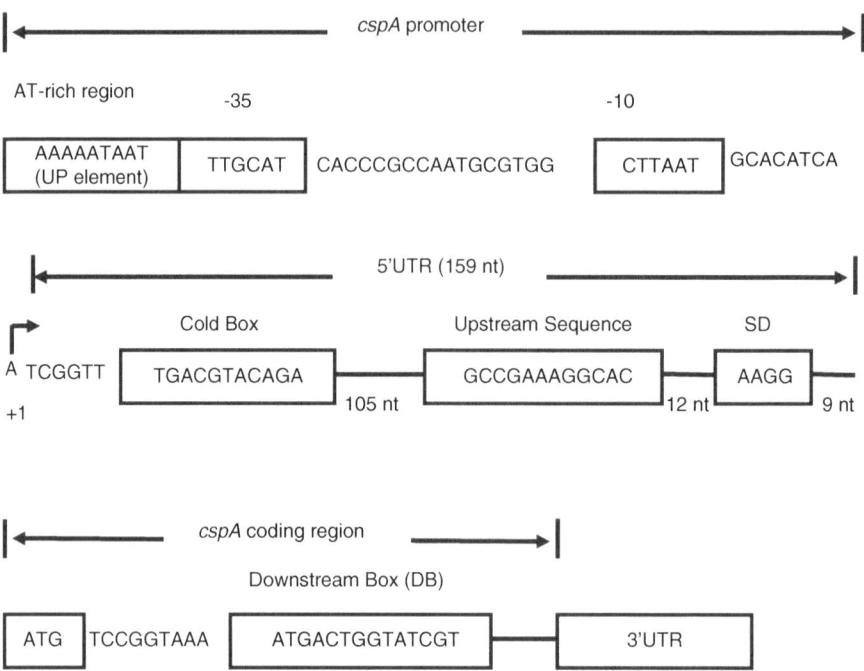

Fig. 5.1 Regulatory elements involved in the transcriptional, posttranscriptional, and translational control of CspA synthesis. (AT-rich region, −35 and −10 hexamers are the transcriptional control elements; cold box is the posttranscriptional control element; upstream and downstream boxes are the translational control elements)

but at all temperatures. One AT-rich region referred to as UP element is located upstream of the −35 hexamer (Fig. 5.1), which could increase the *cspA* promoter strength through promoting the initiation of the transcription. In fact, large amounts of *cspA* transcripts are produced at all physiological temperatures of *E. coli*. A special structure located the *cspA* mRNA 5′ end, a 159 nt long untranslated region (UTR), plays an important role in maintaining the level of *cspA*. The structure makes the transcript of *cspA* gene very short-lived (t1/2 ≈ 10 s) at 37 °C, which impedes its efficient translation.

There are two main regulation modules contributed to synthesis of CspA during temperature downshift. The conformational change in the 5′UTR of *cspA* regulation is regarded as the most important role in production level of CspA; it results in an increase of nearly two orders of magnitude, while the *cspA* core promoter just slightly increases the transcription level. Several reports indicate that a conserved region located the 3′ end of the UTR makes *cspA* transcript easier to obtain by cold modified translation machinery. Moreover, it has been reported that the region downstream located 12 bp after the *cspA* start codon can strengthen *cspA* translation initiation after cold shock (Al-Fageeh and Smales 2009; Giuliodori et al. 2010; Leppek et al. 2018).

The inhibition of CspA synthesis was consistent with the recovery of cell growth. It can be explained using the ribosome adaptation model. The model shows that the Csps, such as RbfA, CsdA, and IF-2, are connected with the free ribosomal subunits and 70S ribosome. Following cold shock, these cold shock proteins accumulate immediately and progressively convert free ribosome subunits and 70S particles into functional, cold-adapted ribosomes and polysomes which are able to translate noncold shock mRNAs. Because *rbfA* mutants produce cold shock proteins constitutively after temperature downshift, these changes occurring at the translational level also have the chance to repress the CspA synthesis. It is well known that *rbfA* cells have no effect on the repression of Csps synthesis at the lag phase end. It is very worthy to obviously enhance the intracellular gene product accumulation under the transcriptional control of *cspA* in shaking bottles and ferments (Joers and Tenson 2016).

5.3.4 Enzymes

Cold-adapted enzymes produced by microorganisms of cold environment (psychrophiles and psychrotrophs) are distinguished from mesophilic and thermophilic enzymes by their ability to catalyze efficiently biochemical reactions at low temperatures (Ohgiya et al. 1999). Due to their potential biotechnological application, they are receiving a great deal of attention. Prospective applications of cold-adapted enzymes include food processing, washing, biomass conversion, and environmental bioremediation. Besides the application in biotechnology, it is important to elucidate the mechanism of molecular adaptation to cold environment.

Early studies on low-temperature adaptation mechanism mostly focused on the isolation and characteristics of cold-adapted enzymes. Cold-adapted enzymes show higher catalytic efficiency and lower the Km value at low temperatures. Because of the high flexibility, the structure allows the enzyme to have a higher substrate-binding capacity and thermal instability. Antarctic *Euphausia superba* possess the krill's ability to thrive in a cold habitat, originally from its capacity to synthesize enzymes that adapt to low temperatures (Olivera-Nappa et al. 2013). Fornbacke and Clarsund (2013) have reported that the cold-adapted trypsin was the major material in metabolic reactions and played a very important role in low-temperature adaptation. In addition, higher random coil ratio and lower steric resistance may also be the main factors to promote cold adaptation. The cold-active xylanases have low-temperature catalytic activity and have great application potential in the food industry. The first crystal structure of cold-active xylanase XynGR40, a number of glycoside hydrolase (GH) family 10, was described (Zheng et al. 2016). The folds into A typical GH10 $(\beta/\alpha)8$ TIM-barrel was folded by the enzyme, as well as E132 and E243 acted as the catalytic residues. By comparing the structure of a thermophilic GH10 xylanase, various parameters that might illuminate the characteristics of cold adaptation were evaluated. The increased revelation of hydrophobic residues, flexibility of substrate-binding residues, loop structure flexibility, and the proportion of special amino acid residues might lead to XynGR40 cold adaptation.

To date, only a handful of cold-adapted enzymes were studied in depth, and their use in industrial processes is limited, mainly due to their limited availability. Most psychrophilic microbes, which could produce cold-active enzymes, grow only slowly with a low expression level of target proteins even under the optimum condition. This brings difficulties in the processes of production and purification of cold-adapted enzymes. Large-scale production of thermolabile proteins for research and biotechnological purposes has been hampered, as they are easily denatured when produced in heterologous hosts at mesothermal conditions. Overproduction system with a psychrophilic bacterium as host is supposed to be ideal for the production of thermolabile proteins.

5.3.5 Increased Membrane Fluidity and Cell Envelope Modification

The maintenance of membrane fluidity is an important mechanism for microorganisms to adapt to low temperatures. Suitable cold microorganisms often enhance the fluidity of the membrane through augmenting the content of unsaturated fatty acids, branched fatty acids, and cis fatty acids. Envelope modifications were recognized as well-known mechanisms. The change of lipid composition in cell membrane is conducive to the formation of short chains and the reduction of lipid saturation. In *P. haloplanktis*, the higher proportion of cell envelope genes presented in the genome was thought to be a special adaptation to cold environment. In the Antarctic bacterium *P. syringae* Lz4w, the changes of the composition and fluidity of LPS were observed at low temperature, with the increase of hydroxy fatty acids and the content of polymyxin B enhanced (Kumar et al. 2002). At $-15\,^{\circ}\mathrm{C}$, *Planococcus halocryophilus* Or1 has a unique cell extracellular membrane, which is characterized by the formation of a hard shell around the cell, during which the saturation of fatty acid increases with the decrease of temperature. The results indicated that *P. halocryophilus* had another mechanism for maintaining membrane fluidity, with the presence of fatty acid desaturase, which was inactive at these temperatures (Mykytczuk et al. 2013).

Exopolysaccharides (EPS) are an important component of bacterial cells and are considered as important cellular function in response to cold and ice. *Pseudoalteromonas* sp., isolated from Arctic marine environments, with mannose as a major component, produces a highly complex exopolysaccharide. These results indicate that the production of EPS with specific properties is shared among bacteria adapted to low temperatures. Moreover, extracellular polymeric substances were associated with increased freeze-thaw survival of *Winogradskyella*, *Colwellia*, and *Shewanella* genera isolated from Antarctic sponge, further demonstrating the protective effect of these compounds at low temperatures (Caruso et al. 2018).

OMVs produced by psychrophilic bacteria reported for putative proteolytic enzymes that have potentiality to degrade high molecular weight molecules, helping bacteria survive in such harsh conditions (Casillo et al. 2019). Casillo et al. (2019) describe the structural features of the carbohydrate backbone of the lipooligosaccharide (LOS) isolated from *Shewanella* sp. HM13, a psychrophilic bacterium isolated from the gut of horse mackerel and that produce large amounts of OMVs.

5.3.6 Secondary Metabolites

Some secondary metabolites synthesized by microorganisms at low temperatures are also important for the low-temperature adaptation of microorganisms. These secondary metabolites include trehalose, extracellular polysaccharides, saccharide, betaine, mannitol, and the like. They act to prevent crystallization, concentrate nutrients, and prevent enzyme cold denaturation. For example, trehalose is presumed to prevent denaturation and aggregation of proteins; on the one hand, extracellular polysaccharides can help microorganisms effectively maintain cell membrane fluidity and preserve water by changing physiological and biochemical indicators of microorganisms, and on the other hand, it can assist microorganisms to concentrate nutrients to protect enzymes (Tian et al. 2017; Raymond-Bouchard et al. 2018).

Some bacteria that grow in cold environments can synthesize polyhydroxyalkanoates (PHAs). PHAs reserving polymers possess significant physiological functions. These polymers accumulate under unbalanced growth conditions. PHAs entitle bacteria greater viability and resistance to environmental stress, as well as ecological relevance. In *S. alaskensis*, proteome analysis showed increased abundance of enzymes associated with PHA synthesis at low temperatures. Among the proteins involved in PHA biosynthesis, phasins are the main multifunctional proteins in the PHA granule-related proteins, which are believed to play a positive role in stress protection and fitness enhancement. The PhaP phasins of *S. alaskensis*, possibly secreted 3-hydroxybutyrate dehydrogenase, were found to increase at low temperatures. Especially, the PHB production is the basis of cryopreservation and cold growth. The accumulation of PHB increases the ability of the bacteria to move around in the biofilms formed under cold conditions and the survival of planktonic cells. It has been indicated that the ability of accumulating PHB might form an adaptive superiority for the colonization of new niches in those habitats. Putrescine synthesis of *Psychrobacter* sp. PAMC 21119 was upregulated, but heme protein synthesis was downregulated (Koh et al. 2017). It was found that putrescine and spermidine are associated with cold stress responses. Furthermore, PHA producers were screened in diverse extreme low-temperature habitats. PHA producers belonging to *Pseudomonas* genera and *Janthinobacterium* genera which were isolated from Antarctic soils possessed higher polymer accumulation at 5 °C–20 °C compared to higher temperatures.

Koh et al. (2017) revealed the responses of *Psychrobacter* sp. PAMC 21119 in cold temperature through transcriptome and proteome analysis and found that the metabolism pathway of acetyl-CoA was upregulated; however, the proteins that

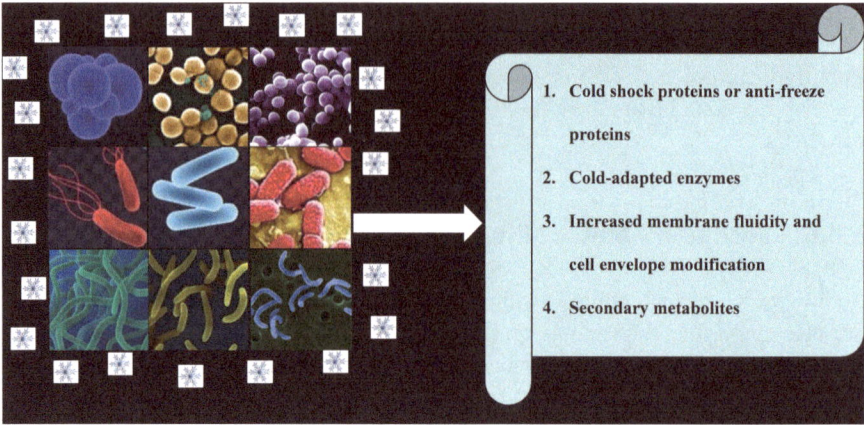

Fig. 5.2 General mechanisms for cold-adapted bacterial community

participated in energy production and transformation were downregulated. The use of glyoxylate shunt can offer intermediate carbon compounds to meet the requirements of biosynthesis. It was observed that the induction of genes encoding enzymes that participated in glyoxylate cycle at 5 °C existed in *Nesterenkonia* sp. AN1 which was isolated from Antarctic soil and was a member of Actinobacteria (Habibu et al. 2016). Methylglyoxal synthase also was found in *Exiguobacterium sibiricum*, isolated from the permafrost by Rodrigues et al. (2008) and it was a significant character of glycolysis downstream of glycolysis, since it is another triose phosphate catabolic pathway. Glyoxalase family proteins are also thought to be the main components of *Planococcus halocryophilus* Or1, nonsporogenic Firmicutes isolated from subzero temperatures of the Arctic high permafrost (Fig. 5.2).

5.4   Conclusion

The understanding of microbial communities is rather complex due to numerous metabolic groups coexisting in the environment and much remaining to be discovered. However, some researches have emphasized the appearance of features associated with coping with cold, such as Csps, compatible solutes, exopolysaccharides, and membrane modifications. Moreover, it was suggested that permafrost communities might employ a pretty high diverse and complex system of biochemical courses, including carbon cycling, organic decomposition, methanogenesis, oxidation cycling, and nitrogen cycling. These results offer new understanding on microbial communities and functions in Arctic habitats affected by climate change. Current evidence suggests the significance of maintaining membrane fluidity, which seems to be a usual character in psychrophilic bacteria, but also highlights the relationship of some envelope components to cope with cold environments.

References

Aakvik T, Degnes KF, Dahlsrud R et al (2009) A plasmid RK2-based broad-host-range cloning vector useful for transfer of metagenomic libraries to a variety of bacterial species. FEMS Microbiol Lett 296:149–158

Adekoya OA, Helland R, Willassen N, Sylte I (2006) Comparative sequence and structure analysis reveal features of cold adaptation of an enzyme in the thermolysin family. Protein 62:435–449. https://doi.org/10.1002/prot.20773

Al-Fageeh MB, Smales CM (2009) Cold-inducible RNA binding protein (CIRP) expression is modulated by alternative mRNAs. RNA NY 15(6):1164–1176. https://doi.org/10.1261/rna.1179109

Anwar MN, Li ZF, Gong Y, Singh RP, Li YZ (2019) Omics studies revealed the factors involved in the formation of colony boundary in *Myxococcus xanthus*. Cell 8(6):530. https://doi.org/10.3390/cells8060530

Bae W, Xia B, Inouye M, Severinov K (2000) *Escherichia coli* CspA-family RNA chaperones are transcription antiterminators. PNAS 97(14):7784–7789. https://doi.org/10.1073/pnas.97.14.7784

Baneyx F, Mujacic M (2002) Cold-inducible promoters for heterologous protein expression. Methods Mol Biol 205. *E. coli* gene expression protocols Edited by: P. E. Vaillancourt © Humana Press Inc., Totowa, NJ

Berny JF, Hennebert G (1991) Viability and stability of yeast cells and filamentous fungus spores during freeze-drying: effects of protectants and cooling rates. Mycol 83(6):805–815. https://doi.org/10.2307/3760439

Caruso C, Rizzo C, Mangano S, Poli A, Di DP et al (2018) Production and biotechnological potential of extracellular polymeric substances from sponge-associated Antarctic bacteria. Appl. Environ Microbial 84(4):e01624–e01617. https://doi.org/10.1128/AEM.01624-17

Casillo A, Di GR, Carillo S, Chen C, Kamasaka K, Kawamoto J, Corsaro MM (2019) Structural elucidation of a novel lipooligosaccharide from the Antarctic bacterium OMVs producer *Shewanella* sp. HM13. Marine drug 17(1):34. https://doi.org/10.3390/md17010034

Cavicchioli R (2016) On the concept of a psychrophile. ISME J 10:793–795

Cedric YS, Michel F, Martine C (2001) Mobilization function of the pBHR1 plasmid, a derivative of the broad-host-range plasmid pBBR1. J Bacteriol 183(6):2101–2110. https://doi.org/10.1128/JB.183.6.2101-2110.2001

Charollais J, Dreyfus M, Iost I (2004) CsdA, a cold-shock RNA helicase from *Escherichia coli*, is involved in the biogenesis of 50S ribosomal subunit. Nucleic Acids Res 32:2751–2759. https://doi.org/10.1093/nar/gkh603

Choo DW, Tatsuo K, Takeshi S, Kenji S, Nobuyoshi E (1988) A cold-adapted lipase of an Alaskan Psychrotroph, *Pseudomonas* sp. strain B11-1: gene cloning and enzyme purification and characterization. Appl Environ Microbiol 64(2):486–491

Cody WL, Wilson JW, Hendrixson DR, McIver KS, Hagman KE et al (2008) Skim milk enhances the preservation of thawed −80 °C bacterial stocks. J Microbial Method 75(1):135–138. https://doi.org/10.1016/j.mimet.2008.05.006

Czajka JJ, Abernathy MH, Benites VT, Baidoo EEK, Deming JW et al (2018) Model metabolic strategy for heterotrophic bacteria in the cold ocean based on *Colwellia psychrerythraea* 34H. Proc Natl Acad Sci U S A 115(49):12507–12512

D'Amico S, Collins T, Marx JC, Feller G, Gerday C (2006) Psychrophilic microorganisms: challenges for life. EMBO Rep 7:385–389

Dong D, Ihara T, Motoshima H, Watanabe K (2005) Crystallization and preliminary X-ray crystallographic studies of a psychrophilic subtilisin-like protease Apa1 from Antarctic *Pseudoalteromonas* sp. strain AS-11. Acta Crystallogr Sect F Struct Biol Cryst Commun 61(3):308–311

Duilio A, Tutino ML, Marino G (2004) Recombinant protein production in Antarctic gram-negative bacteria. In: Balbás P, Lorence A (eds) Recombinant gene expression, Methods in molecular biology, vol 267. Humana Press

Ewert M, Deming JW (2011) Selective retention in saline ice of extracellular polysaccharides produced by the cold-adapted marine bacterium *Colwellia psychrerythraea* strain 34H. Ann Glaciol 52:111–117

Feller G (2013) Psychrophilic enzymes: from folding to function and biotechnology. Scientifica (Cairo) 2013:1–28

Feller G, Gerday C (2003) Psychrophilic enzymes: hot topics in cold adaptation. Nat Rev Microbiol 1:200–208

Fornbacke M, Clarsund M (2013) Cold-adapted proteases as an emerging class of therapeutics. Inf Dis Ther 2(1):15–26. https://doi.org/10.1007/s40121-013-0002-x

Giuliodori AM, Pietro FD, Marzi S, Masquida B, Wagner R, Romby P et al (2010) The cspA mRNA is a thermosensor that modulates translation of the cold-shock protein CspA. Mol Cell 37(1):21–33. https://doi.org/10.1016/j.molcel.2009.11.033

Goldstein J, Pollitt NS, Inouye M (1990) Major cold shock protein of *Escherichia coli*. Proc Natl Acad Sci U S A 87:283–287

Gottesman S (2018) Chilled in translation: adapting to bacterial climate change. Mol Cell 70(2):193–194

Guieysse B, Autem Y, Soares A (2005) Biodegradation of phenol at low temperature using two-phase partitioning bioreactors. Water Sci Technol 52(10–11):97–105

Habibu A, Pieter DM, Don C (2016) The genome of the Antarctic polyextremophile *Nesterenkonia* sp. AN1 reveals adaptive strategies for survival under multiple stress conditions. FEMS Microbiol Ecol 92(4):fiw032. https://doi.org/10.1093/femsec/fiw032

Hashimoto-Gotoh T, Timmis KN (1981) Incompatibility properties of Col E1 and pMB1 deriva-tive plasmids: random replication of multicopy replicons. Cell 23(1):229–238. https://doi.org/10.1016/0092-8674(81)90287-7

Hoyoux A, Jennes I, Dubois P et al (2001) Cold-adapted beta-galactosidase from the Antarctic psychrophile *Pseudoalteromonas haloplanktis*. Appl Environ Microbiol 67:1529–1535

Joers A, Tenson T (2016) Growth resumption from stationary phase reveals memory in *Escherichia coli* cultures. Sci Report 6:24055. https://doi.org/10.1038/srep24055

John PB, Carol MN, John AEG (2003) *Algoriphagus ratkowskyi* gen. nov., sp. nov., *Brumimicrobium glaciale* gen. nov., sp. nov., *Cryomorphaignava* gen. nov., sp. nov. and *Crocinitomix catalas-itica* gen. nov., sp. nov., novel flavobacteria isolated from various polar habitats. Int J Syst Evol Microbiol 53(5):1343–1355

Jones PG, VanBogelen RA, Neidhardt FC (1987) Induction of proteins in response to low tempera-ture in *Escherichia coli*. J Bacteriol 169:2092–2095

Jung SK, Jeong DG, Lee MS, Lee JK, Kim HK et al (2008) Structural basis for the cold adaptation of psychrophilic M37 lipase from *Photobacterium lipolyticum*. Proteins 71:476–484

Keen NT, Tamaki S, Kobayashi D, Trollinger D (1998) Improved broad-host-range plas-mids for DNA cloning in gram-negative bacteria. Gene 70(1):191–197. https://doi.org/10.1016/0378-1119(88)90117-5

Keto-Timonen R, Hietala N, Palonen E, Hakakorpi A, Lindström M et al (2016) Cold shock pro-teins: a minireview with special emphasis on Csp-family of enteropathogenic *Yersinia*. Front Microbial 7:1151. https://doi.org/10.3389/fmicb.2016.01151

Knoblauch C, Jørgensen BB, Harder J (1999) Community size and metabolic rates of psychrophilic sulfate-reducing bacteria in Arctic marine sediments. App Environ Microbiol 65(9):4230–4233

Koh HY, Park H, Lee JH, Han SJ, Sohn YC (2017) Proteomic and transcriptomic investigations on cold-responsive properties of the psychrophilic Antarctic bacterium *Psychrobacter* sp. PAMC 21119 at subzero temperatures. Environ Microbiol 19(2):628–644. https://doi.org/10.1111/1462-2920.13578

Kumar S, Jagannadham MV, Ray MK (2002) Low-temperature-induced changes in composition and fluidity of lipopolysaccharides in the Antarctic psychrotrophic bacterium *Pseudomonas syringae*. J Bact 184(23):6746–6749. https://doi.org/10.1128/JB.184.23.6746-6749.2002

Leppek K, Das R, Barna M (2018) Functional 5' UTR mRNA structures in eukaryotic translation regulation and how to find them. Nat Rev Mol Cell Biol 19(3):158–174. https://doi.org/10.1038/nrm.2017.103

Lipson DA, Schmidt SK, Monson RK (1999) Links between microbial population dynamics and nitrogen availability in an alpine ecosystem. Ecol 80:1623–1631

Lipson DA, Schmidt SK, Monson RK (2000) Carbon availability and temperature control the post-snowmelt decline in alpine soil microbial biomass. Soil Biol Biochem 32:441–448

Margesin R, Fonteyne PA, Redl B (2005) Low-temperature biodegradation of high amounts of phenol by *Rhodococcus* spp. and basidiomycetous yeasts. Res Microbiol 156(1):68–75

Meyer AF, Lipson DA, Martin AP, Schadt CW, Schmidt SK (2004) Molecular and metabolic characterization of cold-tolerant alpine soil *Pseudomonas sensu stricto*. Appl Environ Microbial 70(1):483–489. https://doi.org/10.1128/aem.70.1.483-489.2004

Mocali S, Chiellini S, Fabiani S, Decuzzi A, Pascale S et al (2017) Ecology of cold environments: new insights of bacterial metabolic adaptation through an integrated genomic-phenomic approach. Sci Rep 7(1):1–13

Morita RY (1975) Psychrophilic bacteria. Bacteriol Rev 39:144–167

Mujacic M, Cooper KW, Baneyx F (1999) Cold-inducible cloning vectors for low-temperature protein expression in *Escherichia coli*: application to the production of a toxic and proteolytically sensitive fusion protein. Gene 238(2):325–332. https://doi.org/10.1016/S0378-1119(99)00328-5

Muñoz PA, Márquez SL, González-Nilo FD, Márquez-Miranda V, Blamey JM (2017) Structure and application of antifreeze proteins from Antarctic bacteria. Microb Cell Factories 16(1):138. https://doi.org/10.1186/s12934-017-0737-2

Mykytczuk NC, Foote SJ, Omelon CR, Southam G, Greer CW et al (2013) Bacterial growth at −15 °C; molecular insights from the permafrost bacterium *Planococcus halocryophilus* Or1. ISME J 7(6):1211–1226. https://doi.org/10.1038/ismej.2013.8

NASA (2010, January 21). NASA climatologist Gavin Schmidt discusses the surface temperature record. Accessed 30 Nov 2010.

Ohgiya S, Hoshino T, Okuyama H, Tanaka S, Ishizaki K (1999) Biotechnology of enzymes from cold-adapted microorganisms. In: Margesin R, Schinner F (eds) Biotechnological applications of cold-adapted organisms. Springer, Berlin/Heidelberg

Olivera-Nappa A, Reyes F, Andrews BA, Asenjo JA (2013) Cold adaptation, Ca^{2+} dependency and autolytic stability are related features in a highly active cold-adapted trypsin resistant to autoproteolysis engineered for biotechnological applications. PLoS One 8(8):e72355. https://doi.org/10.1371/journal.pone.0072355

Papa R, Rippa V, Sannia G, Marino G, Duilio A (2007) An effective cold inducible expression system developed in *Pseudoalteromonas haloplanktis* TAC125. J Biotechnol 127(2):199–210. https://doi.org/10.1016/j.jbiotec.2006.07.003

Park JW, Oh YS, Lim JY, Roh DH (2006) Isolation and characterization of cold-adapted strains producing beta-galactosidase. J Microbiol 44(4):396–402

Phadtare S (2004) Recent developments in bacterial cold-shock response. Curr Issues Mol Biol 6:125–136

Phadtare S, Inouye M, Severinov K (2002) The nucleic acid melting activity of *Escherichia coli* CspE is critical for transcription antitermination and cold acclimation of cells. J Biol Chem 277:7239–7245. https://doi.org/10.1074/jbc.M111496200

Rampelotto PH (2010) Resistance of microorganisms to extreme environmental conditions and its contribution to astrobiology. Sustain 2(6):1602–1623

Raymond-Bouchard I, Tremblay J, Altshuler I, Greer CW, Whyte LG (2018) Comparative transcriptomics of cold growth and adaptive features of a eury- and steno-psychrophile. Front Microbial 9:1565. https://doi.org/10.3389/fmicb.2018.01565

Rodrigues DF, Ivanova NHZ, Huebner M, Zhou J, Tiedje JM (2008) Architecture of thermal adaptation in an *Exiguobacterium sibiricum* strain isolated from 3 million year old permafrost: a genome and transcriptome approach. BMC Genomics 9:547. https://doi.org/10.1186/1471-2164-9-547

Santiago M, Ramírez-Sarmiento CA, Zamora RA, Parra LP (2016) Discovery, molecular mechanisms, and industrial applications of cold-active enzymes. Front Microbiol 7:1408. https://doi.org/10.3389/fmicb.2016.01408

Schmidhauser TJ, Helinski DR (1985) Regions of the broad-host-range plasmid RK2 involved in replication and stable maintenance in nine species of gram-negative bacteria. J Bacteriol 164:446–455

Siani L, Papa R, Di DA, Sannia G (2006) Recombinant expression of Toluene o-Xylene Monooxygenase (ToMO) from *Pseudomonas stutzeri* OX1 in the marine Antarctic bacterium *Pseudoalteromonas haloplanktis* TAC125. J Biotechnol 126(3):334–341

Soares A, Guieysse B, Delgado O et al (2003) Aerobic biodegradation of nonylphenol by cold-adapted bacteria. Biotechnol Lett 25:731. https://doi.org/10.1023/A:1023466916678

Sobecky PA, Mincer TJ, Chang MC, Toukdarian A, Helinski DR (1998) Isolation of broad-host-range replicons from marine sediment bacteria. Appl Environ Microbial 64(8):2822–2830

Subhashini DV, Singh RP, Manchanda G (2017) OMICS approaches: tools to unravel microbial systems. Directorate of Knowledge Management in Agriculture, Indian Council of Agricultural Research. ISBN: 9788171641703. https://books.google.co.in/books?id=vSaLtAEACAAJ

Tian Y, Li YL, Zhao FC (2017) Secondary metabolites from polar organisms. M drug 15(3):28. https://doi.org/10.3390/md15030028

Tribelliand PM, López NI (2018) Reporting key features in cold-adapted bacteria. Life 8(1):1–12

Turkiewicz M, Kur J, Białkowska A, Cieśliński H, Kalinowska H, Bielecki S (2003) Antarctic marine bacterium *Pseudoalteromonas* sp. 22b as a source of cold-adapted beta-galactosidase. Biomol Eng 20(4–6):317–324

Tutino ML, Parrilli E, Giaquinto L, Duilio A, Sannia G et al (2002) Secretion of alpha-amylase from *Pseudoalteromonas haloplanktis* TAB23: two different pathways in different hosts. J Bacteriol 184(20):5814–5817. https://doi.org/10.1128/jb.184.20.5814-5817

Wang Z, Wang S, Wu Q (2014) Cold shock protein a plays an important role in the stress adaptation and virulence of *Brucella melitensis*. FEMS Microbiol Lett 354:27–36. https://doi.org/10.1111/1574-6968.12430

Wilson S, Kelley D, Walker V (2006) Ice-active characteristics of soil bacteria selected by ice-affinity. Environ Microbiol 8:816–824. https://doi.org/10.1111/j.1462-2920.2006.01066.x

Xiong H, Song L, Xu Y et al (2007) Characterization of proteolytic bacteria from the Aleutian deep-sea and their proteases. J Ind Microbiol Biotechnol 34:63. https://doi.org/10.1007/s10295-006-0165-5

You YW, Wang TH (2005) Cloning and expression of endoglucanase of marine cold-adapted bacteria *Pseudoalteromonas* sp. MB-1. W. Sheng W. X. B 45(1):142–144

Zheng Y, Li Y, Liu W, Chen CC, Ko TP et al (2016) Structural insight into potential cold adaptation mechanism through a psychrophilic glycoside hydrolase family 10 endo-β-1,4-xylanase. J Struct Biol 193(3):206–211. https://doi.org/10.1016/j.jsb.2015.12.010

Secretome of Microbiota in Extreme Conditions

Mohit S. Mishra, Ravi Kant Singh, Sushma Chauhan, and Priyanka Gupta

Abstract

Microorganisms are highly versatile living beings that have survived nearly 4 billion years of evolutionary change. Microbes have adapted to environmental conditions that range from cold to hot and from anoxic to oxic. Their life forms range from free living to symbiotic. During the harsh environment, bacteria adopt various strategies to acquire multiple systems so as to thrive against adverse conditions. Among them, secreted proteins play a major role and facilitate most of the interactions of the cells with their surrounding environment as well as with neighbouring cells. Bacterial cells secrete the proteins, metabolites and various molecules to the extracellular environments to sustain nutrient depletion conditions, adhesion on substrate (biofilm formation), invasion of host or antagonism, etc. These microbial functions are managed through their secretion system, which are very sophisticated molecular machines that include export/secretion of required proteins or metabolic factors required for the conjugation and survival processes. Hence, the present chapter is aimed to describe the microbes, specially bacteria adapted to survival in extreme temperature. Further, the chapter highlights the secretion system to understand how these bacteria deal with the extreme environments.

Keywords

Secretion system · Genetic assets · Pathway · Plasma membrane · ATP

M. S. Mishra (✉) · R. K. Singh · S. Chauhan
Amity Institute of Biotechnology, Amity University, Raipur, Chhatishgarh, India
e-mail: msmishra@rpr.amity.edu

P. Gupta
Radhika Tai Panndav College of Science, Nagpur, Maharashtra, India

© Springer Nature Singapore Pte Ltd. 2020
R. P. Singh et al. (eds.), *Microbial Versatility in Varied Environments*,
https://doi.org/10.1007/978-981-15-3028-9_6

6.1   Introduction

Since the origin of life on earth, prokaryotic life is abundant on our planet and is evolving in all available niches wherever life is possible. On earth, various conditions, from favourable to extreme, are present, and living beings thrive in every possible way to overcome the swaths of unfavourability such as pH, pressure, temperature, salinity, energy, radiation and nutrients (Michels and Clark 1992; Subhashini et al. 2017). In any location of the earth, where liquid water is or was available for life, microorganisms have confirmed their presence. These statements demonstrate the adaptation capabilities of microbiota in the wide range of condition parameter. Hence, it is necessary to overview the machinery or system required in order to combat the conditions of maxima or extreme.

Among the prokaryotes, extremophilic bacteria are the best part of life to be used as an example for the understanding of mechanism and survival tactics. Extremophiles technically cover those bacteria which thrive under a variety of extreme conditions. Moreover, they represent vital objects of research in various fields and disciplines that reveal the ecological, agricultural, industrial and clinical treasures. Furthermore, extremophiles research has exposed the dark shadow of origin of life on the earth as well as on other terrestrial and celestial bodies. Extremophiles are completely opposite to extremotolerants because they are actually adapted creature of that condition with balanced metabolic, biochemical and cellular functionality which is harsh or extreme for normal or tolerant microflora. Merino et al. (2019) have reviewed the extremophiles and hypothesized them as the richest lifeforms on the earth. Moreover, Knoll (2015) speculated that mechanistic life cycle of extremophiles was the only way for survival on the earth during evolution. The environmental conditions and their physical as well as chemical parameters are generally used to categorize the microbiota in different groups. As an example, different temperature values categorize the thermophiles, i.e. bacteria capable of surviving at temperatures more than 60 °C are hyperthermophiles, while those below 15 C are known as psychrophiles. Similarly, the microbes surviving below pH 7 are acidophiles (pH <3) and above 7 are alkaliphiles (pH >9). Those capable of survival at high salt condition belong to halophiles. Similarly, bacteria in absence of oxygen or in low oxygen conditions are anaerobes or microanaerobes, those in low water are xerophiles, at high pressure are barophiles and in high radiations are radioresistant. Interestingly, some of extremophiles are poly-extremophiles which is due to their multi-resistance properties. Auernik et al. (2008) reviewed the thermoacidophiles, a very important poly-extremophilic microbiota and speculated that they majorly belong to the order *Sulfolobales* and *Thermoplasmatales*. By nature, they survive well in high temperature as well as in low pH. Similarly, high NaCl and pH tolerance was reported in some halophilic species (Green and Mecsas 2016).

Under the variety of extreme environments, the extremophiles have adopted various secretion mechanisms and a variety of secretomes to thrive normally. Studies have revealed that thermophiles have special features for degradation of complex extracellular carbohydrates (Vafiadi et al. 2009; Asada et al. 2009) and ability to secrete antimicrobial compounds (Esikova et al. 2002; Muhammad et al. 2009). Thermostable *Taq* polymerase enzyme of *Thermus aquaticus* is the most vibrant example of novel secretome of extremophiles. The secretome of microbiota and its released proteins in their surroundings or niche where it lives are completely mediated by bacterial cell wall and associated secretion system (Tjalsma et al. 2000). Moreover, cell-to-cell interaction, interaction with substrates, virulence factor and environmentally mediated stress are also governed by secretion system (Zhou et al. 2008). Hence, the following chapter is aimed to highlights the secretion strategies of various extremophile as well as the mechanisms of various types of bacterial secretion systems to understand the ecology of bacteria in extreme and bioprocessing environments.

6.2   Secretion and Secretome of Some Extremophiles

Bacterial secretome are derived from cell surface by the specialized tools called secretion system. The secretions contain the numerous types of proteins that facilitated the microbes to interact with own community and with host, environment and substrate. Colonization of microbiota or adhesion on surfaces via biofilm is also facilitated by secretome. Moreover, breakdown of complex polymers into simpler monomers or their defence mechanism is also mediated by secretome. Overall, the physiology and survival as well as their application of microbiota are completely based on their secretome. Here, we have illustrated some extremophiles and their secretion to understand it clearly.

Acidophiles and Alkaliphiles   Acidophiles are microorganisms which grow generally at pH range from 0 to 5.5 and occur in all three life domains. Few well-known acidophiles include *Cephalosporium* sp., *Ferroplasma acidarmanus*, *Acidithiobacillus ferrooxidans*, *Trichosporon cerebriae*, etc. In natural condition, the acidophiles can be obtained from acidic mines, acidic soils and industrial areas (Rothschild and Mancinelli 2001). The acidophiles have unique ability to maintain an intracellular pH near neutral pHs result from high levels of positive cell surface charge, high internal buffer capacity and overexpression of H^+ export proteins (Pick 1999). In recent years, the genome sequence of few acidophiles (e.g. *Ferroplasma acidarmanus*) unrevealed the physiological adaptation to acidic environments.

Alkaliphiles are microorganisms capable of growing optimally at a pH range of 8.5–11.5. These microorganisms represent archaea, prokaryotes and eukaryotes and are isolated from natural habitats such as alkaline soils, mines, hot springs, soda lakes, industrial areas, etc. The key point for alkaliphilic microbiota is to maintain the intracellular pH in their niche where they live. Krulwich (1998) have stated that alkaliphiles performed active import of H^+ into the cells, cell wall polymers and use of complex membrane composition for survival. *B. pseudofirmus* strain OF4 is the best example to understand the molecular machinery of alkaliphiles. The bioenergetic study of cell surface of strain OF4 revealed that NADH dehydrogenase donates electrons in the cellular membrane via menaquinone (MQ) pool. Further, the electron of reduced MQ moves through the menaquinol:cytochrome c oxidoreductase and cytochrome c oxidase, that pumps the protons to deliver it outside the cell and resultantly the molecular O_2 is reduced. Then, proton-motive force (PMF) catalysed the import of proton by electropotential charges and resultantly Na + goes lesser than proton. This inward movement maintained the low pH intracellularly in alkaliphiles (Preiss et al. 2015).

Halophiles These are a versatile group of nature that optimally grown at or above 0.2 M salt concentration and are capable to adapt the rapid alteration of osmotic stress. The secretome of halophiles respond to stress and altered the secretion for homeostasis. For sustainability in saline environments, halophiles have evolved two types of mechanism. One is the cytoplasmic accumulation of inorganic ions that have balanced the osmotic pressure on environments and another is accumulation of specified organic osmolytes in high concentration, that acts as osmoprotectant for the cell, without any metabolic changes in the bacterial cells (Oren 2002). Halophilic bacterial proteomes are basically acidic in nature and possess the amino acids of hydrophilic nature (Weinisch et al. 2018). Expression of the *ect* gene, transcription of *ectABC* operon and biosynthesis of ectoine and hydroxyectoine also induce the salt stress (Ma et al. 2010).

Psychrophiles These are a group of microorganism that thrive well at >15 °C temperature (Steven et al. 2006). However, the actual temperature limits for psychrophiles are not determined clearly. Although for reproduction and metabolism, temperatures of ~12 °C and ~20 °C have been reported, respectively (Rivkina et al. 2000). Moreover, cold niche negatively affects the physiology as well metabolism of psychrophiles, such as cellular integrity, viscosity, diffusion of solutes, fluidity of membrane, metabolome kinetics, etc. (Rodrigues and Tiedje 2008; Piette et al. 2011). Hence, psychrophiles have adapted specialized machinery to maintain the cellular function normally in cold environments and also evolved the polyextremophilic behaviour in adapted cell such as against the radiations, pH alterations, lower in nutrient (Tehei and Zaccai 2005). With the advent of 'omics' technologies, a boost in the molecular insight of psychrophile adaption to cold environment has been speculated. Till date, about 135 whole genomes are available on database and about 125 are ongoing. The analysis of published genomes has highlighted the presence of plasmids, fatty acid biosynthesis gene clusters and transposable elements that contribute to genome plasticity (Allen et al. 2009). These available transposables or mobile genes and genetic clusters directly mediate the cold adaptation mechanisms in psychrophilic microbiota. For example, the higher abundance and diversity in tRNA of the *Alteromonas* sp. SN2 and *Psychromonas ingrahamii* (Riley et al. 2008), though the G + C content is approximately similar across the genus (Rabus et al. 2004).

Genomic data and bioinformatic interpretations of psychrophiles revealed the significance of genes encoded for cold-shock proteins (CSPs). CSPs are very prominent, small, upregulated nucleic acid binders of proteins. CSPs have been responsible for the regulation of various microbial molecular phenomena such as membrane fluidity, genetic transcription and regulation of expression, protein folding and strong upregulated response to cold shock or exposure (Rodrigues and Tiedje 2008; Margesin and Miteva 2011). This cold shock response of CSP genes induces the RNA helicase enzymes to destabilize the DNA/RNA secondary structures, chaperons, HSPs and mediator genes of sugar metabolism and cellular biogenesis (Chen et al. 2012). Cold environments induce the cellular freezing via formation of cytoplasmic ice crystal in the native microbiota. Resultantly, osmotic balance and cell fluidity gets damaged. Psychrophiles fight against these stresses by producing the antifreeze protein (AFP) that binds the ice crystal of cytoplasm and prevents the crystallization by thermal hysteresis. Exopolysaccharide (EPS) production is another tactic of psychrophiles for cryoprotection. Bordat et al. (2005) have found the excess production of EPSs in Antarctic marine microbes and winter ocean ice. Cornelis and Wolf-Watz (1997) have reviewed that EPSs regulate the cellular surface maintenance of bacteria in cold water and appropriate management of nutrients. Moreover, EPSs facilitate the adherence of cells to surfaces via biofilm formation and synergism with other bacterial communities.

6.3   Process of Bacterial Secretome

Bacterial secretome is defined as the set of proteins that contains extracellular matrix, vesicle (outer and inner) and membrane secretory proteins of cell (Makridakis and Vlahou 2010). This term was coined by Tjalsma et al. (2000) and described the secretion system of bacteria as well as their secretome. Secretion system and secretome belong to the cellular complexity, virulence, transports, receptors, toxin, etc. and mediated the cellular communication, signalling with in situ environments, biofilm formation, biodegradation, nutrient mineralization, host defence, etc. (Walsh 2000; Wooldridge 2009; Dalbey and Kuhn 2012). Interestingly, a bacterial genome has been occupied 10–30% by the gene that encoded the bacterial secretion system (Kudva et al. 2013). Now, the development of advanced spectrometry such as 2-D electrophoresis, MS/MS and NanoLC/MS has increased our knowledge in secretome of microbiota (Berman et al. 2013; Anwar et al. 2019). Secretome can also facilitate the pathogens to directly or indirectly infect the host cellular surfaces and to survive within it (Aizawa 2001). In non-pathogenic and saprophytic organisms, secretome facilitates the adaptation to its habitat by means of the shape of the bacterial cell envelope (Wattiau et al. 1996).

Recent development within the molecular analysis of the protein secretion pathways of bacteria has revealed the 12 types of secretion apparatus as Sec, Tat, T1SS, T2SS, T3SS, T5SS, T6SS, T7SS, SecA2, sortase and injectosome. Sec, Tat and T4SS are common in bacteria, while T1SS, T2SS, T3SS, T4SS, T5SS, T6SS has described in gram-negative bacteria and T7SS, SecA2, sortase and injectosome are

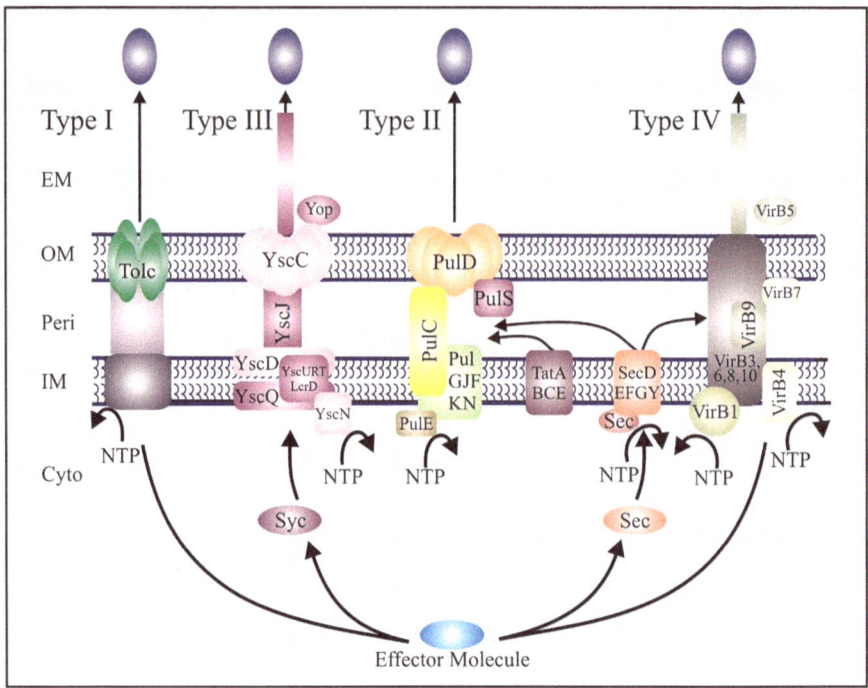

Fig. 6.1 Type I, II, III and IV protein secretion systems

described in gram-positive bacteria (Green and Mecsas 2016). T1SS-T4SS has been illustrated in the Fig. 6.1. Examples and classes of bacterial protein secretion system are displayed in Table 6.1.

Gram-positive bacterial cell wall is arranged in the following manner: a single plasma membrane, the cytoplasmic membrane, observed by way of a thick cellular wall layer, while in gram-negative microorganism, it is a double-membrane system, a cytoplasmic membrane and an outer membrane which sandwich the peptidoglycan and periplasmic area among them. The protein secretion system is found across the inner membrane in both gram-positive and gram-negative organisms and normally follows the Sec-based pathway (regularly referred to as the overall secretory pathway (GSP), despite the fact that other pathways have recently been recognized, the Tat and signal recognition particle (SRP) pathways) (Herskovits et al. 2000, Mori and Ito 2001; Driessen 2001). Different types of secretion systems have been described in detail in the following points.

6.3.1   Type I Secretion System

This secretion system (T1ss) is described widely in gram-negative cells. It governs the secretion of proteins (varied in size) and allowed a single-step transport

Table 6.1 Bacterial protein secretion system

Secretion apparatus	Signal secretion	Secretion steps	Membranes	Gram positive (+) or gram negative (−)	Examples
Sec	N-terminus	1	1	Both	*Listeria monocytogenes Vibrio cholerae*
Tat	N –terminus	1	1	Both	*B. subtilis, E. coli*
T1SS	C- terminus	1	2	Gram (−)	*B. subtilis, Streptomyces lividans*
T2SS	N -terminus	2	1	Gram (−)	*V. cholerae, Klebsiella pneumoniae*
T3SS	N -terminus	1 to 2	2 to 3	Gram (−)	*S. enterica*
T4SS	C- terminus	1	2 to 3	Gram (−)	*A. tumefaciens, Neisseria gonorrhoeae*
T6SS	N -terminus	2	1	Gram (−)	*V. cholerae, S. marcescens*
SecA2	Unknown secretion signal	1	2 to 3	Gram (+)	*Mycobacteria*
Sortase	N -terminus	2	1	Gram (+)	*S. aureus*
Injectosome	N -terminus	2	1	Gram (+)	*Streptococcus pyogenes*
T7SS	C- terminus	1	1 to 3	Gram (+)	*Corynebacteria*

functions from the cytoplasm to the extracellular medium. The secretion of scavenging molecules, outer membrane vesicles proteins, proteases, lipases and different kinds of virulence factors is mediated by T1ss system. T1ss is similar to the ATP-binding cassette (ABC) transporters family and is involved in the transfer of small molecules such as antibiotics and toxins out of the cell cytoplasm (Symmons et al. 2009). The acquisition of iron by HasAp, a heme-binding protein of *P. aeruginosa*, is the best example for T1ss protein secretion (Letoffe et al. 1998).

The secretion signal of secreted protein is basically located at the C-terminal, and it is not cleavable at the time of functioning. The overall machinery is designed by three kinds of proteins. Among it, ABC transporter and membrane fusion protein were found in the inner membrane of bacteria cell and an outer membrane channel forming porins. All of these proteins are localized in the bacterial cell envelope and in the presence of a substrate; these proteins form a 'tunnel channel' which is transverse to the periplasmic space. Morgan et al. (2017) has described the T1ss-mediated translocation of unfolded polypeptides and proved the significance of T1ss system for protein secretion.

6.3.2   Type II Secretion System

Secreted proteins have a major function in the pathogenesis of bacterial infections, along with vital sicknesses of humans, animals and plants. Mostly, type II SS (T2ss) is mostly described in the gram-negative bacterium. It facilitates the translocation of proteins from the periplasm to extracellular environments via outer membrane. It is common in pathogenic bacteria as well as in non-pathogenic bacteria. For example, human pathogens *V. cholerae*, *E. coli*, *P. aeruginosa*, *Klebsiella* sp., *L. pneumophila*, *Yersinia enterocolitica*, etc. mediated the pathogenesis via T2ss (Korotkov et al. 2012). Similar, T2ss machinery was also studied in *A. hydrophila*, an amphibian animal pathogens, by Jiang and Howard (1992).

The T2ss is a classical multi-protein system that contains 12 'centre' proteins, and it was denoted by Anacker et al. (1987) as T2S C, D, E, F, G, H, I, J, K, L, M and O. In recent years, with the development of molecular biology, T2ss structure has been defined well. Presently, it has been divided into four sub-complexes: (i) an OM 'secretin', which is a pentadecamer of the T2S D protein that offers a pore via the membrane; (ii) an IM platform composed of T2S C, F, L and M, with T2S C imparting a connection to the OM secretin; (iii) a cytoplasmic ATPase, which is a hexamer of T2S E that is recruited to the IM platform; and (iv) a periplasm-spanning pseudopilus that is a helical filament of the most important pseudopilin T2S G capped with the minor pseudopilins T2S H, I, J and K. Finally, T2S O, an IM peptidase, cleaves and methylates the pseudopilins for the incorporation it into the pseudopilus (Bachovchin et al. 1990).

6.3.3   Type III Bacterial Secretion System

Many bacterial pathogens along with *salmonella* use tiny molecular syringes so that it can inject proteins into host cellular cytoplasm. This type of transportation is done by type III secretion device. Type III secretion system (T3ss) presents as microscopic needle complexes and comes from inner to the outer membrane of the bacterium. The needle projected is far away from the cell (Aepfelbacher 2004). T3ss device basically contains two rings that can help throughout the secretion across the cellular membrane including the peptidoglycan of bacterial cell. Moreover, the ring present in inner membrane is larger in size, and the ring protein is also identified in several bacteria. The outer membrane ring is composed of the secretin protein family. It is a needle-like structure that associates with the outer membrane ring and projects from the bacterial surface. T3ss is also reported for the involvement in T2ss secretion and also present in the type IV bacterial pili complex. End of the needle system bought the secretion machinery to touch on substrate for initiate the secretion. *Salmonella* sp. uses type III secretion supply a cocktail of at least 13 kinds of protein toxin known as effector protein without delay into host cells. Inside the bacterium, sheperon protein binds to effector protein for the handing of over loosely folded from the needle. ATpase triggers the release of effector protein and helps them to start out their journey through narrow channel inside the needle (Frankel et al. 2001). Direct transport to cytoplasm of the host cellular is right concept because it gets rid of the dilution that arises in the cellular within the gut. These effector proteins interfere with the signal transduction cascade to host reaction. The effector protein intrudes with sign transduction cascades to host response. Once the *salmonella* enters in the host cell, it invades by the vacuoles via the 'phagosome', a normal course of activities (Hapfelmeier et al. 2005). The lysosomes then loose with phagosomes and introduce its digestive content to kill the invader.

6.3.4   Type IV Bacterial Secretion System

Bacterial cells mediate the translocation of DNA (in addition to proteins) into target cells by this unique secretion system type IVss which is found in both gram-negative and gram-positive bacteria and also in some archaea (Alvarez-Martinez and Christie 2009). Conjugation of plasmid DNA has been mediated only by type IVss, and due to these systems, it contributes into the spread of plasmid-borne antibiotic resistance genes, and hence it is the most omnipresent secretion systems in nature. Moreover, TIVss is also involved in bacterial pathogenesis characteristics of few organisms. In *Helicobacter pylori* and *Bordetella pertussis*, it is mediated through the secretion of transforming proteins toxins.

In others, effector proteins are required to support an intracellular lifestyle in bacteria such as *Legionella pneumophila*. Christie et al. (2014) has reported the detailed mechanism and structure of the bacterial type IVss and revealed that it is encoded by the Ti plasmid of *Agrobacterium tumefaciens*, together with the TIVss encoded by the pKM101 and R388 conjugative plasmids of *E. coli*. A total of 12 proteins were notified for the arrangement of IVss which is VirB1, VirB2, VirB3, VirB4, VirB5, VirB6, VirB7, VirB8, VirB9, VirB10, VirB11 and VirD4. The protein categorization in different domains of TIVss has been described in Fig. 6.2.

6.3.5   Type V Bacterial Secretion System

Type V secretion systems (TVss) is a single-membrane-spanning secretion system, which is also known as the autotransporter system. The TVss is mainly for virulence factors but also biofilm formation as well as cell-cell adhesion (Leo et al. 2012). TVss transfers an unfolded autotransporter polypeptide through the IM to the periplasm via SecYEG translocon. The autotransporters are known as 'passenger' domain, which is composed of a secreted domain and that is either semi-unfolded or fully unfolded in the periplasm. For the OM (transmembrane domain), it is also called 'translocator' or 'β-domain'. On the basis proteins' number involved in the secretion process, Green and Mecsas (2016) have separated the TVss into three classes which is autotransporter secretion, two-partner secretion and chaperone-usher secretion.

In autotransporter secretion, it is allowed to secrete themselves by self-contained components (Leyton et al. 2012). These components are described as a translocator

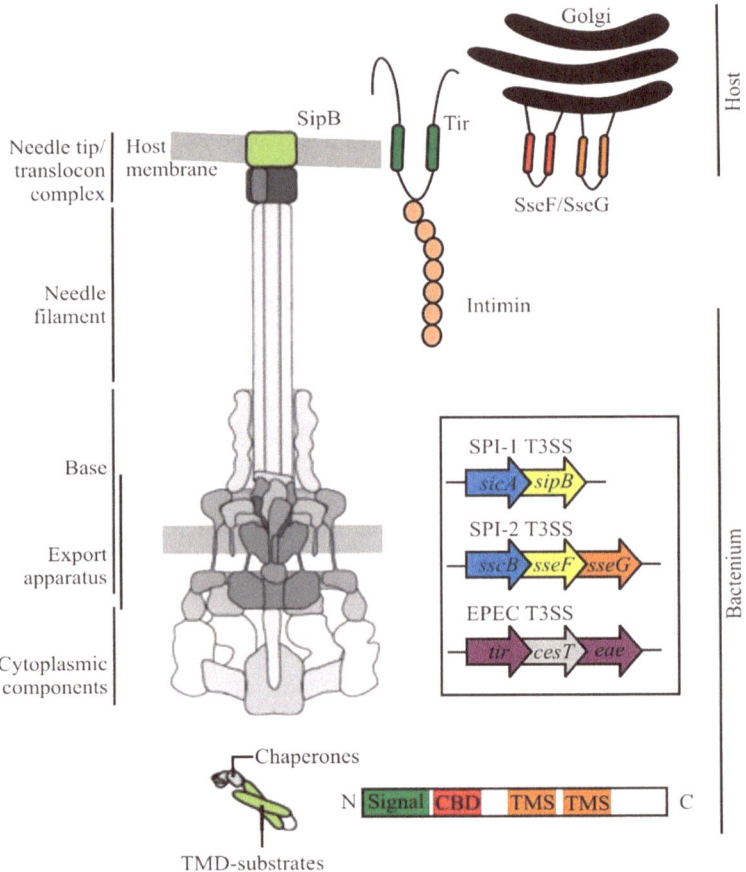

Fig. 6.2 Represent protein categorization in different domains

domain, a linker domain and a protease domain which were responsible for the C-terminus that forms the outer membrane channel, a passenger domain that contains the functional part of the autotransporter protein and cleaves off the passenger domain once it passes through the channel, respectively. Besides the common autotransporter secretion, few rely on the two-partner secretion, a different polypeptide for transport outside of the cells. In this process, a pair of proteins, on for carrying the carries the β-barrel domain and other, serves as the secreted protein, had participates in the secretion process (Henderson et al. 2004).

The chaperone-usher secretion categories are secreted with the usher protein and chaperone. Usher protein forms the β-barrel, a periplasmic protein channel in the outer membrane that is for the facilitation of folding the secreted protein before delivering to the channel (Waksman and Hultgren 2009). This category is commonly used to assemble pilins on the surface of gram-negative bacteria like P pilus of uropathogenic *E. coli*.

6.3.6   Type VI Secretion Systems

The term TVIss is a pivotal secretion system activated during pathogenesis and bacterial competition (Ho et al. 2014), coined firstly by Pukatzki et al. (2006) in the human pathogen *V. cholerae*. However, the actual coded gene (the imp genes) for the TVIss was reported by Bladergroen et al. (2003) in the symbiotic phytobacterium *R. leguminosarum* and reported broadly among *Proteobacteria* (Boyer et al. 2009). TVIss is a cell envelope-spanning machine that is applied to inject toxic effectors into eukaryotic and prokaryotic cells (Ho et al. 2013). Evidences revealed that the TVIss complex is somewhat equivalent to a contractile phage tail that is divided into a tail sheath, an inner tube and a base plate which is anchored to the cell envelope by the membrane complex. The tail sheath extends deep from bacterial outer membrane to cell cytoplasm (Basler et al. 2012). TVIss is composed by several accessory components and 13 essential conserve core components (Zheng and Leung 2007). Generally, all were encoded within the same gene cluster (Taylor et al. 2016). This macromachinery has a membrane and a tail complex. Both are the main complex of TVIss and comprises IM proteins (T4SS homolog) and a tail complex that contains components that are evolutionarily related to contractile bacteriophage tails, respectively (Ma et al. 2009; Leiman et al. 2009).

6.3.7   Type VII Secretion System

Type VII secretion system (TVIIss) is a specialized protein secretion machinery of gram-positive bacterium, described first in *M. tuberculosis* and *M. bovis*. This secretion system is commonly found in phyla of *Actinobacteria* and *Firmicutes*. TVIIss is a ~1.5 MDa protein complex that shares a conserved IM and cytosolic apparatus. Central channel of IM is predicted to be composed by membrane proteins EccB, EccC, EccD and EccE. Among them, EccC possesses an ATPase cytoplasmic domain (Solomonson et al. 2013) and EccD have hydrophobic domains.

Generally, TVIIss is divided into two types of component: (a) FtsK/SpoIIIE protein and (b) EsxA-/EsxB-related protein (Pallen 2002; Sysoeva et al. 2014). The atomic structures of EsxA–EsxB, EsxG–EsxH and PE25–PPE41 have been resolved, but their functional attributes remain uncleared (Houben et al. 2014). Hence, the bioinformatic predictions and homology modelling studies revealed that each type of VIIss machinery secretes numerous substrates and establish the several components based complex and multi-protein machinery.

6.4   Conclusion

Microbes utilize the various methods to survive in the extreme conditions and establish immune system to thwart against it. For that, bacterial secretome and secretion system, across the phospholipid membranes, is foremost and essential component in promoting bacterial virulence, attaching to the substrate, formation of biofilm, to scavenging resources in extreme niche. Moreover, this chapter is illustrating the bacterial secretion system and secretome. It is offering the detailed molecular and functional details of secretion system which is important for the survival of microbiota in the extreme conditions.

References

Aepfelbacher M (2004) Modulation of Rho GTPases by type III secretion system translocated effectors of *Yersinia*. Rev Physiol Biochem Pharmacol 152:65–77

Aizawa SI (2001) Bacterial flagella and type III secretion systems. FEMS Microbiol Lett 202:157–116

Allen MA, Lauro FM, Williams TJ, Burg D, Siddiqui KS et al (2009) The genome sequence of the psychrophilic archaeon, *Methanococcoides burtonii*: the role of genome evolution in cold adaptation. ISME J 3:1012–1035

Alvarez-Martinez CE, Christie PJ (2009) Biological diversity of prokaryotic type IV secretion systems. Microbiol Mol Biol Rev 73:775–808

Anacker RL, Mann RE, Gonzales C (1987) Reactivity of monoclonal antibodies to Rickettsia rickettsii with spotted fever and typhus group rickettsiae. J Clin Microbiol 25:167–171

Anwar MN, Li ZF, Gong Y, Singh RP, Li YZ (2019) Omics studies revealed the factors involved in the formation of Colony boundary in *Myxococcus xanthus*. Cell 8(6):530. https://doi.org/10.3390/cells8060530

Asada Y, Endo S, Inoue Y, Mamiya H, Hara A et al (2009) Biochemical and structural characterization of a short-chain dehydrogenase/reductase of Thermus thermophilus HB8 A hyperthermostable aldose-1-dehydrogenase with broad substrate specificity. Chem Biol Interact 178:117–126

Auernik KS, Cooper CR, Kelly RM (2008) Life in hot acid: pathway analyses in extremely thermoacidophilic archaea. Curr Opin Biotechnol 19(5):445–453

Bachovchin WW, Plaut AG, Flentke GR, Lynch M, Kettner CA (1990) Inhibition of IgA1 proteinases from *Neisseria gonorrhoeae* and *Haemophilus influenzae* by peptide prolyl boronic acids. J Biol Chem 265:3738–3743

Basler M, Pilhofer M, Henderson GP, Jensen GJ, Mekalanos JJ (2012) Type VI secretion requires a dynamic contractile phage tail-like structure. Nature 483:182–186

Berman HM, Coimbatore NB, Di CL, Dutta S, Ghosh S et al (2013) Trendspotting in the protein data bank. FEBS Lett 587:1036–1045

Bladergroen MR, Badelt K, Spaink HP (2003) Infection-blocking genes of a symbiotic strain that are involved in temperature-dependent protein secretion. Mol Plant-Microbe Interact 16(1):53–64

Bordat P, Lerbret A, Demaret J-P, Affouard F, Descamps M (2005) Comparative study of trehalose, sucrose and maltose by molecular modelling. Europhys Lett 65:41–47

Boyer F, Fichant G, Berthod J, Vandenbrouck Y, Attree I (2009) Dissecting the bacterial type VI secretion system by a genome wide *in silico* analysis: what can be learned from available microbial genomic resources? BMC Genomics 10:104

Chen Z, Yu H, Li L, Hu S, Dong X (2012) The genome and transcriptome of a newly described psychrophilic archaeon, *Methanolobus psychrophilus* R15, reveal its cold adaptive characteristics. Environ Microbiol Rep 4:633–641

Christie PJ, Whitaker N, González-Rivera C (2014) Mechanism and structure of the bacterial type IV secretion systems. Biochim Biophys Acta (BBA) Mol Cell Res 1843(8):1578–1591

Cornelis GR, Wolf-Watz H (1997) The *Yersinia* Yopvirulon: a bacterial system for subverting eukaryotic cells. Mol Microbiol 23(5):861–867

Dalbey RE, Kuhn A (2012) Protein traffic in Gram-negative bacteria – how exported and secreted proteins find their way. FEMS Microbiol Rev 36(6):1023–1045

Driessen AJ (2001) SecB, a molecular chaperone with two faces. Trends Microbiol 9:193–196

Esikova TZ, Temirov YV, Sokolov SL, Alakhov YB (2002) Secondary antimicrobial metabolites produced by thermophilic *Bacillus* spp. strains VK2 and VK21. Appl Biochem Microbiol 38:226–231

Frankel GA, Phillips L, Trabulsi S, Knutton GD et al (2001) Intimin and the host cell-is it bound to end in Tir(s)? Trends Microbiol 9:214–218

Green ER, Mecsas J (2016) Bacterial secretion systems: an overview. Microbiol Spectr 4(1). https://doi.org/10.1128/microbiolspec.VMBF-0012-2015

Hapfelmeier S, Stecher B, Barthel M, Kremer M, Muller AJ (2005) The Salmonella pathogenicity island (SPI)-2 and SPI-1 type III secretion systems allow *Salmonella* serovar *Typhimurium* to trigger colitis via MyD88-dependent and MyD88-independent mechanisms. J Imunol 174:1675–1685

Henderson IR, Navarro-Garcia F, Desvaux M, Fernandez RC, Ala'Aldeen D (2004) Type V protein secretion pathway: the autotransporter story. Microbiol Mol Biol Rev 68(4):692–744

Herskovits AA, Bochkareva ES, Bibi E (2000) New prospects in studying the bacterial signal recognition particle pathway. Mol Microbiol 38:927–939

Ho B, Basler M, Mekalanos J (2013) Type 6 secretion system–mediated immunity to type 4 secretion system–mediated gene transfer. Science 342:250–253

Ho BT, Dong TG, Mekalanos JJ (2014) A view to a kill: the bacterial type VI secretion system. Cell Host Microbe 15:9–21

Houben EN, Korotkov KV, Bitter W (2014) Take five- type VII secretion systems of *Mycobacteria*. Biochim Biophys Acta 1843:1707–1716

Jiang B, Howard SP (1992) The Aeromonas hydrophila exeE gene, required both for protein secretion and normal outer membrane biogenesis, is a member of a general secretion pathway. Mol Microbiol 6(10):1351–1361

Knoll AH (2015) Life on a young planet: the first three billion years of evolution on earth. Princeton University Press, Princeton. https://doi.org/10.1515/9781400866045

Korotkov KV, Sandkvist M, Hol WG (2012) The type II secretion system: biogenesis, molecular architecture and mechanism. Nat Rev Microbiol 10(5):336–351. https://doi.org/10.1038/nrmicro2762. Published 2012 April 2

Krulwich TA (1998) Alkaliphilic prokaryotes, the prokaryotes, pp 283–308

Kudva R, Denks K, Kuhn P, Vogt A, Müller M et al (2013) Protein translocation across the inner membrane of Gram-negative bacteria: the Sec and Tat dependent protein transport pathways. Res Microbiol 164:505–534. https://doi.org/10.1016/j.resmic.2013.03.016

Leiman PG et al (2009) Type VI secretion apparatus and phage tail-associated protein complexes share a common evolutionary origin. Proc Natl Acad Sci U S A 106:4154–4159

Leo JC, Grin I, Linke D (2012) Type V secretion: mechanism(s) of autotransport through the bacterial outer membrane. Philos Trans R Soc B 367:1088–1101

Letoffe S, Redeker V, Wandersman C (1998) Isolation and characterization of an extracellular haem-binding protein from *Pseudomonas aeruginosa* that shares function and sequence similarities with the *Serratia marcescens* HasA haemophore. Mol Microbiol 28:1223–1234

Leyton DL, Rossiter AE, Henderson IR (2012) From self sufficiency to dependence: mechanisms and factors important for autotransporter biogenesis. Nat Rev Microbiol 10:213–225. https://doi.org/10.1038/nrmicro2733

Ma LS, Lin JS, Lai EM (2009) An IcmF family protein, ImpLM, is an integral inner membrane protein interacting with ImpKL, and its walker a motif is required for type VI secretion system-mediated Hcp secretion in *Agrobacterium tumefaciens*. J Bacteriol 19:4316–4329

Ma Y, Galinski EA, Grant WD, Oren A, Ventosa A (2010) Halophiles 2010: life in saline environments. Appl Environ Microbiol 76(21):6971–6981. https://doi.org/10.1128/AEM.01868-10

Makridakis M, Vlahou A (2010) Secretome proteomics for discovery of cancer biomarkers. J Proteome 73:2291–2305

Margesin R, Miteva V (2011) Diversity and ecology of psychrophilic microorganisms. Res Microbiol 162:346–361

Merino N, Aronson HS, Bojanova DP, Feyhl-Buska J, Wong ML et al (2019) Living at the extremes: extremophiles and the limits of life in a planetary context. Front Microbiol 10:780

Michels PC, Clark DS (1992) Pressure dependence of enzyme catalysis. In: Adams MWW, Kelly R (eds) Biocatalysis at extreme environments. American Chemical Society Books, Washington, DC, pp 108–121

Morgan JLW, Justin FA, Jochen Z (2017) Structure of a Type-1 secretion system ABC transporter. Structure 25(3):522–529

Mori H, Ito K (2001) The Sec protein-translocation pathway. Trends Microbiol 9:494–500

Muhammad SA, Ahmad S, Hameed A (2009) Antibiotic production by thermophilic bacillus specie Sat-4. Pak J Pharm Sci 22:339–345

Oren A (2002) Diversity of halophilic microorganisms: environments, phylogeny, physiology, and applications. J Ind Microbiol Biotechnol 28:58–63

Pallen MJ (2002) The ESAT-6/WXG100 superfamily and a new Gram-positive secretion system? Trends Microbiol 10:209–212

Pick U (1999) *Dunaliella acidophila*: a most extreme acidophilic alga. In: Enigmatic microorganisms and life in extremophiles, vol 2. Springer, Dordrecht, pp 141–148

Piette A, D'Amico S, Mazzucchelli G, Danchin A, Leprince P et al (2011) Life in the cold: a proteomic study of cold-repressed proteins in the Antarctic bacterium *Pseudoalteromonas haloplanktis* TAC125. Appl Environ Microbiol 77:3881–3883

Preiss L, Hicks DB, Suzuki S, Meier T, Krulwich TA (2015) Alkaliphilic bacteria with impact on industrial applications, concepts of early life forms, and bioenergetics of ATP synthesis. Front Bioeng Biotechnol 3:75. https://doi.org/10.3389/fbioe.2015.00075

Pukatzki S, Ma AT, Sturtevant D, Krastins B, Sarracino D et al (2006) Identification of a conserved bacterial protein secretion system in Vibrio cholerae using the Dictyostelium host model system. Proc Natl Acad Sci U S A 103:1528–1533

Rabus R, Ruepp A, Frickey T, Rattei T, Fartmann B et al (2004) The genome of *Desulfotalea psychrophila*, a sulfate-reducing bacterium from permanently cold Arctic sediments. Environ Microbiol 6:887–902

Riley M, Staley JT, Danchin A, Wang TZ, Brettin TS et al (2008) Genomics of an extreme psychrophile, *Psychromonas ingrahamii*. BMC Genomics 9:210

Rivkina EM, Friedmann EI, McKay CP, Gilichinsky D (2000) Metabolic activity of permafrost bacteria below the freezing point. Appl Environ Microbiol 66(8):3230–3233

Rodrigues DF, Tiedje JM (2008) Coping with our cold planet. Appl Environ Microbiol 74:1677–1686

Rothschild LJ, Mancinelli RL (2001) Life in extreme environment. Nature 409:1092–1101

Solomonson M, Wasney GA, Watanabe N, Gruninger RJ et al (2013) Structure of the mycosin-1 protease from the mycobacterial ESX-1 protein type VII secretion system. J Biol Chem 288:17782–17790

Steven B, Leveille R, Pollard WH, Whyte LG (2006) Microbial ecology and biodiversity in permafrost. Extremophiles 10:259–267

Subhashini DV, Singh RP, Manchanda G (2017) OMICS approaches: tools to unravel microbial systems. Directorate of Knowledge Management in Agriculture, Indian Council of Agricultural Research. ISBN: 9788171641703. https://books.google.co.in/books?id=vSaLtAEACAAJ

Symmons MF, Bokma E, Koronakis E, Hughes C, Koronakis V (2009) The assembled structure of a complete tripartite bacterial multidrug efflux pump. Proc Natl Acad Sci U S A 106(17):7173–7178

Sysoeva TA, Zepeda-Rivera MA, Huppert LA, Burton BM (2014) Dimer recognition and secretion by the ESX secretion system in *Bacillus subtilis*. Proc Natl Acad Sci U S A 111:7653–7658

Taylor NM, Prokhorov NS, Guerrero-Ferreira RC, Shneider MM, Browning C et al (2016) Structure of the T4 baseplate and its function in triggering sheath contraction. Nature 533:346–352

Tehei M, Zaccai G (2005) Adaptation to extreme environments: macro- molecular dynamics in complex systems. Biochim Biophys Acta 1724:404–410

Tjalsma H, Bolhuis A, Jongbloed JDH, Bron S, van Dijl JM (2000) Signal peptide-dependent protein transport in *Bacillus subtilis*: a genome-based survey of the secretome. Microbiol Mol Biol Rev 64:515–547

Vafiadi C, Topakas E, Biely P, Christakopoulos P (2009) Purification, characterization and mass spectrometric sequencing of a thermophilic glucuronoyl esterase from *Sporotrichum* thermophile. FEMS Microbiol Lett 296:178–184

Waksman G, Hultgren SJ (2009) Structural biology of the chaperone-usher pathway of pilus biogenesis. Nat Rev Microbiol 7(11):765–774

Walsh C (2000) Molecular mechanisms that confer antibacterial drug resistance. Nature 406(6797):775–781

Wattiau P, Woestyn S, Cornelis GR (1996) Customized secretion chaperones in pathogenic bacteria. Mol Microbiol 20:255–262

Weinisch L, Kühner S, Roth R, Grimm M, Roth T et al (2018) Identification of osmoadaptive strategies in the halophile, heterotrophic ciliate *Schmidingerothrix salinarum*. PLoS Biol 16(1):e2003892

Wooldridge K (2009) Bacterial secreted proteins: secretory mechanisms and role in pathogenesis. Academic, Norfolk

Zheng J, Leung KY (2007) Dissection of a type VI secretion system in *Edwardsiella tarda*. Mol Microbiol 66:1192–1206

Zhou M, Boekhorst J, Francke C, Siezen RJ (2008) LocateP: Genome-scale subcellular-location predictor for bacterial proteins. BMC Bioinformatics 9:173

Deciphering the Key Factors for Heavy Metal Resistance in Gram-Negative Bacteria

Raghvendra Pratap Singh, Mian Nabeel Anwar,
Dipti Singh, Vivekanand Bahuguna,
Geetanjali Manchanda, and Yingjie Yang

Abstract

Heavy metals (HMs) are versatile elements of nature with five times higher atomic weight and density than water. HMs are ubiquitous in nature due to the industrial, domestic, agricultural, medical and technological applications. These are toxic at trace levels and therefore attract more and more interest for their least bioaccumulation and thus high persistence in the environment. Among HMs, arsenic, cadmium, chromium, lead and mercury rank as priority metals that are of public health significance and ecological concern. Interestingly, bacteria have been found as efficient tool for heavy metal degradation as well as resistance. Several bacteria have been reported for the HM accumulation which has been controlled by the metal resistance gene, carried on genome or in plasmid. In nature, Gram-negative bacteria are dependent on plant-derived simple carbon (C) compounds. In HMs abundant flora and fauna, they survive by different cellular mechanisms like metal sorption, mineralization, uptake and accumulation,

Authors Raghvendra Pratap Singh and Geetanjali Manchanda have been equally contributed for this chapter.

R. P. Singh · V. Bahuguna
Department of Research & Development, Biotechnology, Uttaranchal University, Dehradun, Uttarakhand, India

M. N. Anwar
State Key Laboratory of Microbial Technology, Shandong University, Jinan, China

D. Singh
Department of Microbiology, V.B.S. Purvanchal University, Jaunpur, Uttar Pradesh, India

G. Manchanda
Department of Botany and Environmental Studies, DAV University, Jalandhar, Punjab, India

Y. Yang (✉)
Marine Agriculture Research Center, Tobacco Research Institute of Chinese Academy of Agricultural Sciences, Qingdao, China

© Springer Nature Singapore Pte Ltd. 2020
R. P. Singh et al. (eds.), *Microbial Versatility in Varied Environments*,
https://doi.org/10.1007/978-981-15-3028-9_7

extracellular precipitation, enzymatic mechanisms for oxidation or reduction to a less toxic form and efflux of heavy metals from the cells to adapt in HM stresses. Hence, here we focus on the mechanism of microbial interaction with these heavy metals which can open the new horizon for the exploitation of Gram-negative bacteria and their gene pool as HM remediator agents, biological indicator and plant growth promoters.

Keyword
Metal resistance · Toxicity · Gene · Mechanism · Adaptation

7.1   Introduction

The extensive existence of the anthropogenic-based pollutant and effluent are the important stress factors that have been responsible for several diseases all over the biological kingdom of environments. Among them, the metals and metalloids with a density above $5 \text{ g}^{-1} \text{ cm}^3$, known as heavy metals (HMs), are increasing day by day and are a global threat to living beings and ecological health on earth (Zhou et al. 2015). HMs are high-density nondegradable naturally occurring earth crust compounds, which are much toxic even at a very trivial dosage or concentration. HMs are problematic to environment because of their nonbiodegradability, higher toxicity and bioaccumulation in food chain of living organism. They enter our systems through inhalation, adsorption by cell surface contact within industrial exposure, manufacturing, agriculture and residential settings. HMs are represented by arsenic (As), cadmium (Ca), chromium (Cr), cobalt (Co), lead (Pb), mercury (Hg), nickel (Ni), selenium (Se) and zinc (Zn) which are highly toxic even in trace amounts (Turpeinen et al. 2002; Siddiquee et al. 2015). HMs basically come from naturally as well as anthropogenic sources (Fu and Wang 2011). Some important natural sources of heavy metals are natural rock weathering process, volcanic eruptions, forest fires, sea salt sprays, wind-borne soil particles and biogenic sources. Industrial activities such as leather tanning, energy production, electroplating, oil industries including crude oil and hydrocarbon exploration and utilization, emissions from vehicular traffic gas exhausts, fuel production and downwash from power lines are also the major sources of HMs. However, the growth and development of living beings require some traces of heavy metals like Fe, Cu, Zn, Mo, etc., but the excess of these metals can be harmful for plants as well as concerned food chain for animals (Wintz and Fox 2002). For example, during plant growth, if they accumulate the HMs in more concentration, the plant growth and cellular metabolism, absorption as well as transportation of vital component will be inhibited (Xu and Shi 2000). Robin et al. (2012) have reported the harmful effects of HMs to plant growth and development, which are responsible for various diseases in animal. Studies have revealed that HM pollution is a serious global environmental problem which is adversely affecting the composition and activity of soils and its microbial communities (Xie et al. 2016).

HM pollutants consistently get deposited in the major sink of nature, "the soils". In the nature, the cycle of elements and metals is managed by soil bacteria, fungi, actinomycetes, algae and other microorganisms, and they are responsible for the decomposition of material elements and nutrient conversion via various biochemical reactions in the soil. Soil microbes are more sensitive to soil conditions than large animals or plants, and hence they serve as an indicator for soil environmental quality (Dian 2018). But the HM pollution can alter the soil microbiota and their activity such as soil enzyme activity, composition of soil microbial community and structure, plant growth, etc. (Sadler et al. 1967; Giller et al. 1998; Rajapaksha et al. 2004; Singh et al. 2017). Gülser and Erdoğan (2008) studied the effects of HM pollution on microbial enzyme activities and basal soil respiration of soils. Mills and Colwell (1977) found that HMs are detrimental to microorganisms even at the low concentrations. Ahmad et al. (2005) reported microbial diversity losses in soil by metal toxicity and which was validated by microcosm test. The study revealed that Pb, Mn and Ni were highly toxic, followed by Cd, Hg, Cr and Cu and Zn were the least. However, the toxicity of HMs was concentration as well as time dependent.

During the long-term exposures and history of HM contamination, several microorganisms have followed the resistance strategies and developed the detoxification or assimilation or bioremediation mechanisms to counter the toxic effects (Azarbad et al. 2016; Tipayno et al. 2018; Yang et al. 2019). The term "resistance to HMs" refers to the mechanism of detoxification of toxic metals by bacteria (Gadd 1992). Khan et al. (2016) isolated the HM-resistant Gram-negative bacteria *Salmonella enterica* 43Ca and found that the resistance was in order of $Pb^{2+} > Cd^{2+} > As^{3+} > Zn^{2+} > Cr^{6+} > Cu^{2+} > Hg^{2+}$. The metal resistance mechanisms allow the microbial populations to survive and maintain the functional sustainability of their communities. Moreover, the HM resistance in bacteria allow them to be employed as potential eco-friendly and cost-effective bioremediation tools for HMs which transform the toxic HMs into a less harmful state (Abbas et al. 2014; Ma et al. 2016; Ndeddy and Babalola 2016). Some of the metal resistance factors such as bioaccumulation (Jin et al. 2018), reduction (Nies 1999), biosorption (Quintelas et al. 2011), siderophore production (Singh et al. 2019; Prajakta et al. 2019) and the formation of biofilms (Von Bodman et al. 2003) have been explored for the control of metal pollution with a view to promote mitigation of the environmental impacts.

Gram-negative bacteria are attractive model microorganisms for the laboratory because of their fast growth, easy manipulation, genetic stability in large cultures and well-studied secretion system. Gram-negative bacteria have been studied extensively for the metal resistance. Moreover, Gram-negative microbes and their secreted products are also utilized at commercial scale which has made it more biotechnologically relevant for researchers, who have focused significantly on the homeostasis and resistance of Gram-negative bacteria in the presence of HMs. Therefore, it can open the opportunity for microbial mediated alleviation of HM stress and decreased accumulation of metals in agriculture. Hence, the critical evaluation of mechanistic system of Gram-negative bacteria is needed. So, the present chapter is aimed to display the resistance capabilities and mechanism of Gram-negative bacteria to cope with toxic concentrations of HMs generally considered to be environmental pollutants.

7.2   HM Resistance in Gram-Negative Bacteria: Molecular and Ecological Prospective

In the course of evolution, bacteria adapted to the increased content of HM ions in places of ore deposits. Its adaptation ensured the appearance in bacteria of various mechanisms for the protection of sensitive components from the action of heavy metal ions such as the type and number of pathways for the transport of metal ions into the cell; the localization of resistance genes on the chromosome, plasmid or transposon; and the role of these ions in normal cell metabolism (Choudhury and Srivastava 2001; Subhashini et al. 2017). Resistance to HMs in Gram-negative bacteria is described here according to previous research. Basically, the protection mechanisms are driven by some fundamental procedures, such as biosorption, intracellular sequestration, extracellular sequestration, extracellular barrier for preventing the metal entry into the microbial cell, methylation of metals and reduction of heavy metal ions by the microbial cell (Gomathy and Sabarinathan 2010; Chandrangsu et al. 2017). An illustration of HM resistance in Gram-negative bacteria is shown in Fig. 7.1.

The extracellular barrier is an important mechanism to prevent these ions from going into the cell. The membrane of the cell, cell wall or capsule can avoid entry of the metal ions within the cell. Different taxonomic groups of bacteria can bind metal ions by polarized groups of the cell wall or capsule (phosphate, carboxyl, hydroxyl and amino groups) (Taniguchi et al. 2000). Sorption is considered as a passive process, and during this process, bacterial cells of dead bacteria bind with the ions of metal. It was shown that bacterial cells killed by heating possessed a similar or higher sorption capacity as viable cells.

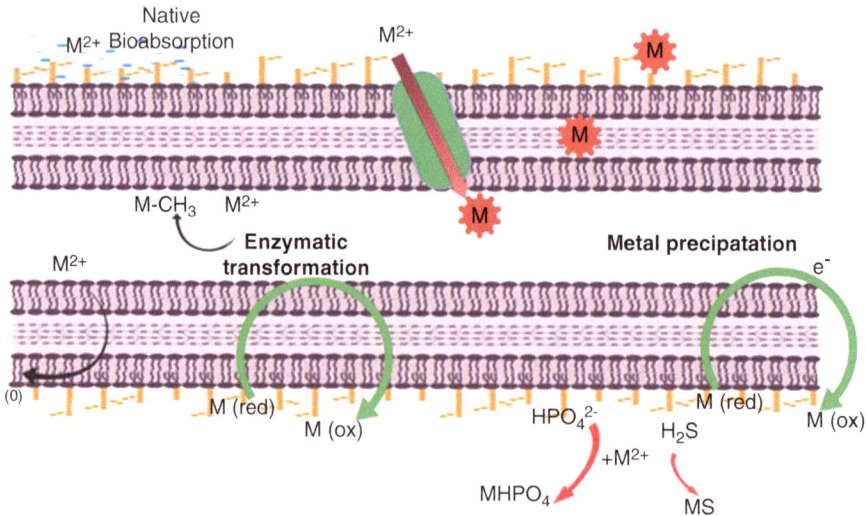

Fig. 7.1 Metal resistance system in Gram-negative bacteria

A passive sorption of ions of heavy metal was observed in nonviable cells of *P. putida* at high levels, while some studies revealed that the accumulation of metal ions by viable bacterial cells occurs in two stages – rapid non-specific sorption on the surface of the cell wall and, far along, long-term gathering of these ions of metal in the cytoplasm (Gadd 1990, McEldowney 2000). HM ions can also be bound by bacterial capsular polymers, mainly carboxyl groups of polysaccharides. The ability to bind metal ions by extracellular biopolymers was observed in *Marinobacter* sp. and *Acinetobacter* sp. (Bhaskar and Bhosle 2006). Interestingly, in *P. aeruginosa* biofilm cells had a significantly higher resistance to HMs (Cu, Pb and Zn) (Teitzel and Parsek 2003).

Active transport or, in other words, efflux represents the most extensive system for the resistance of bacteria to metal ions. By active transport, bacteria can remove metal ions from the cell. Efflux systems can be encoded by both chromosomal (Franke et al. 2001) and plasmid genetic determinants (Cervantes and Gutierrz-Corona 1994).

HM resistance in Gram-negative bacteria is mostly contributed by lipopolysaccharide of cell walls, a significant component of the outer membrane. The outer layers of cells probably determine how much of a metal penetrates the cytoplasm. HM resistance in Gram-negative bacteria also relates to secretion system that deliver multiple effector proteins into cells or into the extracellular milieu. Wang et al. (2015) have revealed that T6SS4 of *Y. pseudotuberculosis* have a prominent role in transporting the zinc ions (Zn^{2+}) from the environment into bacterial cells to mitigate the detrimental hydroxyl radicals induced by multiple stressors and prevent the cells from having oxidative damage. Similarly, an H3-T6SS secreted effector TseF (PA2374) of *P. aeruginosa* is involved in iron uptake by interacting with outer membrane vesicles (OMVs) and the *Pseudomonas* quinolone signal (PQS) system (Lin et al. 2017).

7.2.1   Lead (Pb)

Pb is not known to be of any biological significance but is toxic at very low concentrations (Bruins et al. 2000). However, the bioavailable fraction of Pb(II), to which microbes are exposed, may be rather low (Kotuby-Amacher et al. 1992). In the *Micrococcus luteus* and *Azotobacter* sp., the cell wall and several functional macromolecules are involved in binding Pb(II). The ions were revealed to be presented in the cell wall and cell membrane, and the least portion was found in the cytoplasmic fraction (Tornabene and Edwards 1972; Tornabene and Peterson 1975). Jarosławiecka and Piotrowska-Seget (2014) have reviewed *Cupriavidus metallidurans* CH34 for its unique mechanism combining efflux and lead precipitation. Pb(II) toxicity can alter the conformation of nucleic acids and proteins, inhibit the enzyme activity and disrupt the membrane functions and oxidative phosphorylation, as well as osmotic balance disruption (Bruins et al. 2000). Hasnain et al. (1993) have isolated the Pb(II)-resistant Gram-negative bacteria (*P. marginalis*, *P. vesicularis* and *Enterobacter* sp.) from metal-contaminated soils, industrial wastes and plants growing on metal-contaminated soils.

7.2.2 Cadmium (Cd)

In the hazardous substance list, Cd is also in the priority list and is classified at the seventh place by the Agency for Toxic Substances and Disease Registry (ATSDR) in 2017 (ATSDR 2017). Several studies have marked Cd as a highly toxic element even at low concentrations (Figueira et al. 2005, Lima et al. 2006, Aksoy et al. 2014, Vinodini et al. 2015). Tremaroli et al. (2009) have reported that the exposure to metals changes bacterial metabolism and altered growth pattern in time-dependent manner (Khan et al. 2016). Fazeli et al. (2010) have found that Gram-negative bacteria were less sensitive to cadmium compared to the Gram-positive bacteria. During the experiment, 23–50 mg kg^{-1} cadmium (drinking water) was given to healthy mice, and after 45 days intestinal microflora was aseptically collected and bacterial count was performed. In comparison with the control, it was revealed that bacteria of genera *Bacillus cereus*, *Lactobacillus* spp., *Clostridium* spp., *E. coli*, *Klebsiella* spp., *Pseudomonas* spp., *Enterococcus* spp. and *Proteus* spp. were presented which could be due to their possible ability to uptake the Cd ions. In *E. coli*, Cd effects by extending the lag phase of cultures, though normal proliferation was detected at the end of the lag phase (Jaiganesh et al. 2012). Among the microbial plethora, plant-associated soil bacteria are of great interest because of their potentiality in nitrogen fixation, agriculture and industrial production, crop protection and plant-mediated biodegradation (Singh et al. 2016, Singh et al. 2019, Prajakta et al. 2019, Yang et al. 2019). High Cd resistance in soilborne bacterium *Enterobacter* sp. strain EG16 was found in multimetal-polluted site of Dabao Mountain, Guangdong, China. It tolerates high level of Cd^{2+} concentrations (MIC, >250 mg). Strain EG16 accumulated the 31% of the total Cd by surface biosorption (Chen et al. 2016). Similarly, *Rhizobium* sp. strain E20–8, isolated from *Pisum sativum* root nodules, grown at non-contaminated field in Southern Portugal was reported as Cd tolerant (Matos et al. 2019).

7.2.3 Mercury (Hg)

In HMs, Hg is in any form poisonous, is persistent in nature and has been ranked third by the US Government Agency (1999) for Toxic Substances and Disease Registry. These most toxic elements or substances are continuing to be dumped and spilled into the soil, water and atmosphere and consumed by living beings (Clifton 2007). Currently, Hg resistance in bacteria is in its fifth decade. Unfortunatelly, the bacterial metal resistant mechanism is leads to transformation of its toxic target at large scale (Barkay et al. 2003). The Gram-negative bacteria consist of a regulatory gene (merR), an operator/promoter region and at least three structural genes merT, merP and merA as Hg-resistant elements (Etesami 2017). Regulatory genes are basically encoded for three components – a membrane transport protein, a Hg^{2+}-binding protein for periplasmic and an enzyme subunit for mercuric reductase and participated as central player for Hg resistance. In Gram-negative bacteria, the *mer* operons, a fairly high GC content, averaging 61% overall ((Liebert et al. 2000), have been studied extensively for the bacterial Hg resistance. However, *mer* operons are highly homologous in most of Gram-negative bacteria (Trajanovska et al. 1997). Hg(II) in Gram-negative bacteria competes with MerT's cytosolic cysteines as a dithiol derivative which is a substrate form for MerA. MerC is also the most common in Gram-negative bacteria, but the presence of MerF and MerE in sequence data is also recorded. Resistance to inorganic and organic mercurial in bacteria drives with MerB. Recently, MerG, a speciously periplasmic protein, has been reported in some Gram-negative bacteria. Moreover, in several Gram-negative bacteria, an additional protein, MerD, appears which antagonizes MerR's activation of *mer* operon transcription. Parkhill and Brown (1990) have reported the presence of MerO, an 18-bp hyphenated dyad with 7-bp palindromes flanking a 4-bp AT-rich centre, in most Gram-negative bacteria.

Several studies have revealed the Hg-resistant Gram-negative bacteria. Pérez-Valdespino et al. (2013) isolated the *Aeromonas* strains from diarrhoea sample and showed that Hg resistance occurs via mercuric ion reduction and indicated the presence of high variable *mer* operons in *Aeromonas*. Hg-resistant *Aeromonas* strains, *A. hydrophila*, *A. caviae*, *A. veronii*, *A. aquariorum* and *A. media* were characterized in Mexico (Aguilera-Arreola et al. 2007). Hg-tolerant *P. aeruginosa* has isolated from hospital sewage of Brazil (Lima de Silva et al. 2012), surface river water of Pole Khan and Pole Petroshimi stations, Iran (Mirzaei et al. 2013). Gram-negative bacteria from water samples were reported to be resistant at 110–200 μg/ml (Shakoori and Muneer 2002, Alam and Imran 2017). *Beijerinckia* and *Azotobacter* sp., N_2-fixing bacteria, have been reported for their ability to remove Hg (Ray et al. 1989).

7.2.4 Chromium (Cr)

Cr is also speculated as one of the 17 extremely hazardous chemicals by the US Environmental Protection Agency (USEPA) (Marsh and McInerney 2001). Cr(VI) compounds are not only toxic for humans but also responsible for the alteration of bacterial diversity in soil ecosystem (Turpeinen et al. 2004; Viti 2006). Moreover, the bacterial growth rate declines (Garbisu et al. 1998), or the lag phase is extended with uncoupling of energy (Nepple et al. 2000), as the chromate concentration is progressively increased (Chardin et al. 2002). The toxicity of Cr(VI) in bacteria also affected the morphological symmetry and filamentous growth, altered the gene expression, activated the SOS response to counteract the oxidative stress and resulted in the induction of prophage-related genes (Ackerley et al. 2006). Chourey et al. (2006) have revealed that Cr stress also negatively affected the cell division, DNA metabolism and gene regulation, chemotaxis and protein transport system, biosynthesis and degradation of murein, membrane response and environmental stress protection mechanisms.

Cr resistance in bacteria has been generally observed in chromium-contaminated habitats such as soil, wastewater, industrial effluents, etc. (Pal et al. 2005). *Pseudomonas* sp., a Gram-negative bacterial strain, was first reported as Cr(VI) resistant and has the ability to reduce Cr(VI) (Romanenko and Korenkov 1977). Cr detoxification was observed by the reduction of Cr(VI) to Cr(III), through Cr(V) and Cr(IV) intermediates through *P. aeruginosa* (Bopp and Ehrlich 1988a). Interestingly, chromium (III) is less toxic than chromium (VI) (approximately 1000 times) due to their impermeability to cell membrane. Several studies have already been revealed the resistance of Gram-negative bacteria to Cr such as Cr-resistant and Cr-reducing bacteria *Serratia marcescens* (Campos et al. 2005) *Acinetobacter* and *Ochrobactrum* isolated from the activated sludge of a wastewater treatment plant, Portugal (Francisco et al. 2002), *E. casseliflavus* (Saranraj et al. 2010) *E. gallinarum* with the ability to reduce the chromate to 100% at a concentration of 200 mg/l (Sayel et al. 2012). The removal of Cr(VI) from aqueous solution using kaolin-supported bacterial biofilms was evaluated by Khyle et al. (2018). They revealed that the adsorptive capacity of Gram-negative *E. coli* is higher than the Gram-positive *S. epidermidis*. Thacker and Madamwar (2005) have isolated and identified the Cr-resistant (>300 ppm) *Ochrobactrum* sp. DM1 from chemical industry sites and speculated 30 kDa inducible protein for chromium reduction.

Cr resistance in bacterial cell was clearly illustrated by Ahemad (2014). He reported that Cr enters into the bacterial cell via sulphate transporter (encoded by chromosomal DNA), which is due to the homology between Cr and sulphate. Bacterial cells resist to chromate toxicity by exorcise the intracellular chromates outside through efflux systems (encoded by plasmid DNA). Aerobic and anaerobic reduction of Cr(VI) ion into Cr(III) ion involves soluble reductase which requires NAD(P) and electron transport pathway by cytochrome b or c along the respiratory chains in the inner membrane, respectively. While, Cr(VI) ion redox cycle produced the Cr(V) ion by the production of reactive oxygen species (ROS) in oxidative stresses.

7.2.5 Copper (Cu)

Cu is an essential element required in traces for cellular process and participates as component of proteins. But, during changes in cuprous and cupric, it generates the reactive toxic radicals (Ridge et al. 2008). Hiniker et al. (2005) reported that free copper causes the cellular sulfhydryl pool depletion and decreases the cellular viability. Hence, the bacterial system has evolved the mechanism to control the intracellular copper level and save itself from Cu cation toxicity (Waldron and Robinson 2009).

Biocidal action of Cu ions is due to the electrostatic attraction with cell membrane and is affected more in Gram-negative bacteria than Gram-positive bacteria (Vergara-Figueroa et al. 2019). The excess of Cu in Gram-negative bacteria is mainly controlled by the cue regulon which is composed by CueR, included with P1B-1-type ATPase coding genes (a periplasmic multicopper oxidase (MCO)), for sensing the Cu(I) ions' presence in cytoplasm and small metal chaperones of cytoplasm (Outten et al. 2000, Espariz et al. 2007, Zhang and Rainey 2008). Cu(I) export from cytoplasm to the periplasm is done by ATPase (Rensing et al. 2000), and Cu(I) is converted into a less toxic Cu(II) by oxygen-mediated oxidation governed by MCO (Singh et al. 2004). Majorly, periplasmic Cu homeostasis in aerobic condition is maintained by the above described mechanisms in Gram-negative bacteria. In *E. coli*, CueR regulon is composed by the ATPase (copA) and the MCO, while in *P. fluorescens* SBW25 it is by the ATPase and the Cu chaperone-coding genes, respectively (Outten et al. 2001; Zhang and Rainey 2008).

Cu-resistant *P. syringae* was reported by Cha and Cooksey (1991), which has the ability to produce Cu-inducible proteins CopA, CopB and CopC that are responsible for binding bacterial cell and copper ions. Moreover, the long-term exposure of Cu also stimulates the genetic determinants for Cu adaptation. Five isolates from three Gram-negative genera, *Sphingomonas*, *Stenotrophomonas* and *Arthrobacter*, were procured and all were resistant (3.1–4.7 mM) to Cu (Altimira et al. 2012). The Gram-negative Enterobacteriaceae (*E. coli* and *S. enterica*) encoded the homologous Cue-responsive regulon system (Samanovic et al. 2012). Gram-negative strains of *E. coli*, *Enterobacter* spp., *Klebsiella pneumoniae* and *Pseudomonas aeruginosa* have been isolated from the hospital environment of AGH University of Science and Technology, Kraków, and all were more sensitive to copper (Różańska et al. 2018).

7.2.6   Arsenic (As)

Arsenic (As) is also often present in the environment and is very toxic for most microorganisms. For the resistance, some microbial strains have genetic determinants, e.g. bacterial plasmids. In Gram-negative bacteria, it encodes specific efflux pumps that have the ability to extrude the As from cell cytoplasm. The efflux pump consists of a two-component ATPase complex (ArsA and ArsB) and is integrated with membrane subunit (Rosen and Liu 2009; Yang et al. 2012). arsBC gene pair is commonly found in Gram-negative bacterial chromosome. Several studies have reported the various genes and gene clusters associated with plasmids as well as chromosomes, for example, As-resistant arsRBC or arsRDABC gene cluster associated with plasmids in *E. coli* and *A. multivorans* AIU 301 (Suzuki et al. 1998), ars operon variants in the marine strain *P. fluorescens* MSP3 (Chen et al. 2016), As genomic island of SinA plasmid in *Sinorhizobium* sp. and Thiomonas sp. (Freel et al. 2015). The diversity analysis of the arsenic-contaminated old tin mine area in Thailand has procured 262 isolates, and among them, A. koreensis and β-proteobacteria were found as the dominant species of the soil. Interestingly, Areonmit et al. (2010) have revealed that majority of the As-resistant isolates were Gram-negative bacteria. Harmin et al. (2018) have isolated the As-resistant *E. asburiae*, *S. paucimobilis*, *Pantoea* spp., *Rhizobium rhizogenes* and *R. radiobacter* (MIC of >1500 mg/L of As).

7.2.7   Iron (Fe)

In HM resistance, iron uptake system is very significant in bacteria because Fe-mediated gene expression is controlled by the "global" transcriptional regulator *fur*, which is conserved in many bacterial genera. More than 90 genes in *E. coli* and

87 in *P. aeruginosa* are known to be regulated by *fur*. Studies revealed that Fur also regulates a varied range of metabolic functions in bacteria, such as respiration, chemotaxis, the tricarboxylic acid cycle, glycolysis, amino acid biosynthesis, DNA synthesis and sugar metabolism, protecting the cell from oxidative damage and redox stress conditions. Typically, genes that are involved in Fe uptake are expressed only when Fe is deficient (Guerinot and Yi 1994). Some of the bacteria like *B. japonicum* is best exemplified to know the regulatory function of fur protein (Singh et al. 2011). In *B. japonicum*, the symbiont of soybeans, an outer membrane protein that is made in response to Fe starvation is similar to the hydroxamate receptor, FhuA, of *E. coli* (James et al. 2008). When Fe is replete, Fur, which interacts with Fe, binds to DNA sequences (*fur* boxes) that overlap the target promoters, repressing their transcription (Guerinot and Yi 1994, Braun et al. 1998). In the absence of Fe, Fur no longer binds, allowing transcription to occur. Fur also affects the transcription of genes concerned with traits as varied as toxin production, superoxide dismutase or acid tolerance (Tsolis et al. 1995). At some promoters, Fur is a positive regulator, in response to the cell's Fe status (Tsolis et al. 1995). Viable *fur* alleles have also been isolated in *P. aeruginosa*, based on the fact that fur mutant strains are manganese resistant (Prince et al. 1993).

7.3   Biotechnological Prospective of Heavy Metal Resistance

Interactions of microorganisms with HMs are vital for various biotechnological interests. Tolerance to HMs is generally present in bacteria due to the horizontal gene transfer (Ianeva 2009). Naturally occurring microorganisms are capable of reducing and detoxifying heavy metal contamination from industrial effluents. This adaptation occurs due to the development of the cellular protection mechanism system against toxic metal ions of microorganisms. Naturally occurring bacteria such as *Gemella* sp. and *Micrococcus* sp. showed biodegradation capacity to metals like cadmium (Cd), chromium (Cr) and lead (Pb) where *Hafnia* sp. showed resistance to cadmium (Cd) (Marzan et al. 2017). Several studies have revealed that Gram-negative bacteria are likely to be more tolerant to HMs than Gram-positive (Silva et al. 2012). HM resistance in biotechnology has great importance for creating the value-added product by adding metal resistance to a microorganism for the facilitation of biotechnological process, biomining of expensive metals and bioremediation of metal-contaminated environments. Bacteria might be established in the sewage plant or plasmids with a broad host range of replication, and it could be cloned into the bacterial community for economic and social uses. For HM resistance *E. coli*, *Pseudomonas* sp., etc. could be a better system to understand how bacteria manage metal homeostasis via several mechanisms and timing of metal sequestration.

References

Abbas HS, Ismail MI, Mostafa MT, Sulaymon HA (2014) Biosorption of heavy metals: a review. J Chem Sci Technol v3:74–102

Ackerley DF, Barak Y, Lynch SV, Curtin J, Matin A (2006) Effects of chromate stress on *Escherichia coli* K-12. J Bacteriol 188:3371–3381

Aguilera-Arreola MG, Hernández-Rodríguez C, Zúñiga G, Figueras MJ, Garduño RA et al (2007) Virulence potential and genetic diversity of *Aeromonas caviae*, *Aeromonas veronii*, and *Aeromonas hydrophila* clinical strains from Mexico and Spain: a comparative study. Can J Microbiol 53:877–887

Ahemad M (2014) Bacterial mechanisms for Cr(VI) resistance and reduction: an overview and recent advances. Folia Microbiol (Praha) 59(4):321–332

Ahmad I, Hayat S, Ahmad A, Inam A, Samiullah (2005) Effect of heavy metal on survival of certain groups of indigenous soil microbial population. J Appl Sci Environ Manage 9(1):115–121

Aksoy E, Salazar J, Koiwa H (2014) Cadmium determinant 1 is a putative heavy-metal transporter in *Arabidopsis thaliana*. FASEB J 28(617):4

Alam M, Imran M (2017) Metal tolerance analysis of gram-negative bacteria from hospital effluents of Northern India. J Appl Pharm Sci 7(4):174–180

Altimira F, Yáez C, Bravo G, González M, Rojas L et al (2012) Characterization of copper-resistant bacteria and bacterial communities from copper-polluted agricultural soils of Central Chile. BMC Microbiol 12:193. https://doi.org/10.1186/1471-2180-12-193

Areonmit PJ, Sajjaphan K, Sadowsky MJ (2010) Structure and diversity of arsenic-resistant bacteria in an old tin mine area of Thailand. J Microbiol Biotechnol 20(1):169–178

ATSDR (2017) Substance priority list. Available from: https://www.atsdr.cdc.gov/ SPL

Azarbad H, Van GCAM, Niklińska M, Laskowski R, Röling WFM et al (2016) Resilience of soil microbial communities to metals and additional stressors: DNA-based approaches for assessing stress-on-stress responses. Int J Mol Sci 17:20

Barkay T, Susan M, Miller Anne Summers O (2003) Bacterial mercury resistance from atoms to ecosystems. FEMS Microbiol Rev 27(2–3):355–384

Bhaskar PV, Bhosle NB (2006) Bacterial extracellular polymeric substance (EPS): a carrier of heavy metals in the marine food-chain. Environ Int 32(2):191–198

Bopp LH, Ehrlich HL (1988a) Chromate resistance and reduction in *Pseudomonas fluorescens* strain LB300. Arch Microbiol 150:426–431

Bopp LH, Ehrlich HL (1988b) Chromate resistance and reduction in *Pseudomonas fluorescens* strain LB300. Arch Microbiol 150:426–431

Braun V (1998) Pumping iron through cell membranes. Science 282:2202–2203

Bruins MR, Kapil S, Oehme FW (2000) Microbial resistance to metals in the environment. Ecotoxicol Environ Saf 45:198–207

Campos VL, Moraga R, Yánez J, Zaror CA, Mondaca MA (2005) Chromate reduction by *Serratia marcescens* isolated from tannery effluent. Bull Environ Contam Toxicol 75(2):400–406

Cervantes C, Gutierrz-Corona F (1994) Copper resistance mechanisms in bacteria and fungi. FEMS Microbiol Rev 14(2):121–138

Cha JS, Cooksey DA (1991) Copper resistance in *Pseudomonas syringae* mediated by periplasmic and outer membrane proteins. PNAS U S A 88(20):8915–8919

Chandrangsu P, Rensing C, Helmann JD (2017) Metal homeostasis and resistance in bacteria. Nat Rev Microbiol 15(6):338–350

Chardin B, Dolla A, Chaspoul F, Fardeau ML, Gallice P et al (2002) Bioremediation of chromate: thermodynamic analysis of the effects of Cr(VI) on sulfate-reducing bacteria. Appl Microbiol Biotechnol 60:352–360

Chen J, Yoshinaga M, Garbinski LD, Rosen BP (2016) Synergistic interaction of glyceraldehydes-3-phosphate dehydrogenase and ArsJ, a novel organoarsenical efflux permease, confers arsenate resistance. Mol Microbiol 100:945–953. https://doi.org/10.1111/mmi.13371

Choudhury R, Srivastava S (2001) Zinc resistance mechanisms in bacteria. Curr Sci 81(7):768–775

Chourey K, Thompson MR, Morrell-Falvey J, VerBerkmoes NC, Brown SD et al (2006) Global molecular and morphological effects of 24-h chromium(VI) exposure on *Shewanella oneidensis* MR-1. Appl Environ Microbiol 72:6331–6344

Clifton JC (2007) Mercury exposure and public health. Pediatr Clin N Am 54(2):237–269

Dian C (2018) Effects of heavy metals on soil microbial community. IOP Conf Ser Earth Environ Sci 113:12009. https://doi.org/10.1088/1755-1315/113/1/012009

Espariz M, Checa SK, Audero ME, Pontel LB, Soncini FC (2007) Dissecting the *Salmonella* response to copper. Microbiology 153:2989–2997

Etesami H (2017) Bacterial mediated alleviation of heavy metal stress and decreased accumulation of metals in plant tissues. Mechanisms and future prospects. Ecotoxicol Environ Saf 147:175–191

Fazeli M, Hassanzadeh P, Alaei S (2010) Cadmium chloride exhibits a profound toxic effect on bacterial microflora of the mice gastrointestinal tract. Hum Exp Toxicol 30(2):152–159

Figueira EMDAP, Gusmão LAI, Pereira SIA (2005) Cadmium tolerance plasticity in *Rhizobium leguminosarum* bv. viciae: glutathione as a detoxifying agent. Can J Microbiol 51:7–14

Francisco R, Alpoim MC, Morais PV (2002) Diversity of chromium-resistant and -reducing bacteria in a chromium-contaminated activated sludge. J Appl Microbiol 92(5):837–843

Franke S, Grass G, Nies DH (2001) The product of the *ybd*E gene of the *Escherichia coli* chromosome is involved in detoxification of silver ions. Microbiology 147(4):965–972

Freel KC, Krueger MC, Farasin J, Brochier-Armanet C, Barbe V et al (2015) Adaptation in toxic environments: arsenic genomic islands in the bacterial genus *Thiomonas*. PLoS One 10:e0139011. https://doi.org/10.1371/journal.pone.0139011

Fu F, Wang Q (2011) Removal of heavy metal ions from wastewaters: a review. J Environ Manag 92:407–418

Gadd GM (1990) Heavy metal accumulation by bacteria and other microorganisms. Experientia 46:834–840. https://doi.org/10.1007/BF01935534

Gadd GM (1992) Metals and microorganisms: a problem of definition. FEMS Microbiol Lett 100:197–204

Garbisu C, Alkorta I, Llama MJ, Serra JL (1998) Aerobic chromate reduction by *Bacillus subtilis*. Biodegradation 9:133–141

Giller KE, Witter E, McGrath SP (1998) Toxicity of heavy metals to microorganisms and microbial processes in agricultural soil: a review. Soil Biol Biochem 30:1389–1414

Gomathy M, Sabarinathan KG (2010) Microbial mechanisms of heavy metal tolerance- a review. Agric Rev 31(2):133–138

Guerinot ML, Yi Y (1994) Iron: nutritious, noxious and not readily available. Plant Physiol 104:815–820

Gülser F, Erdoğan E (2008) The effects of heavy metal pollution on enzyme activities and basal soil respiration of roadside soils. Environ Monit Assess 145(1):127–133

Harmin T, Abdullah S, Idris SRS, Mushrifah A, Nurina B (2018) Arsenic resistance and biosorption by isolated rhizobacteria from the roots of Ludwigia octovalvis. Int J Microbiol 2018:1–10

Hasnain, Yasmin S, Yasmin A (1993) The effects of lead resistant pseudomonads on the growth of Triticum aestivum seedlings under lead stress. Environ Pollut 81:179–184

Hiniker A, Collet JF, Bardwell JC (2005) Copper stress causes an in vivo requirement for the *Escherichia coli* disulfide isomerase DsbC. J Biol Chem 280:33785–33791

Ianeva OD (2009) Mechanisms of bacteria resistance to heavy metals. Mikrobiol Z 71(6):54–65

Jaiganesh T, Rani JDV, Girigoswami A (2012) Spectroscopically characterized cadmium sulfide quantum dots lengthening the lag phase of *Escherichia coli* growth. Spectrochim Acta A 92:29–32

James KJ, Hancock MA, Moreau V, Molina F, Coulton JW (2008) TonB induces conformational changes in surface-exposed loops of FhuA, outer membrane receptor of *Escherichia coli*. Protein Sci 17:1679–1688

Jarosławiecka A, Piotrowska-Seget Z (2014) Lead resistance in micro-organisms. Microbiol (United Kingdom) 160(1):12–25

Jin Y, Luan Y, Ning Y, Wang L (2018) Effects and mechanisms of microbial remediation of heavy metals in soil: a critical review. Appl Sci 8:1336

Khan Z, Rehman A, Hussain SZ, Nisar MA (2016) Cadmium resistance and uptake by bacterium, *Salmonella enterica* 43C, isolated from industrial effluent. AMB Exp 6:54. https://doi.org/10.1186/s13568-016-0225-9

Khyle GQ, Bonifacio D, Cybelle MF, Meng-Wei W (2018) Removal of chromium(VI) and zinc(II) from aqueous solution using kaolin-supported bacterial biofilms of Gram-negative *E. coli* and gram-positive *Staphylococcus epidermidis*. Sustain Environ Res 28(5):206–213

Kotuby-Amacher J, Gambrell RP, Amacher MC (1992) The distribution and environmental chemistry of lead in soil at an abandoned battery reclamation site. Eng Aspects Metal Waste Manag:1–24

Liebert CA, Watson A, Summers O (2000) The quality of *merC*, a module of the *mer* mosaic. J Mol Evol 51:607–622

Lima AIG, Corticeiro SC, Figueira EMDAP (2006) Glutathione-mediated cadmium sequestration in Rhizobium leguminosarum. Enzym Microb Technol 39:763–769

Lima de Silva AA, de Carvalho MA, de Souza SA, Dias PM, Filho d S et al (2012) Heavy metal tolerance (Cr, Ag and Hg) in bacteria isolated from sewage. Brazil J Microbiol 43(4):1620–1631

Lin J, Zhang W, Cheng J, Yang X, Zhu K et al (2017) A *Pseudomonas* T6SS effector recruits PQS-containing outer membrane vesicles for iron acquisition. Nat Commun 8:14888. https://doi.org/10.1038/ncomms14888

Ma Y, Rajkumar M, Zhang C, Freitas H (2016) Beneficial role of bacterial endophytes in heavy metal phytoremediation. J Environ Manag 174:14–25

Marsh TL, McInerney MJ (2001) Relationship of hydrogen bioavailability to chromate reduction in aquifer sediments. Appl Environ Microbiol 67(4):1517–1521

Marzan LW, Hossain M, Mina SA, Akter Y, Chowdhury AMMA (2017) Isolation and biochemical characterization of heavy-metal resistant bacteria from tannery effluent in Chittagong city, Bangladesh: bioremediation viewpoint. Egypt J Aquat Res 43:65–74

Matos D, Sa C, Cardoso P, Pires A, Rocha SM et al (2019) The role of volatiles in Rhizobium tolerance to cadmium: effects of aldehydes and alcohols on growth and biochemical endpoints. Ecotoxicol Environ Saf 186:109759

McEldowney S (2000) The impact of surface attachment on cadmium accumulation by *Pseudomonas fluorescens* H2. FEMS Microbiol Ecol 33(2):121–128

Mills AL, Colwell RR (1977) Microbiological effects of metal ions in Chesapeake Bay water and sediment. Bull Environ Contam Toxicol 18:99–103

Mirzaei N, Rastegari H, Kargar M (2013) Antibiotic resistance pattern among gram- negative mercury resistant bacteria isolated from contaminated environments, jundishapur. J Microbiol 6(10):e8085

Ndeddy ARJ, Babalola OO (2016) Effect of bacterial inoculation of strains of *Pseudomonas aeruginosa*, *alcaligenes faecalis* and *Bacillus subtilis* on germination, growth and heavy metal (cd, Cr, and Ni) uptake of *Brassica juncea*. Int J Phytorem 18(2):200–209

Nepple BB, Kessi J, Bachofen R (2000) Chromate reduction by *Rhodobacter sphaeroides*. J Ind Microbiol Biotechnol 25:198–203

Nies DH (1999) Microbial heavy-metal resistance. Appl Microbiol Biotechnol 51(6):730–750. https://doi.org/10.1007/s002530051457

Outten FW, Outten CE, Hale J, O'Halloran TV (2000) Transcriptional activation of an *Escherichia coli* copper efflux regulon by the chromosomal MerR homologue, CueR. J Biol Chem 275:31024–31029

Outten FW, Huffman DL, Hale JA, O'Halloran TV (2001) The independent cue and cus systems confer copper tolerance during aerobic and anaerobic growth in *Escherichia coli*. J Biol Chem 276:30670–30677

Pal A, Dutta S, Mukherjee PK, Paul AK (2005) Occurrence of heavy metal-resistance in microflora from serpentine soil of Andaman. J Basic Microbiol 45(3):207–218

Parkhill J, Brown NL (1990) Site-specific insertion and deletion mutants in the *mer* promoter-operator region of Tn*501*, the 19bp spacer is essential for normal induction of the promoter by MerR. Nucleic Acids Res 18:5157–5162

Pérez-Valdespino A, Celestino-Mancera M, Villegas-Rodríguez VL, Curiel-Quesada E (2013) Characterization of mercury-resistant clinical Aeromonas species. Brazil J Microbiol 44(4):1279–1283. https://doi.org/10.1590/S1517-83822013000400036

Prajakta BM, Suvarna PP, Singh RP, Rai AR (2019) Potential biocontrol and superlative plant growth promoting activity of indigenous *Bacillus mojavensis* PB-35(R11) of soybean (*Glycine max*) rhizosphere. SN Appl Sci 1:1143. https://doi.org/10.1007/s42452-019-1149-1

Prince RW, Cox CD, Vasil ML (1993) Coordinate regulation of siderophore and exotoxin A production: molecular cloning and sequencing of the *Pseudomonas aeruginosa* fur gene. J Bacteriol 175:2589–2598

Quintelas C, da Silva VB, Silva B, Figueiredo H, Tavares T (2011) Optimization of production of extracellular polymeric substances by *Arthrobacter viscosus* and their interaction with a 13X zeolite for the biosorption of Cr(VI). Environ Technol 32:1541–1549

Rajapaksha RMCP, Tobor-Kapłon MA, Bååth E (2004) Metal toxicity affects fungal and bacterial activities in soil differently. Appl Environ Microbiol 70(5):2966–2973

Ray S, Gachhui R, Pahan K, Chaudhury J, Mandal A (1989) Detoxification of mercury and organo-mercurials by nitrogen-fixing soil bacteria. J Biosci 14(2):173–182

Rensing C, Fan B, Sharma R, Mitra B, Rosen BP (2000) CopA: an *Escherichia coli* Cu(I)-translocating P-type ATPase. PNAS U S A 97:652–656

Ridge PG, Zhang Y, Gladyshev VN (2008) Comparative genomic analyses of copper transporters and cuproproteomes reveal evolutionary dynamics of copper utilization and its link to oxygen. PLoS One 3:e1378. https://doi.org/10.1371/journal.pone.0001378

Robin RS, Muduli PR, Vardhan KV, Ganguly D, Abhilash KR et al (2012) Heavy metal contamination and risk assessment in the marine environment of Arabian Sea, along the southwest coast of India. A J Chem 2(4):191–208. https://doi.org/10.5923/j.chemistry.20120204.03

Romanenko VI, Korenkov VN (1977) A pure culture of bacterial cells assimilating chromates and bichromates as hydrogen acceptors when grown under anaerobic conditions. Mikrobiolo 46:414–417

Rosen BP, Liu Z (2009) Transport pathways for arsenic and selenium: a minireview. Environ Int 35:512–515

Różańska A, Chmielarczyk A, Romaniszyn D, Majka G, Bulanda M (2018) Antimicrobial effect of copper alloys on *Acinetobacter* species isolated from infections and hospital environment. Antimicrob Resist Infect Control 7:10. https://doi.org/10.1186/s13756-018-0300-x

Sadler WR, Trudinger PA, Mineral D (1967) The inhibition of microorganisms by heavy metals. 2(3):158–168. https://doi.org/10.1007/BF00201912

Samanovic MI, Ding C, Thiele DJ, Darwin KH (2012) Copper in microbial pathogenesis: meddling with the metal. Cell Host Microbe 11(2):106–115

Saranraj P, Stella D, Reetha D, Mythili K (2010) Bioadsorption of chromium resistant *Enterococcus casseliflavus* isolated from tannery effluents. J Ecobiotechnol 2(7):17–22

Sayel H, Bahafid W, Joutey NT, Derraz K, Benbrahim KF (2012) Cr (VI) reduction by *Enterococcus gallinarum* isolated from tannery waste-contaminated soil. Ann Microbiol 62(3):1269–1277

Shakoori AR, Muneer B (2002) Copper-resistant bacteria from industrial effluents and their role in remediation of heavy metals in wastewater. Folia Microbiologica (Praha) 47:43–50

Siddiquee S, Rovina K, Azad SA (2015) Heavy metal contaminants removal from wastewater using the potential filamentous fungi biomass: a review. J Micro Biochem Technol 07(06):384–393

Silva AADL, Agostinho ARDC, Márcia ALDS, Sérgio ALTD, Patrícia MDSF et al (2012) Heavy metal tolerance (Cr, Ag and Hg) in bacteria isolated from sewage. Braz J Microbiol 43(4):1620–1631

Singh SK, Grass G, Rensing C, Montfort WR (2004) Cuprous oxidase activity of CueO from *Escherichia coli*. J Bacteriol 186:7815–7817

Singh RP, Singh RN, Srivastava AK, Kumar S, Dubey RC et al (2011) Structural analysis and 3D-modelling of fur protein from *Bradyrhizobium japonicum*. J Appl Sci Environ Sani 6(3):357–366

Singh RP, Manchanda G, Singh RN, Srivastava AK, Dubey RC (2016) Selection of alkalotolerant and symbiotically efficient chickpea nodulating rhizobia from North-West Indo Gangetic Plains. J Basic Microbiol 56:14–25. https://doi.org/10.1002/jobm.201500267

Singh RP, Manchanda G, Li ZF, Rai AR (2017) Insight of proteomics and genomics in environmental bioremediation. In: Bhakta JN (ed) Handbook of research on inventive bioremediation techniques. IGI Global, Hershey. https://doi.org/10.4018/978-1-5225-2325-3

Singh RP, Manchanda G, Maurya IK, Maheshwari NK, Tiwari PK et al (2019) *Streptomyces* from rotten wheat straw endowed the high plant growth potential traits and agro-active compounds. Biocatal Agric Biotechnol 17:507–513. https://doi.org/10.1016/j.bcab.2019.01.014

Subhashini DV, Singh RP, Manchanda G (2017) OMICS approaches: tools to unravel microbial systems. Directorate of Knowledge Management in Agriculture, Indian Council of Agricultural Research. ISBN: 9788171641703. https://books.google.co.in/books?id=vSaLtAEACAAJ

Suzuki K, Wakao N, Kimura T, Sakka K, Ohmiya K (1998) Expression and regulation ofthe arsenic resistance operon of Acidiphilium multivorum AIU 301 plasmid pKW301 in Escherichia coli. Appl Environ Microbiol 64:411–418

Taniguchi J, Hemmi H, Tanahashi K, Amano N, Nakayama T et al (2000) Zinc biosorption by a zinc-resistant bacterium, *Brevibacterium* sp. strain HZM-1. Appl Microbiol Biotechnol 54(4):581–588

Teitzel GM, Parsek MR (2003) Heavy metal resistance of biofilm and planktonic *Pseudomonas aeruginosa*. Appl Environ Microbiol 69(4):2313–2320

Thacker U, Madamwar D (2005) Reduction of toxic chromium and partial localization of chromium reductase activity in bacterial isolate DM1. W J Microbiol Biotechnol 21:891–899

Tipayno SC, Truu J, Samaddar S, Truu M, Preem JK (2018) The bacterial community structure and functional prof*ile in t*he heavy metal contaminated paddy soils, surrounding a nonferrous smelter in South Korea. Ecol Evol 8:6157–6168. https://doi.org/10.1002/ece3.4170

Tornabene TG, Edwards HW (1972) Microbial uptake of lead. Science 176:1334–1335

Tornabene TG, Peterson SL (1975) Interaction of lead and bacterial lipids. Appl Microbiol 29:680–684

Trajanovska S, Britz ML, Bhave M (1997) Detection of heavy metal ion resistance genes in gram-positive and gram-negative bacteria isolated from a lead-contaminated site. Biodegradation 8:113–124

Tremaroli V, Workentine ML, Weljie AM, Vogel HJ, Ceri H et al (2009) Metabolomic investigation of the bacterial response to a metal challenge. Appl Environ Microbiol 75:719–728

Tsolis RM, Bäumler AJ, Heffron F (1995) Role of Salmonella typhimurium Mn-superoxide dismutase (SodA) in protection against early killing by J774 macrophages. Infect Immun 63(5):1739–1744

Turpeinen R, Kairesalo T, Haggblom M (2002) Microbial activity community structure in arsenic, chromium and copper contaminated soils. J Environ Microbiol 35(6):998–1002

Turpeinen R, Kairesalo T, Haggblom M (2004) Microbial community structure and activity in arsenic, chromium and copper contaminated soils. FEMS Microbiol Ecol 47:39–50

US Department of Health and Human Services, Public Health Service (1999) Toxicological profile for mercury. US Department of Health and Human Services, Atlanta, pp 1–600

Vergara-Figueroa J, Alejandro-Martín S, Pesenti H, Cerda F, Fernández-Pérez A et al (2019) Obtaining nanoparticles of Chilean natural zeolite and its ion exchange with copper salt (Cu^{2+}) for antibacterial applications. Material 12(13):E2202. https://doi.org/10.3390/ma12132202

Vinodini N, Chatterjee PK, Chatterjee P, Chakraborti S, Nayanatara A et al (2015) Protective role of aqueous leaf extract of *Moringa oleifera* on blood parameters in cadmium exposed adult wistar albino rats. Int J Curr Res Acad Rev 3:192–199

Viti C (2006) Response of microbial communities to different doses of chromate in soil microcosms. J Appl Soil Ecol 34:125–139 von Bodman SB, Bauer WD, Coplin DL (2003) Quorum sensing in plant-pathogenic bacteria. Annu Rev Phytopathol 41:455–482

Waldron KJ, Robinson NJ (2009) How do bacterial cells ensure that metalloproteins get the correct metal? Nat Rev Microbiol 7:25–35

Wang T, Si M, Song Y, Zhu W, Gao F et al (2015) Type VI secretion system transports Zn^{2+} to combat multiple stresses and host immunity. PLoS Pathog 11:e1005020. https://doi.org/10.1371/journal.ppat.1005020

Wintz HT, Fox VC (2002) Functional genomics and gene regulation in biometals research. Biochem Soc Trans 30:766–768

Xie Y, Fan J, Zhu W, Amombo E, Lou Y et al (2016) Effect of heavy metals pollution on soil microbial diversity and bermudagrass genetic variation. Front Plant Sci 7:755. https://doi.org/10.3389/fpls.2016.00755

Xu Q, Shi G (2000) The toxic effects of single Cd and interaction of Cd with Zn on some physiological index of [*Oenanthe javanica* (Blume) DC]. J Nanjing Normal Uni 23(4):97–100

Yang HC, Fu HL, Lin YF, Rosen BP (2012) Pathways of arsenic uptake and efflux. Curr Top Membr 69:325–358

Yang YJ, Singh RP, Lan X, Zhang CS, Sheng DH et al (2019) Synergistic effect of *Pseudomonas putida* II-2 and *Achromobacter* sp. QC36 for the effective biodegradation of the herbicide quinclorac. Ecotoxicol Environ Saf. https://doi.org/10.1016/j.ecoenv.2019.109826

Zhang XX, Rainey PB (2008) Regulation of copper homeostasis in *Pseudomonas fluorescens* SBW25. Environ Microbiol 10:3284–3294

Zhou Y, Xu YB, Xu JX, Zhang XH, Xu SH et al (2015) Combined toxic effects of heavy metals and antibiotics on a *Pseudomonas fluorescens* strain ZY2 isolated from swine wastewater. Int J Mol Sci 16:2839–2850

Bioactive Compounds from Extremophiles

8

Indresh Kumar Maurya, Rahul Dilawari, Dipti Singh, and Raghvendra Pratap Singh

Abstract

Extremophiles are microorganisms that grow and survive in harsh environmental conditions (e.g., extreme temperature and pressure, variable pH, high salinity, radiation, toxic waste, and metal concentrations). Extremophilic microorganisms have developed diverse strategies in order to survive in harsh conditions and produced different types of bioactive molecules such as extremolytes, extremozymes, and cryoprotectant. These biomolecules possess extraordinary properties such as salt tolerance, thermostability, and pH adaptivity and represent unique attributes under unembellished conditions, which can be compared to existing industrial procedures. These bioactive molecules have great potential for application in various biotechnological processes including agriculture, pharmaceutical, and food industries. In recent years, due to innovative molecular biology tools, genomics, bioinformatics, data mining, and culturing approaches have given unique prospects to explore novel biomolecules of extremophiles. So, this chapter discusses the sources, properties, extraction techniques, and varieties of bioactive compounds from different extremophiles and their different industrial applications for human welfare.

Indresh Kumar Maurya and Rahul Dilawari have equally contributed to this chapter.

I. K. Maurya (✉)
Department of Microbial Biotechnology, Panjab University, Chandigarh, India

Centre of Infectious Diseases (CID), NIPER, S.A.S. Nagar (Mohali), Punjab, India

R. Dilawari
CSIR-Institute of Microbial Technology (CSIR-IMTECH), Chandigarh, India

D. Singh
Department of Microbiology, V.B.S. Purvanchal University, Jaunpur, Uttar Pradesh, India

R. P. Singh
Department of Research & Development, Biotechnology, Uttaranchal University, Dehradun, Uttarakhand, India

© Springer Nature Singapore Pte Ltd. 2020
R. P. Singh et al. (eds.), *Microbial Versatility in Varied Environments*,
https://doi.org/10.1007/978-981-15-3028-9_8

Keywords
Extremophiles · Biomolecules · Industrial applications · Harsh conditions

8.1   Introduction

Earth is full of various desolate and uninhabitable environments which are unsuitable for human beings to survive, but a variety of microbes thrive there. These microbes are called extremophiles and are known to produce myriad of products including enzymes (extremozymes), bioactive compounds (extremolytes), novel chemotypes, pharmacophores, and cryoprotectants (Stierle and Stierle 2014; Subhashini et al. 2017). These extremophiles are of variety including bacteria, actinomycetes, and fungi of different origins like halophilic, acidophilic, alkalophilic, and psychrophilic. Natural sources include hot water bodies, hydrothermal vents (high temperature), low-temperature areas in arctic regions, oceans, sediments of marine origin, extreme low-temperature deserts, glaciers, temporal forests area, and plants from extreme environments. Nature provides unique chemical scaffold for elaboration by employing chemical- and biochemical-based combinatorial approaches which would ultimately lead to unique and improved versions of agents optimized on the basis of their chemical and pharmacological properties. There are continuous destruction and threat to biodiversity of terrestrial and marine ecosystems along with scarcity of new chemical compounds in pharmaceutical industries, pushing a compelling thought to scientific community that there is an urgent need of new approaches in terms of expanded multidisciplinary experimentations and novel techniques for isolation, characterization, and applications. Improvement of techniques such as transcriptomics, proteomics, and metabolomics leads to discovery of several biologically active compounds isolated from bacteria, yeasts, and virus-infected cells that are important for improving the human health (Mazzoli et al. 2017). This would definitely be the best sustainable approach to utilize natural resources of novel leads for the development of bioactive compounds (Cragg and Newman 2005). Bioactive compounds include antibiotics; compounds with neurotoxic, cytotoxic, antimitotic, antiviral, and antineoplastic activities; enzyme inhibitors; and novel drugs (Singh et al. 2006). The cells of extremophiles are full of diverse changes in comparison to their normal counterparts. These changes include saturated/unsaturated fatty acids (branched/single chain) in cell wall providing differential environment and structural rigidity, modified DNA sequences, and cytosolic proteins helping cells to thrive extreme conditions. The specific altered cellular changes, chemical entity differences, and genetic material attributes are discussed in this chapter along with their bioactive compounds.

8.2 Bioactive Compounds from Extremophiles at a Glance

Microbes thriving in extreme environments have been documented to produce a variety of bioactive compounds. These products have been isolated and characterized well from archaea, eubacteria, fungi, etc. The extremophiles producing bioactive compounds include thermophiles, psychrophiles, acidophiles, alkaliphiles, xerophile, metallotolerant, and radioresistant extremophiles in particular (Fig. 8.1).

8.3 Gene Expression System of Extremophiles

Cultivable microbial extremophiles can be manipulated to increase products of interests, but it may be time-consuming and laborious (Gomez-Escribano and Bibb 2012). Heterologous gene expression in a variety of DNA vectors (plasmids, cosmids, fosmids) and myriad of hosts like bacteria, *E. coli*, yeasts, and insect cells arrived as main strategy used to improve expression, characterization, and modification of biosynthetic pathways for new products (Yang et al. 2019). Lots of challenges are associated with hosts including inclusion bodies formation and lack of posttranslational modifications. These problems have been rectified by introducing or deletion of biosynthetic pathways. Culture-based methods like changes in fermentation components and co-culturing microbes have been started to be exploited for the production of unknown bioactive compounds.

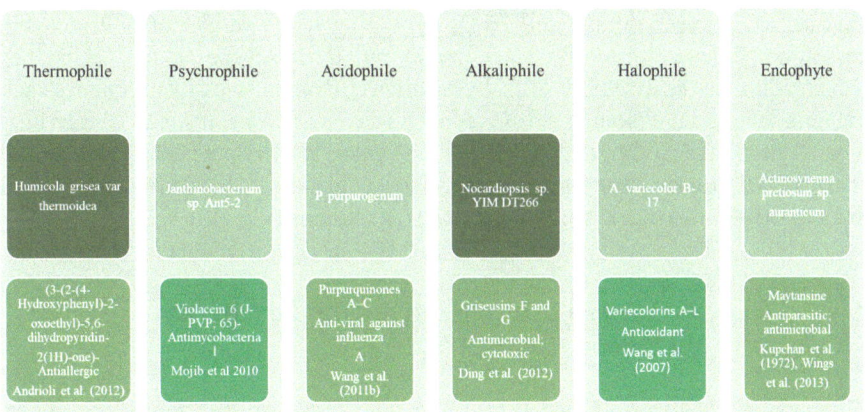

Fig. 8.1 Bioactive compounds produced by different extremophiles

8.4 Genetic and Metabolic Engineering of Extremophiles

Microbes have been engineered using recombinant DNA techniques along with genomic sequencing that helped a lot in manipulation and characterization of extremophilic genes for the last few decades using expression vectors, site-directed mutagenesis, CRISPR, etc. (Yang et al. 2018). There are lots of bioinformatic tools to predict novel natural and unnatural products like antiSMASH (Blin et al. 2014) and SpirPep for bioactive peptides discovery from a genome-wide database (Anekthanakul et al. 2018). Few examples include fluorometabolite fluorosalinosporamide A1 and salinosporamide X7. Further, the fluorosalinosporamide A1 was formed by mutating chlorinase gene salL in *Salinispora tropica*. Further, the fermentation media were supplemented with synthetic 5-fluoro-5-deoxyribose (Eustáquio and Moore Eustáquio and Moore 2008).

8.4.1 Heterologous Expression

For optimal and enhanced titer of foreign gene of interest, a series of constitutive promoters are required for the manipulation and integration of gene into host chromosome expression vector. Generally, the host microorganisms produce limited amount of own's metabolites for minimal biosynthesis interference of bioactive compounds. *S. coelicolor* produces its own secondary metabolites which interfere to heterologous expression of bioactive compounds. *E. coli* lacks endogenous secondary metabolic pathways (Giddings et al. 2015). Some of the best-known examples include the production and overexpression of moonlighting 300KDa type I PKS [6-deoxyerythronolide B synthase (DEBS)] in *E. coli* (Wang and Pfeifer 2008) which produces 6-deoxyerythronolideB (6-DEB), the aglycone variant of erythromycin antibiotic (Wiley et al. 1957). There are various modern gene cloning techniques which facilitate the assembly of DNA with fewer cloning steps like sequence- and ligation-independent cloning (SLIC) (Li and Elledge 2007), Gibson assembly ligation (Gibson et al. 2009), circular polymerase extension cloning (CPEC) (Quan and Tian 2014), Golden Gate assembly method (Engler and Marillonnet 2014), and transformation-associated recombination (TAR) (Larionov et al. 1996).

8.4.2 Gene Activation Using Additives and Cocultivation with Microbes

Commonly, when there is a need for active gene expression without genetic engineering or sequence expression, various factors such as pH, temperature, metal, antibiotics, and flavonoids can be used to boost gene expression levels. The addition of scandium and/or lanthanum in fermentation media of *S. sioyaensis* activated silent or poorly expressed nine genes in *S. coelicolor* A3 (2) from 2.5- to 12-folds (Tanaka et al. 2010). Addition of histone deacetylase (HDAC) inhibitors like sodium butyrate, MS-275, and valproic acid induced the expression of genes for the production of bioactive compounds including pigmented secondary metabolites, such as actinorhodin and undecylprodigiosinin *S. coelicolor* grown on minimal medium (Giddings and Newman 2013). Addition of $CaBr_2$ or inorganic fluoride to fungus *Phoma herbarum*, which was extracted from *Gloiopeltis tenax* (marine red alga) in Korea (Cheeokpo, Tongnyeong), resulted in the production of halogenated benzo-quinones bromochlorogentisylquinones A and B (Nenkep et al. 2010). Addition of acetyl-CoA to the fermentation media can increase the amount of secondary metabolite production. Activation of involved in secondary metabolites, like non-ribosomal peptides and polyketides (Giddings and Newman 2013), can be induced by eubacterial along with fungal (Schroeckh et al. 2009; Oh et al. 2007) and fungal-fungal interaction (Zhu and Lin 2006).

8.5 Metagenomics Approach to Identify New Bioactive Molecules

In natural environment, there are plenty of microbes and extremophiles which are not cultivable, but they may capable of lots of secondary metabolites production. It has been estimated that approximately 10^5 non-culturable species of eubacteria are present per gram of soil and ubiquitous myriad of non-culturable extremophiles are thriving in solfataric hot springs, glacier soil, and glacier ice by DNA rejoining kinetics (Torsvik et al. 1990, 1998), 16S rRNA sequencing, and fluorescent in situ hybridization (Langer-Safer et al. 1982). Metagenomic sequencing can help in identification of new compounds from unculturable microorganisms. In order to uncover the metabolic diversity of environmental samples and identify new biosynthetic clusters of interest, experts have devised sequence plus function-based metagenomic methods. These both techniques involve recombinant cloning of DNA from culture-free living cells and constructing metagenomic libraries transformed in a bacterial host for further sequencing purposes.

8.5.1 Sequence-Based Metagenomic Analyses

This approach uses DNA probe sequence tags from conserved regions of genes to find out homolog sequences from metagenomic sample that could not be otherwise searched or found. For example, non-purified DNA sample isolated from different samples were used by utilizing PCR-based primers covering cyanobactin sequences to recuperate gene bunches for production of novel natural molecules like ribosomal peptides. This gene fragment was latterly cloned and heterologous expressed in *E. coli* for industrial-scale production. Further, the DNA was extracted from flecking colonies of *Didemnum molle ascidians* from the regions of tropical pacific (Donia et al. 2011). With gene-targeted metagenomics (GT-metagenomics), a combination of PCR-based targeting and pyrosequencing, here DNA sequence information, is used to make DNA probes which can help in the recovery of full-length genes from clades of interest (Iwai et al. 2010).

8.5.2 Function-Based Metagenomic Analyses

This technique uses phenotypic traits like pigment production, antibiotic activity, enzymatic activity, and morphological changes commonly associated with secondary metabolite production. On the basis of known phenotypic trait, a variety of common metabolites of the same nature can be identified and isolated further for industrial usage. There are challenges associated with function-based metagenomics approach related to heterologous expression, environmental DNA size, library enrichment, toxic gene association encoded within DNA, and low detection frequency to point few of them (Penesyan et al. 2013).

8.6 Novel Biomolecules from Phylogenetic Diverse Microbial Communities

Taxonomic and phylogenetically similar sequences can help in identification and derivatization of novel bioactive compounds present in extremophilic environment. Ribosomal RNA samples can be analyzed typically by using samples from 16S or 18S rDNA gene library. Later on, PCR amplification using degenerate primers and sequencing is done to identify existing or homolog of eubacterial or fungal populations. Proteins like heat shock protein 70 (Hsp70), RNA polymerase B, recA/radA, and elongation factors Tu and G are reported as phylogenetic markers (Venter et al. 2004). Moreover, rRNA-based classification databases have been developed like SILVA database (Pruesse et al. 2007), Greengenes (DeSantis et al. 2006), and Ribosomal Database Project II (RDP II) (Cole et al. 2003).

8.7 Bioactive Compounds from Extremophiles and Their Applications

Various acidophiles from archaeal, bacterial, fungal, algal, and protozoal origin are thriving in extreme acidic solfataric fields, hot springs, marine vents, sulfuric pools and geysers, coal mines, and metallic ores (Sharma et al. 2011). Various species of following genera fall under this category: *Acidianus* sp., *Desulphurolobus* sp., *Stygiolobus* sp., *Sulfolobus* sp., *Sulphurisphaera* sp., *Sufurococcus* sp., *Thermoplasma* sp., and *Picrophilus* sp. (Bertoldo et al. 2004). They can grow even at pH 0.7 (Schleper et al. 1995). They have established specialized metabolites, genetic acquisitions, and macromolecular structural and functional characteristics for pH maintenance (Crossman et al. 2004, Baker-Austin and Dopson 2007; Rai et al. 2012).

Some of it grown optimally at pH >9 and extreme alkaliphiles with pH >10. Archaea, bacteria, and fungal alkaliphiles are distributed worldwide like soda lakes, hypersaline lakes, industrial effluents, some parts of insect guts, estuaries, and soil alkaline microenvironments (Hicks et al. 2010; Singh et al. 2016). The cells maintain alkaline pH using Na⁺ as the efflux substrate and antiporters with Li⁺/H⁺ antiport (Padan et al. 2005; Slonczewski et al. 2009). Alkaline proteases and cellulases are used as biological detergents, and cyclodextrin by alkaline cyclomaltodextrin glucan transferase is used in foodstuffs (Horikoshi 1999).

Commonly, the halophilic microorganisms need more than 0.2 M salt concentration for optimal growth and withstand osmotic stress of extracellular media are participates as a best asset for various metabolites. They are classified as slight halophiles if growing in range of 2–5% NaCl, moderate halophiles if growing in range of 5–20% NaCl, and extreme halophiles if growing in range of 20–30% NaCl (Dassarma and Dassarma 2015). They have optimized survival via changes in various cellular components like biased amino acid composition in proteins for stability and optimal activity at high ionic potency. Such proteins have surface exposed acidic amino acids like glutamate and aspartate on their surface. These proteins are negatively charged and thus bind hydrated ions ultimately reducing hydrophobicity and decreasing aggregation propensity with elevated salt concentration (Mevarech et al. 2000; Da Costa et al. 1998; Danson and Hough 1997; Madern et al. 2000; Gomes and Steiner 2004).

Bioactive metabolites of psychrophiles have procured from cold environments extremophiles and distributed in polar regions, mountains, glaciers, ocean deep area, upper layer of atmosphere, and subterranean ecosystems (Deming 2002; Gomes and Steiner 2004). These organisms have the capability to grow in such cold extreme habitat. These microorganisms have the capability to grow in such cold extreme habitat because of increased polyunsaturated and fatty acids (branched) content which helps in fluidity and nutrient transportation, cold shock protein synthesis, and cryoprotectant formation, cold shock protein synthesis, and cryoprotectants (Gomes and Steiner 2004; Feller 2013). Psychrophilic proteins have low ionic, hydrogen bonds along with fewer hydrophobic groups and higher surface charged particles (Margesin et al. 2003; Feller and Gerday 2003; Georlette et al. 2004). They have unsurmountable role in the food and beverage industry because the products are to be stored at low temperature with capacity of low spoilage, taste, and nutritional maintenance at low temperatures. The extremophiles of this category include species of various genera from gram-negative bacteria such as *Moraxella* sp., *Psychrobacter* sp., *Psychroflexus* sp., *Polaribacter* sp., *Moritella* sp., *Vibrio* sp., and *Pseudomonas* sp.; gram-positive bacteria such as *Arthrobacter* sp., *Bacillus* sp., *and Micrococcus* sp.; archaea such as *Methanogenium* sp., *Methanococcoides* sp., yeast of *Candida*, and *Cryptococcus* genera; and fungi from *Cladosporium* and *Penicillium* and genera (Feller and Gerday 2003; Margesin et al. 2003; Georlette et al. 2004).

Thermophiles are extensively studied extremophiles for bioactive metabolites and are categorized as moderate (50–60 °C), outrageous (60–80 °C), and hyperthermophiles (80–110 °C) (Gomes and Steiner 2004). They are extensively dispersed in variety of species: *Bacillus* sp., *Clostridium* sp., *Thermoanaerobacter* sp., *Thermus* sp., *Fervidobacterium* sp., *Rhodothermus* sp., *Thermotoga* sp., and *Aquifer* sp. Such extremophiles are distributed worldwide in hot springs, volcanoes, volcanic islands, fumaroles, deep-sea hydrothermal vents, and even geysers. Altered changes at cellular level include saturated fatty acid in cell wall promoting hydrophobic environment and structural rigidity to cell (Sterner and Liebl 2001; Kumar and Nussinov 2001; Paiardini et al. 2003; Gomes and Steiner 2004). The presence of reverse DNA gyrase producing positive super coiling in the genomic DNA leads to increased melting point (López-García 1999). The presence of specialized chaperones and thermostable proteins possessing enhanced electrostatic, disulfide, and hydrophobic interactions leads to increased thermostability (Gomes and Steiner 2004; Kumar and Nussinov 2001; Pebone et al. 2008; Ladenstein and Ren 2006). Figure 8.2 illustrated the bioactive features of extremophiles.

Microorganisms from low water environments (xerophiles) are capable to thrive and resist the high desiccation, i.e., water activity below 0.8. Endolithic and halophilic microbes come under xerotolerant. Cyanobacteria *Nostoc commune* was recovered from dry area after 13 years, from herbarium storage after 55 years (Shirkey et al. 2003), from dry storage conditions of herbarium after 87 years (Lipman 1941), and after 107 years from dry soil sample (Cameron and Blank 1966). There are reports of synthesis of extracellular polysaccharides (EPS) which withstand dry environment (De Philippis and Vincenzini 1998), cell components stabilization by buildup of compatible solutes like trehalose (Welsh 2000); hyperexpression of genes related to osmosis, salt tolerance, and cold stress; osmoprotection metabolisms, potassium transportation system, and heat shock proteins; and underexpression of genes involved in photosynthesis, nitrogen transport, synthesis of RNA polymerase, and ribosomal proteins (Katoh et al. 2004).

Expressions of enzymes such as catalases, peroxidases, and superoxide dismutase are increased so as to neutralize ROS due to desiccation (Shirkey et al. 2000; Burton and Ingold 1984). Trehalose is reported to induce expression of chaperone proteins (Higo et al. 2006).

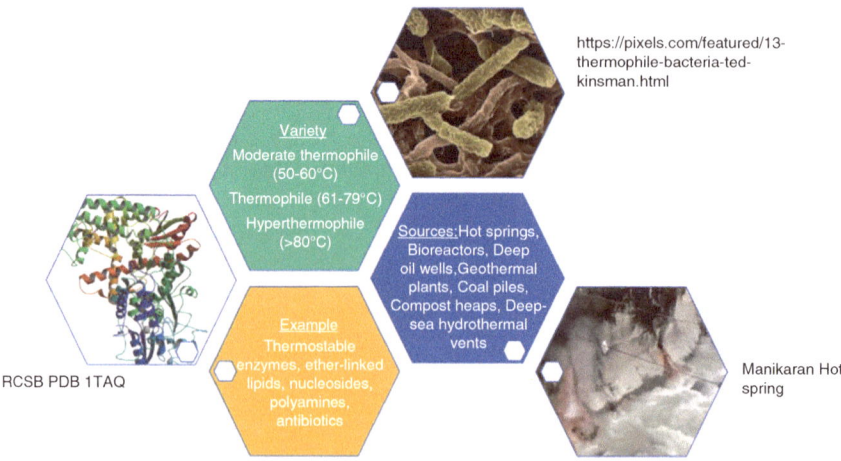

Fig. 8.2 Thermophiles and their properties

Barophiles/piezophile microorganisms have the capacity to thrive at high-pressure area especially greater than 0.1 MPa, or they require high pressure for growth. They thrive in deeper regions of sea, hydrothermal vents, trenches, sediments, and water samples from depths (Certes 1884). *Sh. benthica* and *M. yayanosii* have different strains which are extremely barophilic with optimal growth at 50–80 MPa (Kato et al. 1998). *S. violacea* and *Photobacterium profundum* and *M. japonica* strains come in category of moderate barophilic bacteria thriving at pressures of 10–50 MPa (Kato et al. 1995). *Sporosarcina* sp. strains belong to the category of barotolerant that are capable of growing at a pressure of 0.1 MPa (Kato et al. 1995). These microbes have adapted to such extreme environments because of various modifications like secretion of polyunsaturated fatty acid (PUPA) and eicosapentaenoic acid (EPA) (Nogi et al. 1998a, b; DeLong et al. 1997; Kato et al. 1998). Such microbes have gene expression under the control of pressure-regulated promoter sequences (Nakasone et al. 1998). There are reports of pressure-inducible proteins (Jaenicke et al. 1988); elevated levels of heat shock protein GroES (Gross et al. 1993); repression in outer membrane proteins like Omp-C, Omp-F, and Lam-B (Nakashima et al. 1995, Sato et al. 1996); and increased level of Omp-H (Bartlett et al. 1993). These extremophiles can be exploited for the industrial and high-pressure fermenters because genes and proteins are accustomed to high-pressure conditions, and also the barophilic origin proteases and glucanases can be used for detergents and DNA polymerases in PCR amplification. There is need for exploration of mechanisms behind high-pressure adaptations and biotechnological applications. The research is being extended to mesophilic piezophiles apart from commonly explored psychrophilic and thermophilic piezophiles. Yeast *S. cerevisiae* is transformed to piezophile by alterations in its genome along with the introduction of genes that regulates high-pressure-based cultivation in yeast. High-pressure growth in *S. cerevisiae* is attributed due to the presence of tryptophan permease gene TAT2 (Abe and Horikoshi 2000). Mesophilic organisms with peizophilic applications may help in industrial applications. In the same way, novel antibiotics may be produced from mutants grown under high-pressure conditions. Clubbing the essential information of pressure effects on biochemical/biophysical reactions along with facts of recent advances in rDT and molecular biology techniques can help in potentiating new industrial applications of piezophiles for eco-friendly and energy-saving processes.

Microorganisms that can withstand and survive ionizing and UV radiations are known as radiophiles, and they rarely produce bioactive compounds. Examples include *D. radiophilus* (Yun and Lee 2004), *Deinococcus radiodurans* (Sandigursky et al. 2004), and *T. radiotolerans* (Jolivet et al. 2004), and cyanobacteria like *Nostoc muscorum* , *Microcoleus vaginatus* (Singh 2018), and *Acinetobacter radioresistens* (Jawad et al. 1998). They thrive in various radioactive places like dry climate soil, nuclear reactors like *D. radiodurans*, and Mars Analog Antarctic Dry Valleys (Musilova et al. 2015). They have various adaptation to withstand radiation stresses like Nudix hydrolase enzyme superfamily and homologous proteins of plant dehydration confrontation–linked proteins contributing to dangerous radiations and dehydration resistance of *Deinococcus* sp. (Makarova et al. 2001). PprA protein helps in DNA ligation after DNA fragmentation due to radiation exposure reported in *D. radiodurans* (Narumi et al. 2004), increased Mn/Fe ratio in protection of *D. radiodurans* cellular proteins from oxidative mutilation (Daly 2009), and accumulation of MnII. It helps in resistance toward gamma radiation exposure (Daly et al. 2004), production of mycosporine-like amino acids (MAAs) in cyanobacteria due to exposure of solar UV-B radiations (Sinha et al. 2001), etc. As far as applications of radiophile are concerned, they are helpful in managing nuclear waste-polluted surroundings (Brim et al. 2003; Appukuttan et al. 2006).

The extremophiles grow in the presence of heavy metals (metallophiles) inhabiting industrial sediments, soil, and waste effluents containing heavy metals (Mergeay et al. 2003). These extremophiles possess heavy metal resistance because of the presence of mega plasmids conferring genes for resistance by efflux mechanisms (Gomes and Steiner 2004). Metallophiles eliminate toxic metal via change in redox state of metals aiding in bioleaching of contaminants and precipitation (Wani et al. 2007; Lovley and Coates 1997; Pal and Rai 2010). These extremophiles can be exploited as bioremediation/bioleaching for toxic compound removal from various industrial waste effluents.

8.8   Commercial Applications of Bioactive Molecules from Extremophiles

The bioactive molecules (such as antimicrobials, antivirals and immunomodulatory) from extremophilic origin are best for industrial importance due to tremendous activity and have no matching part in terrestrial higher eukaryotes. The extremophilic bioactive compounds are the focal point of extraordinary enthusiasm for their myriad applications in a few fields running from the nourishment and pharmaceutical ventures to the transformation of biomass to vitality and to the bioremediation of dirtied regions. These compounds are produced by different prokaryotic and eukaryotic species living in extraordinary natural surroundings like cold or hypersaline conditions, aqueous vents, or contaminated locales. Along these lines, these bioactive compounds have accustomed to cruel conditions and are dynamic in the conditions run of the mill of numerous modern procedures like extraordinary temperatures and pH esteems, high saline fixations, and nearness of metals and of natural solvents. In fact, numerous instances of the use of bioactive mixes can be found including the green transformation of various types of biomass to biofuels (ethanol, diesel, hydrogen) and the bioremediation of dirtied locales in outcome of incidental oil slicks or arrival of contaminants by modern exercises. Table 8.1 represents the various types of bioactive metabolites from beneficial extremophiles. The incredible assortment of helpful bioactive mixes is still underexplored. Thus, the distinguishing proof of new bioactive compound delivering species is a key issue for present-day biotechnology that takes favorable circumstances from the abuse of these novel mixes. The recognizable proof of new species could be emphatically pushed by the blend of various pursuit approaches like metagenomic and remote detecting. Extraordinary conditions are promising wellsprings of bioactive

Table 8.1 List of extremophiles and their bioactive metabolites

Species	Bioactive compounds	Optimum pH	Features
Acidophiles			
T. acidophilum	Glucoamylase	2.0	Serour and Antranikian (2002)
S. acidocaldarius	Proteases	2.0	Oda et al. (1987)
F. acidiphilum	Esterases	2.0–3.5	Golyshina and Timmis (2005)
S. solfataricus	Endonuclease	3.0	Limauro et al. (2001)
A. acidocaldarius	Cyclomaltodextrinase	5.5	Matzke et al. (2000)
T. ferrooxidans	Type I Cu rusticyanin	1.0–3.0	Romonsellez et al. (2006)
Alkaliphiles			
S. albidoflavus	Protease	9.0	Proteolysis
B. cereus	Lipase	8.5	Emulsification
P. citrinum	Xylanase	8.0	Xylan degradation
Bacillus sp. DLB 9	Amylase	9.0	Carbohydrate degradation
Bacillus sp. strain YN-2000	2-Phenylamine	9.5	Amino acid
Bacillus sp. strains A-40-2	Carotenes	10.2	Coloring agent
Paecilomyces lilacinus (1907-II, VIII)	Phenazine	10.5	Antibiotics
Nocardiopsis dassonvillei (strain OPC-15)	Antibiotics	10.0	Antibiotics
Halophiles			
Bacillus sp. (BG-CS10)	Cellulase	2.5 M NaCl	Cellulosic
Virgibacillus sp. (EMB 13)	Protease	0–15% NaCl	Proteolysis
Virgibacillus sp. (EMB 13)	Xylanase	12.5% NaCl	Xylan debranching
T. saccharolyticum	Lipase	0–20% NaCl	Lipid emulsification
Chromohalobacter sp. (LY7–8)	Betaine	–	Salt tolerance
H. boliviensis	Ecotain	–	Moisturizer component
Psychrophiles			
M. foliorum GA2	Amylase	20 °C	Carbohydrate degradation
Pseudomonas aeruginosa NCIM 2036	Lipase	5 °C	Lipid oxidation
Bacillus sp. N2a (BNC)	Catalase	25 °C	H_2O_2 production
Pseudomonas sp. MM15	Endoglucanase	30 °C	Waste degradation
Pseudoalteromonas halosplanktis	Feruloyl esterase	20 °C	Feruloylated oligosaccharides debranching

(continued)

Table 8.1 (continued)

Species	Bioactive compounds	Optimum pH	Features
Diatoms	Exopolymeric substances	Below 0 °C	Cryoprotectant
Thermophiles			
Humicola grisea	Cellulase	75 °C	Cellulose degradation
Thermomyces lanuginosus	Phytase	70 °C	Phytic acid degradation
Thermosyntropha lipolytica	Lipase	96 °C	Lipid emulsification
Pyrococcus furiosus	Amylase	>10 °C	Carbohydrate degradation
Bacillus strain HUTBS62	Protease	80 °C	Proteolysis
Phormidium sp.	Bioactive compound	38 °C	Antimicrobial, antifungal
Lb. delbrueckii sp. *Bulgaricus* sp.	2,3-pentanedione	38 °C	Aromatic nature
Radiophile			
Deinococcus radiodurans	Lipase	–	Short-chain lipid esterification, surfactant stable
Halobacterium and *Rubrobacter*	Bacterioruberin	–	Skin cancer prevention
Deinococcus radiodurans	Deinoxanthin	–	Apoptosis of cancer cells (chemoprotective)
Deinococcus radiodurans R1	Superoxide dismutase (Fe/Mn)	–	Reactive oxygen species removal
Metallophiles			
Bacillus jeotgali	–	–	Cadmium and zinc
Geobacillus thermodenitrificans	–	–	Iron, chromium, cobalt, copper, zinc, cadmium, silver
Marine bacteria	–	–	Mercury
Salmonella sp.	–	–	Silver ions
Geobacter sp.	–	–	Iron reduction
Bacterial endophytes	–	–	Lead

compound producers, as metagenomic techniques can be effectively chosen for their potential biotechnology applications.

8.9   Future Perspective

The extremophiles including psychrophiles, thermophiles, acidophiles, alkaliphiles, halophiles, piezophiles, radiophiles, and metalophiles have animated serious endeavors to comprehend the functional adjustments to thrive in extraordinary conditions and to test the potential biotechnological and mechanical uses. This is especially valid for their chemicals (e.g., extremozymes), chemically dynamic under different ranges of temperature, saltiness, pH, weight, radiations, and metals. Numerous fascinating compounds and extraordinary metabolites are confined, filtered, and described from extremophilic microorganisms, and they potentiate modern employments. Notwithstanding, a few specialized troubles have avoided the large-scale modern utilization of extremophilic biomolecules, the most significant being the accessibility of these mixes. Newer improvements in the development and creation of extremophiles yet in addition improvements related to the rDT and articulation of their qualities in heterologous expression hosts will increment the quantity of compound-driven changes in nourishment, pharmaceutical, and other modern applications. Further, due to advancement of proper subatomic apparatuses just as better understanding into structure-work standards, another battery of biomolecules will end up accessible to meet the developing biotechnological enthusiasm for these extremophilic items.

References

Abe F, Horikoshi K (2000) Tryptophan permease gene TAT2 confers high-pressure growth in *Saccharomyces cerevisiae*. Mol Cell Biol 20(21):8093–8102

Anekthanakul K, Hongsthong A, Senachak J, Ruengjitchatchawalya M (2018) SpirPep: an *in silico* digestion-based platform to assist bioactive peptides discovery from a genome-wide database. BMC Bioinf 19(1):149

Appukuttan D, Rao AS, Apte SK (2006) Engineering of *Deinococcus radiodurans* R1 for bioprecipitation of uranium from dilute nuclear waste. Appl Environ Microbiol 72(12):7873–7878

Baker-Austin C, Dopson M (2007) Life in acid: pH homeostasis in acidophiles. Trends Microbiol 15:165–171

Bartlett DH, Chi E, Wright ME (1993) Sequence of the *omp*H gene from the deep-sea bacterium Photobacterium SS9. Gene 131(1):125–128

Bertoldo C, Dock C, Antranikian G (2004) Thermoacidophilic microorganisms and their novel biocatalysts. Eng Life Sci 4:521–532

Blin K, Kazempour D, Wohlleben W, Weber T (2014) Improved lanthipeptide detection and prediction for antiSMASH. PLoS One 9(2):e89420

Brim H, Venkateswaran A, Kostandarithes HM, Fredrickson JK, Daly MJ (2003) Engineering *Deinococcus geothermalis* for bioremediation of high-temperature radioactive waste environments. Appl Environ Microbiol 69(8):4575–4582

Burton GW, Ingold KU (1984) Beta-carotene: an unusual type of lipid antioxidant. Science 224(4649):569–573

Cameron RE, Blank GB (1966) Project Mercury (U.S.). Desert algae: soil crusts and diaphanous substrata as algal habitats. Jet Propulsion Laboratory, California Institute of Technology, Pasadena

Certes A (1884) Of the action of high pressure on the phenomena of putrefaction and on the vitality of microorganisms of fresh and sea water. Compt Rend 99:385–388

Cole JR, Chai B, Marsh TL, Farris RJ, Wang Q et al (2003) The ribosomal database project (RDP-II): previewing a new autoaligner that allows regular updates and the new prokaryotic taxonomy. Nucleic Acids Res 31:442–443

Cragg GM, Newman DJ (2005) Biodiversity: a continuing source of novel drug leads. Pure Appl Chem 77(1):7–24

Crossman L, Holden M, Pain A, Parkhill J (2004) Genomes beyond compare. Nat Rev Microbiol 2:616–617

Da Costa MS, Santos H, Galinski EA (1998) An overview of the role and diversity of compatible solutes in Bacteria and Archaea. Adv Biochem Eng Biotechnoloy 61:117

Daly MJ (2009) A new perspective on radiation resistance based on Deinococcus radiodurans. Nat Rev Microbiol 7(3):237

Daly MJ, Gaidamakova EK, Matrosova VY, Vasilenko A, Zhai M et al (2004) Accumulation of Mn (II) in *Deinococcus radiodurans* facilitates gamma-radiation resistance. Science 306(5698):1025–1028

Danson MJ, Hough DW (1997) The structural basis of protein Halophilicity. Comp Biochem Physiol A Physiol 117(3):307–312

Dassarma S, Dassarma P (2015) Halophiles and their enzymes: negativity put to good use. Curr Opin Microbiol 25:120–126

De Philippis R, Vincenzini M (1998) Exocellular polysaccharides from cyanobacteria and their possible applications. FEMS Microbiol Rev 22(3):151–175

DeLong EF, Franks DG, Yayanos AA (1997) Evolutionary relationships of cultivated psychrophilic and barophilic deep-sea bacteria. Appl Environ Microbiol 63(5):2105–2108

Deming JW (2002) Psychrophiles and polar regions. Curr Opin Microbiol 5(3):301–309

DeSantis TZ, Hugenholtz P, Larsen N, Rojas M, Brodie EL et al (2006) Greengenes, a chimerachecked 16S rRNA gene database and workbench compatible with ARB. Appl Environ Microbiol 72:5069–5072

Donia MS, Ruffner DE, Cao S, Schmidt EW (2011) Accessing the hidden majority of marine natural products through metagenomics. Chem Biochem 12:1230–1236

Engler C, Marillonnet S (2014) Golden gate cloning. In: DNA cloning and assembly methods. Humana Press, Totowa, pp 119–131

Eustáquio AS, Moore BS (2008) Mutasynthesis of fluorosalinosporamide, a potent and reversible inhibitor of the proteasome. Angew Chem Int Ed 47(21):3936–3938

Feller G (2013) Psychrophilic enzymes: from folding to function and biotechnology. Science 2013:512840. https://doi.org/10.1155/2013/512840

Feller G, Gerday C (2003) Psychrophilic enzymes: hot topics in cold adaptation. Nat Rev Microbiol 1(3):200

Georlette D, Blaise V, Collins T, Damico S, Gratia E et al (2004) Some like it cold: biocatalysis at low temperatures. FEMS Microbiol Rev 28(1):25–42

Gibson D, Young L, Chuang R, Clyde AH et al (2009) Enzymatic assembly of DNA molecules up to several hundred kilobases. Nat Method 6:343–345. https://doi.org/10.1038/nmeth.1318Giddings

Giddings LA, Newman D (2013) Microbial natural products: molecular blueprints for antitumor drugs. J Ind Microbiol Biotechnol 40:1181–1210

Golyshina OV, Timmis KN (2005) Ferroplasma and relatives, recently discovered cell wall-lacking archaea making a living in extremely acid, heavy metal-rich environments. Environ Microbiol 7(9):1277–1288

Gomes JI, Steiner W (2004) The biocatalytic potential of extremophiles and extremozymes. Food Technol Biotechnol 42(4):223–225

Gomez-Escribano JP, Bibb MJ (2012) *Streptomyces coelicolor* as an expression host for heterologous gene clusters. Methods Enzymol 517:279–300. Academic Press

Gross DS, Adams CC, Lee S, Stentz B (1993) A critical role for heat shock transcription factor in establishing a nucleosome-free region over the TATA-initiation site of the yeast HSP82 heat shock gene. EMBO J 12(10):3931–3945

Hicks DB, Liu J, Fujisawa M, Krulwich TA (2010) F1F0-ATP synthases of alkaliphilic bacteria: lessons from their adaptations. Biochimica et Biophysica Acta (BBA)-Bioenergetics 1797(8):1362–1377

Higo A, Katoh H, Ohmori K, Ikeuchi M, Ohmori M (2006) The role of a gene cluster for trehalose metabolism in dehydration tolerance of the filamentous cyanobacterium *Anabaena* sp. PCC 7120. Microbiology 152(4):979–987

Horikoshi K (1999) Alkaliphiles: some applications of their products for biotechnology. Microbiol Mol Biol Rev 63(4):735–750

Iwai S, Chai B, Sul WJ, Cole JR, Hashsham SA et al (2010) Gene-targeted-metagenomics reveals extensive diversity of aromatic dioxygenase genes in the environment. ISME J 4(2):279–285

Jaenicke R, Bernhardt G, Lüdemann HD, Stetter KO (1988) Pressure-induced alterations in the protein pattern of the thermophilic archaebacterium *Methanococcus thermolithotrophicus*. Appl Environ Microbiol 54(10):2375–2380

Jawad A, Snelling AM, Heritage J, Hawkey PM (1998) Exceptional desiccation tolerance of *Acinetobacter radioresistens*. J Hosp Infect 39(3):235–240

Jolivet E, Corre E, Haridon S, Forterre P, Prieur D (2004) *Thermococcus marinus* sp. nov. and *Thermococcus radiotolerans* sp. nov., two hyperthermophilic archaea from deep-sea hydrothermal vents that resist ionizing radiation. Extremophiles 8(3):219–227

Kato C, Suzuki S, Hata S, Ito T, Horikoshi K (1995) The properties of a protease activated by high pressure from *Sporosarcina* sp. strain DSK25 isolated from deep-sea sediment. JAMSTEC R 32:7–13

Kato C, Li L, Nogi Y, Nakamura Y, Tamaoka J, Horikoshi K (1998) Extremely barophilic bacteria isolated from the Mariana trench, challenger deep, at a depth of 11,000 meters. Appl Environ Microbiol 64(4):1510–1513

Katoh H, Asthana RK, Ohmori M (2004) Gene expression in the cyanobacterium *Anabaena* sp. PCC7120 under desiccation. Microbial Ecol 47(2):164–174

Kumar S, Nussinov R (2001) How do thermophilic proteins deal with heat? Cell Mol Life Sci 58(9):1216–1233

Ladenstein R, Ren B (2006) Protein disulfides and protein disulfide oxidoreductases in hyperthermophiles. FEBS J 273(18):4170–4185

Langer-Safer PR, Levine M, Ward DC (1982) Immunological method for mapping genes on Drosophila polytene chromosomes. Proc Nat Acad Sci U S A 79:4381–4385

Larionov V, Kouprina N, Graves J, Chen XN, Korenberg JR et al (1996) Specific cloning of human DNA as yeast artificial chromosomes by transformation-associated recombination. Proc Nat Acad Sci 93(1):491–496

Li MZ, Elledge SJ (2007) SLIC sub-cloning using T4 DNA polymerase treated inserts with RecA. Protocol Exchange

Limauro D, Cannio R, Fioriella G (2001) Identification and molecular characterization of an endoglucanase gene, celS, from the extremely thermophilic archaeon *Sulfolobus solfataricus*. Extremophiles 5(4):213–219

Lipman CB (1941) The successful revival of Nostoc commune from a herbarium specimen eighty-seven years old. Bull Torrey Bot Club 1:664–666

López-García P (1999) DNA supercoiling and temperature adaptation: a clue to early diversification of life? J Mol Evol 49(4):439–452

Lovley DR, Coates JD (1997) Bioremediation of metal contamination. Curr Opin Biotechnol 8(3):285–289

Madern D, Ebel C, Zaccai G (2000) Halophilic adaptation of enzymes. Extremophiles 4(2):91–98

Makarova KS, Aravind L, Wolf YI, Tatusov RL, Minton KW et al (2001) Genome of the extremely radiation-resistant bacterium *Deinococcus radiodurans* viewed from the perspective of comparative genomics. Microbiol Mol Biol Rev 65(1):44–79

Margesin R, Sproer C, Schumann P, Schinner F (2003) *Pedobacter cryoconitis* sp. nov., a facultative psychrophile from alpine glacier cryoconite. Int J Syst Evol Microbiol 53(5):1291–1296

Matzke J, Herrmann A, Schneider E, Bakker EP (2000) Gene cloning, nucleotide sequence and biochemical properties of a cytoplasmic cyclomaltodextrinase (neopullulanase) from *Alicyclobacillus acidocaldarius*, reclassification of a group of enzymes. FEMS Microbiol Lett 183:55–61

Mazzoli R, Riedel K, Pessione E (2017) Bioactive compounds from microbes. Front Microbiol 8:392

Mergeay M, Monchy S, Vallaeys T, Auquier V, Benotmane A et al (2003) *Ralstonia metallidurans*, a bacterium specifically adapted to toxic metals: towards a catalogue of metal-responsive genes. FEMS Microbiol Rev 27(2–3):385–410

Mevarech M, Frolow F, Gloss LM (2000) Halophilic enzymes: proteins with a grain of salt. Biophys Chem 86(2–3):155–164

Musilova M, Wright G, Ward JM, Dartnell LR (2015) Isolation of radiation-resistant bacteria from Mars analog Antarctic Dry Valleys by preselection, and the correlation between radiation and desiccation resistance. Astrobiology 15(12):1076–1090

Nakashima K, Horikoshi K, Mizuno T (1995) Effect of hydrostatic pressure on the synthesis of outer membrane proteins in *Escherichia coli*. Biosci Biotechnol Biochem 59(1):130–132

Nakasone K, Ikegami A, Kato C, Usami R, Horikoshi K (1998) Mechanisms of gene expression controlled by pressure in deep-sea microorganisms. Extremophiles 2(3):149–154

Narumi I, Satoh K, Cui S, Funayama T, Kitayama S et al (2004) PprA: a novel protein from *Deinococcus radiodurans* that stimulates DNA ligation. Mol Microbiol 54(1):278–285

Nenkep VN, Yun K, Li Y, Choi HD, Kang JS et al (2010) New production of haloquinones, bromochlorogentisylquinones A and B, by a halide salt from a marine isolate of the fungus Phoma herbarum. J Antibiot 63:199–201

Nogi Y, Kato C, Horikoshi K (1998a) Taxonomic studies of deep-sea barophilic *Shewanella* strains and description of *Shewanella violacea* sp. nov. Arch Microbiol 170(5):331–338

Nogi Y, Masui N, Kato C (1998b) *Photobacterium profundum* sp. nov., a new, moderately barophilic bacterial species isolated from a deep-sea sediment. Extremophiles 2(1):1–8

Oda K, Nakazima T, Terashita T, Suziki KA, Murao S (1987) Purification and properties of an S-PI (Pepstatin Ac) insensitive carboxyl proteinase from a *Xanthomonas* sp. bacterium. Agric Biol Chem 51:3073:3080

Oh DC, Kauffman CA, Jensen PR, Fenical W (2007) Induced production of emericellamides A and B from the marine-derived fungus *Emericella* sp. in competing co-culture. J Nat Prod 70:515–520

Padan E, Bibi E, Ito M, Krulwich TA (2005) Alkaline pH homeostasis in bacteria: new insights. Biochim Biophys Acta 1717(2):67–88

Paiardini A, Gianese G, Bossa F, Pascarella S (2003) Structural plasticity of thermophilic serine hydroxymethyltransferases. Proteins Struct Funct Bioinf 50(1):122–134

Pal R, Rai JPN (2010) Phytochelatins: peptides involved in heavy metal detoxification. Appl Biochem Biotechnol 160(3):945–963

Pebone E, Limauro D, Bartolucci S (2008) The machinery for oxidative protein folding in Thermophiles. Antioxid Redox Signal 10(1):157–169

Penesyan A, Ballestriero F, Daim M, Kjelleberg S, Thomas T et al (2013) Assessing the effectiveness of functional genetic screens for the identification of bioactive metabolites. Mar Drugs 11:40–49

Pruesse E, Quast C, Knittel K, Fuchs BM, Ludwig W et al (2007) SILVA: a comprehensive online resource for quality checked and aligned ribosomal RNA sequence data compatible with ARB. Nucl Acids Res 35:7188–7196

Quan J, Tian J (2014) Circular polymerase extension cloning. In: DNA cloning and assembly methods. Humana Press, Totowa, pp 103–117

Rai AR, Singh RP, Sivastava AK, Dubey RC (2012) Structure prediction and evolution of a halo-acid dehalogenase of Burkholderia mallei. Bioinfo 8(22):1111–1113. https://doi.org/10.6026/97320630081111

Romonsellez F, Orell A, Jerez CA (2006) Copper tolerance of the thermoacidophile archaeon *Sulfolobus metallicus*: possible role of polyphosphate metabolism. Microbiology 152:59–66

Sandigursky M, Sandigursky S, Sonati P, Daly MJ, Franklin WA (2004) Multiple uracil-DNA glycosylase activities in *Deinococcus radiodurans*. DNA Repair 3(2):163–169

Sato T, Nakamura Y, Nakashima KK, Kato C, Horikoshi K (1996) High pressure represses expression of the operon in Escherichia coli. FEMS Microbiol Lett 135(1):111–116

Schleper C, Puehler G, Holz I, Gambacorta A, Janekovic D et al (1995) *Picrophilus* gen. nov., fam. nov.: a novel aerobic, heterotrophic, thermoacidophilic genus and family comprising archaea capable of growth around pH 0. J Bacteriol 177:7050–7059

Schroeckh V, Scherlach K, Nützmann HW, Shelest E, Schmidt HW et al (2009) Intimate bacterial-fungal interaction triggers biosynthesis of archetypal polyketides in *Aspergillus nidulans*. Proc Nat Acad Sci U S A 106:14558–14563

Serour E, Antranikian G (2002) Novel thermoactive glucoamylases from the thermoacidophilic Archaea *Thermoplasma acidophilum*, *Picrophilus torridus* and *Picrophilus oshimae*. Antonie Van Leeuwenhoek 81(1/4):73–83

Sharma A, Kawarabayasi Y, Satyanarayana T (2011) Acidophilic bacteria and archaea: acid stable biocatalysts and their potential applications. Extremophiles 16(1):1–19

Shirkey B, Kovarcik DP, Wright DJ, Wilmoth G, Prickett TF et al (2000) Active Fe-containing superoxide dismutase and abundant sodF mRNA in Nostoc commune (cyanobacteria) after years of desiccation. J Bacteriol 182(1):189–197

Shirkey B, McMaster NJ, Smith SC, Wright DJ, Rodriguez H et al (2003) Genomic DNA of Nostoc commune (cyanobacteria) becomes covalently modified during long-term (decades) desiccation but is protected from oxidative damage and degradation. Nucleic Acids Res 31(12):2995–3005

Singh H (2018) Desiccation and radiation stress tolerance in cyanobacteria. J Basic Microbiol 58(10):813–826

Singh LS, Baruah I, Bora TC (2006) Actinomycetes of Loktak habitat: isolation and screening for antimicrobial activities. Biotechnology 5(2):217–221

Singh RP, Manchanda G, Singh RN, Srivastava AK, Dubey RC (2016) Selection of alkalotolerant and symbiotically efficient chickpea nodulating rhizobia from North-West Indo Gangetic Plains. J Basic Microbiol 56:4–25

Sinha RP, Klisch M, Gröniger A, Hader DP (2001) Responses of aquatic algae and cyanobacteria to solar UV-B. In: Responses of plants to UV-B radiation. Springer, Dordrecht, pp 219–236

Slonczewski JL, Fujisawa M, Dopson M, Krulwich TA (2009) Cytoplasmic pH measurement and homeostasis in bacteria and archaea. Adv Microb Physiol 55:1–79

Sterner RH, Liebl W (2001) Thermophilic adaptation of proteins. Crit Rev Biochem Mol Biol 36(1):39–106

Stierle AA, Stierle DB (2014) Bioactive secondary metabolites from acid mine waste extremophiles. Nat Prod Commun 9(7):1037–1044

Subhashini DV, Singh RP, Manchanda G (2017) OMICS approaches: tools to unravel microbial systems. Directorate of Knowledge Management in Agriculture, Indian Council of Agricultural Research. ISBN:9788171641703. https://books.google.co.in/books?id=vSaLtAEACAAJ

Tanaka Y, Hosaka T, Ochi K (2010) Rare earth elements activate the secondary metabolite–biosynthetic gene clusters in *Streptomyces coelicolor* A3 (2). J Antibiot 63(8):477

Torsvik V, Goksøyr J, Daae FL (1990) High diversity in DNA of soil bacteria. Appl Environ Microbiol 56:782–787

Torsvik V, Daae FL, Sandaa RA, Ovreas L (1998) Novel techniques for analysing microbial diversity in natural and perturbed environments. J Biotechnol 64:53–62

Venter JC, Remington K, Heidelberg JF, Halpern AL, Rusch D et al (2004) Environmental genome shotgun sequencing of the Sargasso Sea. Science 304:66–74

Wang Y, Pfeifer BA (2008) 6-Deoxyerythronolide B production through chromosomal localization of the deoxyerythronolide B synthase genes in E. coli. Metab Eng 10(1):33–38

Wani PA, Khan MS, Zaidi A (2007) Effect of metal tolerant plant growth promoting *Bradyrhizobium* sp.(*vigna*) on growth, symbiosis, seed yield and metal uptake by greengram plants. Chemosphere 70(1):36–45

Welsh DT (2000) Ecological significance of compatible solute accumulation by micro-organisms: from single cells to global climate. FEMS Microbiol Rev 24(3):263–290

Wiley PF, Sigal JMV, Weaver O, Monahan R, Gerzon K (1957) Erythromycin. XI. 1 structure of erythromycin B2. J Am Chem Soc 79(22):6070–6074

Yang YJ, Singh RP, Lan X, Zhang CS, Sheng DH et al (2018) Genome editing in model myxobacteria *Myxococcus xanthus* DK1622 by site-specific Cre/loxP recombination system. Biomol Ther 8(4):137. https://doi.org/10.3390/biom8040137

Yang YJ, Singh RP, Lan X, Zhang CS, Sheng DH et al (2019) Whole transcriptome analysis and gene deletion to understand the chloramphenicol resistance mechanism and develop a new bacterial tool in *Myxococcus xanthus*. Microb Cell Fact 18(1):123. https://doi.org/10.1186/s12934-019-1172-3

Yun YS, Lee YN (2004) Purification and some properties of superoxide dismutase from *Deinococcus radiophilus*, the UV-resistant bacterium. Extremophiles 8(3):237–242

Zhu F, Lin Y (2006) Marinamide, a novel alkaloid and its methyl ester produced by the application of mixed fermentation technique to two mangrove endophytic fungi from the South China Sea. Chin Sci Bull 51:1426–1430

Metallotolerant Bacteria: Insights into Bacteria Thriving in Metal-Contaminated Areas

9

Dina Barman, Dhruva K. Jha, and Kaushik Bhattacharjee

Abstract

The overall condition of the environment is inevitably linked to nature of life on the Earth. However, due to industrial revolution, the global upsurge of accumulation of toxic metals has increased enormously which is posing a serious problem to human health. In such environment, where survival of indigenous microorganisms is difficult, metallotolerant bacteria are able to thrive by tolerating high levels of heavy metals. To cope with this extreme condition, they employ diverse mechanisms to overcome the toxic effects of metals and metalloids with alteration of different genes and proteins, and these mechanisms also help their possible commercial exploitation. Hence, it is essential to understand their unique metabolic capacity or physical structure which encourages thriving in these metal-rich environments. This chapter also sheds light on evolutionary strategies that facilitate the metallotolerant bacteria to adapt to the environment and associated ecophysiological aspects.

Keywords

Metallotolerant bacteria · Bioaccumulation · Biotransformation

Abbreviations

APX   ascorbate peroxidase
CAT   catalase
CDF   cation diffusion facilitators

D. Barman · D. K. Jha
Microbial Ecology Laboratory, Department of Botany, Gauhati University, Guwahati, India

K. Bhattacharjee (✉)
Division of Life Sciences, Institute of Advanced Study in Science and Technology, Guwahati, India

© Springer Nature Singapore Pte Ltd. 2020
R. P. Singh et al. (eds.), *Microbial Versatility in Varied Environments*,
https://doi.org/10.1007/978-981-15-3028-9_9

EPS	exopolysaccharide
FT-IR	Fourier-transform infrared
GC-MS	gas chromatography-mass spectrometry
GST	glutathione S-transferase
HGT	horizontal gene transfer
IAA	indole-3-acetic acid
LC-MS	liquid chromatography-mass spectrometry
MFP	membrane fusion protein
MIP	major intrinsic protein
MTs	metallothioneins
NMR	nuclear magnetic resonance
NTPs	nucleoside triphosphates
OMF	outer membrane factors
PGPB	plant growth-promoting bacteria
PMF	proton motive force
POD	peroxidase
RND	resistance-nodulation-cell division
SOD	superoxide dismutase
SRB	sulfur-reducing bacteria

9.1 Introduction

In the present era, the consequences of heavy metal accumulation in our planet are increasing day by day due to various natural processes (bioweathering of metal-containing minerals, volcanic emissions, forest fires, deep-sea vents, and geysers) and human anthropogenic activities (mining, surface finishing, energy and fuel producing, fertilizer, pesticide, metallurgy, etc.). Both of these have contributed significantly toward the increase in metal contamination (Ayangbenro and Babalola 2017; Romaniuk et al. 2018). Though most of the metals play an important role in various life processes, they become toxic at their high concentrations (Romaniuk et al. 2018). However, some of the heavy metals including mercury (Hg), lead (Pb), chromium (Cr), arsenic (As), cadmium (Cd), uranium (U), selenium (Se), silver (Ag), gold (Au), and nickel (Ni) have negligible biological role and are toxic even in trace amount (Gupta et al. 2016). Toxic effect of heavy metals in various environmental niches is mainly influenced by change of pH and temperature. At acidic pH, the bioavailability of heavy metal increases due to the presence of more protons (H^+) available to saturate the metal-binding sites, thus decreasing the attraction between adsorbent and metal cations. Under basic conditions, protons form other species by replacing metals ions, such as hydroxo-metal complexes which lead to the formation of metal complexes, of which some are soluble (Cd, Ni, Zn) and some are insoluble (Cr and Fe) (Olaniran et al. 2013).

Metals present in the environment are in nondegradable form and cannot be broken down by chemical or biological processes. Due to this, they become persistent in the environment for long time and thus adversely influence the microbial community in environment. It causes serious damage to the ecosystem and soil fertility and negatively impact human health by causing diseases like chronic lung disease, cancer, neurodegeneration, and diabetes (Oyetibo et al. 2015; Fashola et al. 2016; Gupta et al. 2016; Ayangbenro and Babalola 2017). To mitigate the higher concentration of heavy metal ions, various physicochemical approaches including chemical precipitation, chemical oxidation and reduction, ion exchange, filtration, and electrochemical treatment have been used. However, due to various harmful impacts of these physicochemical approaches on the environment, the researchers have explored microbial world to ameliorate metal toxicity (Giovanella et al. 2017).

Among the different organisms on the Earth, microbial world is highly diverse, and some of them possess higher adaptability to tolerate a range of extreme environmental conditions (Gupta et al. 2016; Subhashini et al. 2017). Microbes can respond to the fluctuation of environmental conditions by changing their genetic makeup, by transferring genetic elements and by using many other adaptability mechanisms (Ryan et al. 2009). Metallotolerant bacteria are one of the extremotolerants or extremoresistants that can tolerate high levels of heavy metal concentration by means of their intrinsic properties or have unique metabolic capacities and/or physical structures to tolerate and survive in these extreme conditions (Tse and Ma 2016). These bacteria can survive in toxic metal-rich environment by expressing metal-resistant genes which are further used by bacteria to remediate those metal-contaminated areas (Das et al. 2016). They exert different mechanisms to resist the heavy metal toxicity which includes extracellular barrier, efflux of toxic ions from bacterial cells, incorporation of heavy metals into complexes by metal-binding proteins, and enzymatic transformations of metals (Singh et al. 2011; Romaniuk et al. 2018). A variety of genes and proteins are likely to be involved in such mechanisms which can be understood by transcriptomics, proteomics, genomics, and metabolomics studies. Genes responsible for tolerance to the metal toxicity/detoxification are reported by various researchers (Das et al. 2016). The genes responsible for conferring resistance to cadmium toxicity have been identified as *cadB* and *cadD*, which can protect bacterial cell by binding cadmium at cell membrane (Perry and Silver 1982; Robinson et al. 1990; Crupper et al. 1999). Similarly, *cueO* gene in *E.coli* can detoxify copper toxicity by oxidizing Cu(I) to less toxic Cu(II) (Yu et al. 2014). The *pbrD* gene, encoding a Pb(II)-binding protein, can reduce the toxic effect of Pb (Borremans et al. 2001). The bacterial Ni/Co transporter (NiCoT) gene encodes transporter proteins which mediate energy-dependent uptake of Co and Ni ions into the cell facilitating bioaccumulation (Gogada et al. 2015). *merA* and *merB* are some of the genes which are expressed in response to toxicity of mercury (Schaefer et al. 2011; Dash et al. 2014).

In the present chapter, we have summarized various mechanisms employed by metallotolerant bacteria in order to successfully thrive under various metal-rich environments.

9.2   Strategies to Study Metallotolerant Bacteria

Metallotolerant bacteria can be isolated from metal-contaminated sites since bacteria isolated from these sites are more resistance to a range of metals in comparison to other non-contaminated sites (Sarma et al. 2016). They can be isolated in laboratory with or without enrichment technique. Generally, soil, sediment, sewage, or water collected from different metal-contaminated sites should be serially diluted in saline water (0.85%) or low phosphate buffer followed by mixing vigorously at 120 rpm for 2 h at 30 °C. After that, the diluted samples are spread plated on suitable agar medium followed by incubation at appropriate temperature for allocated time (Romaniuk et al. 2018). The heavy metal-resistant bacteria can be also selectively isolated by serial dilution method using appropriate agar medium by incorporating the desired metal ions (Marzan et al. 2017). Enrichment and isolation of metallotolerant bacteria can be performed by suspending the metal-contaminating environmental sample in a low-nutrient broth supplemented with the desired metal and allowed to incubate in an incubator shaker at 120 rpm and at 30 °C for 24 h. After tenfold serial dilutions of this enrichment culture, the desired metal tolerant bacteria can be isolated by plating on standard media (Fig. 9.1) (Sarma et al. 2016).

The selection of appropriate media for isolation of metallotolerant bacteria in laboratory condition has significant importance. Different researchers have used

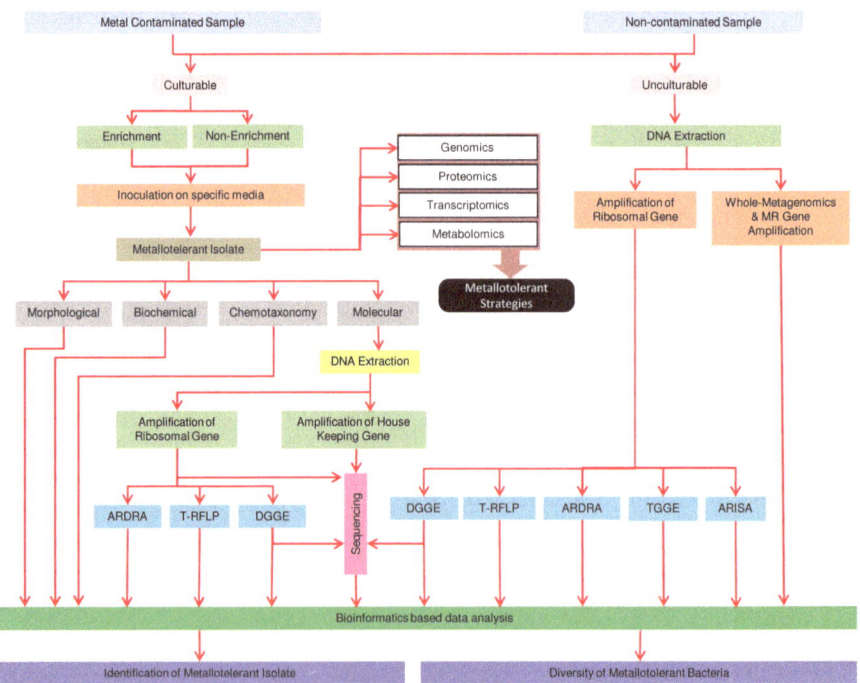

Fig. 9.1 Strategies to study metallotolerant bacteria

different media for isolation of metallotolerant bacteria. A Cr(VI)-tolerant bacterium *Bacillus dabaoshanensis* sp. nov. was isolated from paddy soil altered with sludge compost in Dabaoshan Mine, China (Cui et al. 2015), by inoculating the diluted soil sample with mineral salts medium. Similarly, Abbas et al. (2015) isolated bacteria from tannery effluent (water and sludge sample) collected from Leather Pak Road, Pakistan, on tryptic soy agar amended with various concentrations of heavy metals. Sarma et al. (2016) isolated uranium tolerant bacteria from sediment samples collected from water bodies of three different locations of the uranium rich mining site of Domiasiat in India followed by incubation on tryptone soy agar. Rodriguez-Sanchez et al. (2017) collected soil from lead-contaminated land of Guadalupe, Mexico, and isolated lead tolerant bacteria onto Luria-Bertani (LB) agar plates. Romaniuk et al. (2018) used both R2A and LB media for isolation of metal-resistant bacteria. Oliveira et al. (2009) employed combined carbon medium, a semisolid nitrogen-free medium to isolate arsenic tolerant heterotrophic nitrogen-fixing microorganisms from soil sample collected from both contaminated and non-contaminated soil of Portugal.

Similar to other bacterial populations, metallotolerant bacteria can also be characterized based on morphological, biochemical, and physiological approaches as recommended by Holt et al. (1994). Chemotaxonomic analyses including whole-cell sugar pattern, peptidoglycan type, fatty acid pattern, major menaquinone, and phospholipid type are also useful in characterizing bacteria (Rainey et al. 1996; Maidak et al. 1999). Though morphological, biochemical, physiological, and chemotypic characterization of bacteria can identify genera, sometimes it is not adequate in itself to differentiate between species. The advent of molecular criteria for the characterization of bacteria has provided taxonomists with a set of reliable and reproducible tools for studying the systematics. Molecular characterization of bacteria is performed by 16S rRNA gene amplification of genomic DNA (Shi et al. 2014). Percentage of G + C content of DNA and DNA/DNA-hybridization techniques are also useful tools for the identification of microbes. To characterize taxa at and below the rank of species, the DNA:DNA relatedness, molecular fingerprinting, and phenotypic techniques are methods of choice (Rossello-Mora and Amann 2001).

It is well known that with culture-dependent approach, only 1% of total microbial population can be characterized due to their unknown growth requirements (Vartoukian et al. 2010). In this regard, metagenomics provides a correct scenario regarding the presence of various microorganisms present in different metal-rich environments (Sharma et al. 2008). It also helps to identify the type of microorganism prevalent in metal-rich conditions which enable us to understand the biogeochemical cycles and habits of metal tolerant bacteria (Sharma et al. 2008). For metagenomic analysis, extraction of DNA is the first and foremost requirement followed by cloning DNA into a suitable vector, transforming the clones into a host bacterium, and screening the resulting transformants (Handelsman 2004; Zhang et al. 2017). Metagenomics also provides valuable information about the presence of metal-resistant genes prevalent in microbes thriving in extreme environmental condition (He et al. 2010). One such example is the study of Mina stream sediment, Brazil, which is the world's largest mining regions and is exceptionally rich in iron and gold ores (Costa et al. 2015). A total of 30,738 operational taxonomic units (OTUs) comprising of 52 bacterial phyla particularly belonging to *Proteobacteria*, *Bacteroidetes*, *Acidobacteria*, *Gemmatimonadetes*, *Cyanobacteria*, and *Spirochaetes* with higher abundance of *Proteobacteria* were found from this mining region. Functional annotations also disclose the presence of higher diversity of several metal resistance genes, which indicate that the bacterial community is able to adapt successfully to metal-contaminated environments. Another good example is the complete metagenomics of the acid mine drainage of the Richmond mine. The bacterial community was dominated by *Leptospirillum*, *Sulfobacillus*, and *Acidimicrobium* which predominantly have metal-resistant genes (Handelsman 2004). Similarly, Tyson et al. (2004) explored the microbial community of acid mine drainage (AMD) biofilm in California and which dominated by the genera *Leptospirillum*, *Ferroplasma*, *Sulfobacillus*, and *Acidimicrobium*.

9.3 Community Structure of Metallotolerant Bacteria in Various Metal-Rich Environments

Among the various heavy metal ions, some of them including iron (Fe), zinc (Zn), copper (Cu), and cobalt (Co) are required for metabolic activities of microorganism at low concentrations, but with increasing concentration, they become lethal to living cells. There are some other categories of metal ions such as Hg, As, Ag, Cd, Cr, and Pb which have no roles in biological activities and become toxic even at low concentration (Rodriguez-Sanchez et al. 2017). The presence of high concentration of heavy metals affects the diversity of microbial communities (Romaniuk et al. 2018). The metal-rich environment is usually found across the world. This type of environment is created due to both natural processes and anthropogenic activities (Romaniuk et al. 2018). Among the natural processes, weathering of metal-containing rocks results in the accumulation of heavy metals and metalloids in soil. Serpentine soil is one type of soil which is originated from serpentine rocks containing silica and high concentrations of heavy metals Fe, Mg, Cr, Ni, and Co. A total of 11 different metal tolerating bacteria isolated from this type of soil of Marmara and Aegean regions of Turkey were found to tolerate Ni, Pb, Cd, and Zn in the range of 50–2000 mgL^{-1} (Turgay et al. 2012). However, it is important to note that extreme polar environments are less influenced by anthropogenic activities. The presence of heavy metals in those areas is mainly due to natural process such as biogeochemical weathering of terrigenous sources and global atmospheric pollution. The physiochemical analysis of soil collected from King George Island of Antarctica revealed the presence of high concentrations of heavy metals mainly copper, mercury, and zinc. About 200 bacterial strains were isolated from these regions and found to have many heavy metal resistance genes including *arsB*, *copA*, *czcA*, and *merA* in 62 different strains (Romaniuk et al. 2018).

The massive accumulation of heavy metal generally results from various anthropogenic activities such as large-scale burning of fossil fuels, mining, and industrial processes. Mexico is among one of the major mining countries and extensive producer of Ag and Pb in the world. The mining activities generate huge amount of mine spoils rich in high amount of metal content. This type of soil though generally inhibit the establishment of plants, but some of the plants have been found in these areas may be due to the presence of endophytic bacteria which can detoxify the heavy metal content. The endophytic bacteria belonging to Firmicutes, Actinobacteria, and Proteobacteria isolated from *Prosopis laevigata* and *Sphaeralcea angustifolia* collected from Villa de la Paz in the state of San Luis Potosí, Mexico, showed resistance to heavy metals like Pb, Zn, Cu, As(III), and As(V) with minimum inhibitory concentration (MIC) of 1.1, 3.1, 4.3, 11.0, and 94.3 mM, respectively (Román-Ponce et al. 2016). Uranium contamination is another among the major concerns due to its toxicity to the environment and human health. Domiasiat in the state of Meghalaya, India, is reported as uranium rich mining site with an average ore grade of 0.1% U_3O_8 (Sarma et al. 2016). The bacteria isolated from sediment samples collected from Domiasiat belong to *Serratia nematodiphila* and *S. marcescens* subsp. *sakuensis* and are found to tolerate uranium (U), Cd, Cu, Zn, and Pb (Sarma et al. 2016). Similarly, Shreedhar et al. (2014) isolated bacteria belong to Firmicutes, Proteobacteria, Actinobacteria, and Bacteroidetes having capacity to tolerate uranium from monazite sand of Someshwara beach coast, India. This site is reported to have high radiation because of the deposition of monazite sand containing the actinide element, thorium (Th). Thus, these bacteria provide scope for bioremediation of radionuclide-contaminated sites. Fly ash generated from combustion of coal is also contributing significantly to environmental contamination due to the presence of high content of heavy metals. Bacterial strains isolated from DVC-Mejia Thermal Power Station located at Durlavpur, India, belong to *Bacillus*, *Micrococcus*, *Kytococcus*, and *Staphylococcus* genera and were found to be tolerant to As, with MIC ranging from 14–30 mM for As(III) to 36–72 mM for As(V). Hence, these bacteria are important candidate for As bioremediation in case of fly ash (Roychowdhury et al. 2018).

Industrial processes also have led to the accumulation of metals in the environment which constitutes a major hazard to soil, water, and animal-human health. The soil samples collected from As-polluted areas due to industrial effluents from Estarreja region, Portugal, showed the occurrence of many gram-positive and gram-negative arsenic-tolerant bacteria belonging to Firmicutes, Actinobacteria, and Proteobacteria (Oliveira et al. 2009). Mustapha and Halimoon (2015) also isolated heavy-metal-tolerant bacteria from the outlet of electroplating industry in Klang, Malaysia. In addition to the release of toxic metals from industries, the daily release of sewage from household, agriculture, and health sectors also increases the load of toxic metals in the environment. The banks of Kestopur canal, which runs through the northern fringes of Kolkata, India, is contaminated with heavy metals due to influx of effluents from domestic activities, industries, as well as health sectors. The soil sample collected from this area contained heavy-metal-tolerant bacteria belonging to *Exiguobacterium* and *Bacillus*, which are found to tolerate Cr, Pb, Co, Ni, and Fe with MIC ranging from 7–9 mM, 5–8 mM, 0.6–0.9 mM, 0.6–0.8 mM, to 7–9 mM, respectively (Gupta et al. 2012). Although metallotolerant bacteria are widespread in metal-rich soils of the world, interestingly some of the metallotolerant bacteria are also present in non-contaminated soil, which supports that metal resistance capacity in those bacteria is present intrinsically (Oliveira et al. 2009).

9.4   Metallotolerance in Bacteria

Metallotolerant bacteria have unique feature to thrive in metal-rich environments due to their small size, high surface area to volume ratio, ability to transfer genetic traits, and adaptability (Das et al. 2016). They can convert heavy metals into non-toxic forms by various mechanisms including exclusion by permeability barrier, effluxing metal ions, oxidizing metals, enzymatic conversion of metals, intracellular and extracellular metal sequestration, and producing metal chelators like metallo-thioneins and biosurfactants (Igiri et al. 2018). Nonetheless, a complete understanding of the mechanism behind it is yet a major challenge.

9.5   Underlying Mechanisms of Metallotolerance

9.5.1   Extracellular Barrier as a Way for Averting Metal into Cell

Bacterial extracellular membrane, including the plasma membrane, cell wall, and capsule, acts as a barrier for the toxic metal to enter into the cell. Bacterial capsule, mainly carboxyl groups of polysaccharides, acts as a barrier and resists the entry of toxic metal into the cell by adsorbing metal ions (Fig. 9.2). For example, extracellular biopolymers of *Enterobacter chloacae* and *Marinobacter* sp. can accumulate metal ions on the surface of the cell (Iyer et al. 2005; Bhaskar and Bhosle 2006). Similar to the capsule, the plasma membrane also resists the entry of metal ions into cell. In the case of gram-positive bacteria, peptidoglycan layer of cell surface is thick and composed of alanine, glutamic acid, meso-di-aminopimelic acid, polymer of glycerol, and teichoic acid. Contrarily, in gram-negative bacteria, the peptidoglycan layer is composed of enzymes, glycoproteins, lipopolysaccharides, lipoproteins, and phospholipids. All the lipids, proteins, and polysaccharides of the plasma membrane act as active sites for binding of metals on microbial cell surface (Ayangbenro and Babalola 2017). On binding the heavy metals to bacterial cell, bacteria can transform them from one oxidation state to another and thus reduce their toxicity (Ayangbenro and Babalola 2017). The changes in the permeability of the plasma membrane can also inhibit the entry of metal ions into the cell. All the extracellular surfaces of microbial cell are negatively charged containing many ionizable groups (carboxyl, amino, phosphate, and hydroxyl groups) which provide a platform to adsorb the positively charged heavy metals (Diep et al. 2018). Bacteria can trap heavy metal ions and then adsorb metals onto the binding sites of the cell wall by electrostatic interaction, ion exchange, precipitation, redox process, and surface complexation (Ayangbenro and Babalola 2017) which lead to prohibition of transport of metal ions into the cytoplasm.

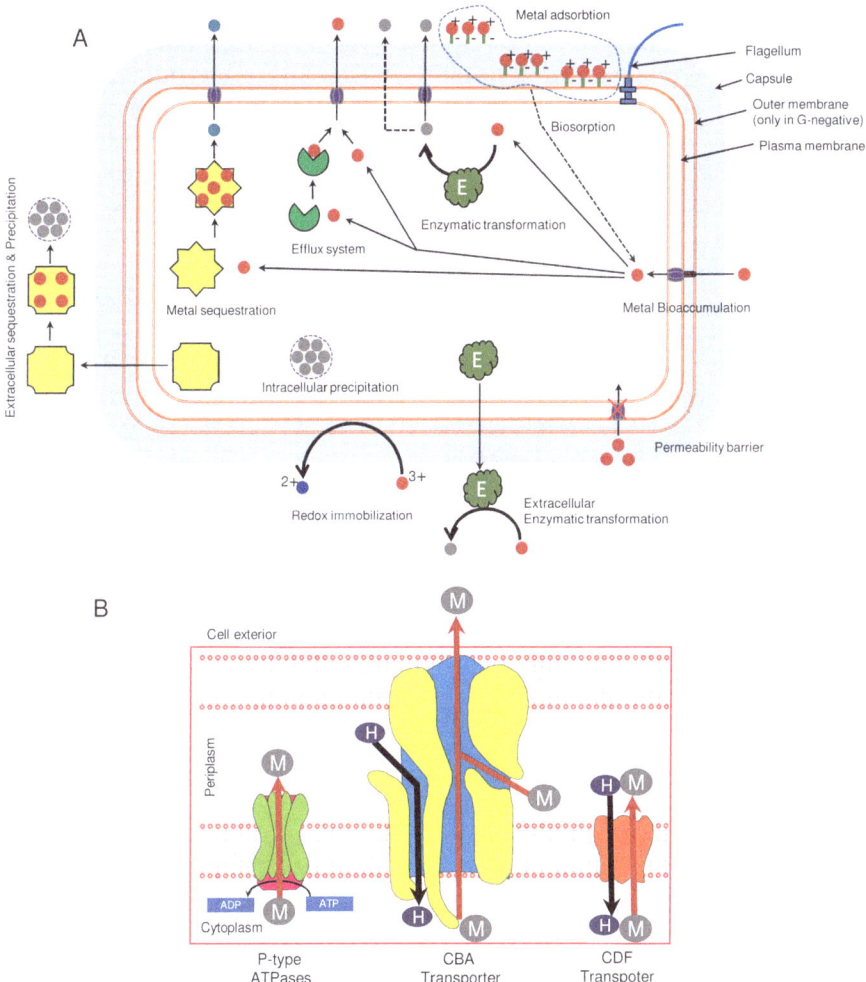

Fig. 9.2 (**a**) A generalized illustration of mechanisms involved in providing tolerance to toxic metals in bacteria. (**b**) Schematic illustration of major types of bacterial efflux system transporter families in heavy metal resistance. (Adapted from Prabhakaran et al. 2016)

9.5.2 Metal Uptake – Biosorption and Bioaccumulation

In bacteria, two different types of heavy metals uptake mechanisms prevail: biosorption and bioaccumulation. Biosorption is the adsorption of metals to the surface layer of bacteria using physicochemical interactions (electrostatic forces, ion exchange), complexation, or chelation and an energy-independent passive metabolic process (Fig. 9.2a) (Diep et al. 2018). On the other hand, bioaccumulation is a metabolically dependent active metal uptake into the living cell across the cell membrane. This process usually involves adsorption of heavy metal ions at the bacterial cell wall or cell membrane through various physicochemical interactions and then transports into the periplasmic space and translocation through the lipid bilayer into the cytoplasm (i.e., metal import system) (Diep et al. 2018). In this manner, the metals are contained and accumulated within the bacterial cell (Chojnacka 2010). The bioaccumulation of metal ions normally relies upon the translocation of metals from the periplasmic space into the cytoplasm of bacteria through the inner membrane importers which are from three major transporter classes: channels, secondary carriers, and primary active transporters (Diep et al. 2018).

Channels are single component α-helical proteins which facilitate passive diffusion of metal ions in step with their concentration gradient across the inner membrane. They are largely energy-independent, which means that they do not need the proton motive force (PMF) or nucleoside triphosphates (NTPs) like ATP and GTP to translocate their substrates (Saier 2016). Some researchers reported a transporter channel, which belongs to the major intrinsic protein (MIP) superfamily, facilitating diffusion of As and Hg ions in *Escherichia coli*, *Corynebacterium diphtheriae*, and *Streptomyces coelicolor* (Singh et al. 2008; Villadangos et al. 2014). These ion channels appear to be the most effective selection for the uptake of metal ions since there is less energy burden on the cell because of their zero-energy requirement (Diep et al. 2018).

Secondary carriers are single component membrane transport proteins which are involved in the movement of ions across the inner membrane. They are of three different types, i.e., uniporters, symporters, and antiporters (Saier 2016). Uniporter can transport single type of metal ions across a cell membrane based on diffusion gradient. On the contrary, symporters and antiporters are types of cotransporters that transport different types of metal ions across a cell membrane. However, symporters involved in movement of ions in the same direction in relation to each other, and antiporters involved in movement of different ions in both the directions (i.e., in and out) based on electrochemical and concentration gradient. Uniporters translocate positively charged metal ions across the inner membrane which depends on varying the charge difference facilitated by PMF, whereas symporters translocate metal ions by utilizing the protons generated due to the charge difference as a co-substrate during uptake (Diep et al. 2018). For example, symporters are used by *Staphylococcus aureus* and *Helicobacter pylori* to transport Ni, Co, and As^{4+} (Deng et al. 2013).

Furthermore, primary active transporters are multicomponent membrane transport proteins containing transmembrane component for the translocation pathway, a cytoplasmic energy-coupling ATPase component that can hydrolyze phosphoanhydride bond (i.e., in NTPs like ATP and GTP) for translocation of metal ions. Primary active transporters MntA and CdtB in *Lactobacillus plantarum* and *CopA* in *Enterobacter hirae* have been used for Cd uptake and Cu uptake respectively, are belong to the P-type ATPase superfamily (Diep et al. 2018).

Another important class of transporter is porins, which are β-barrel transmembrane protein channels, present across the outer membrane of gram-negative bacteria. These porin channels transport the metal ions from the outer surface to the periplasmic space and later to the cytoplasm by different other transporters present in inner lipid bilayer (Saier 2016). Schauer et al. (2007) reported about the FrpB4 channels present in *Helicobacter pylori* and their involvement in Ni uptake. The *ropAe* gene encodes a porin-like protein involved in copper transit in *Rhizobium etli* CFN42 (González-Sánchez et al. 2018). A potential role of multiple porins belonging to the OprD family has also been suggested in *P. aeruginosa* in the uptake of Cu (Teitzel et al. 2006).

Microbial exopolysaccharide (EPS) is also one of the significant components that can take part in metal biosorption. They are released by bacterial cells as self-defense against harsh environmental condition. EPS is a negatively charged polymer, and this negative charge is imparted due to the presence of different active and ionizable functional groups such as acetamido group of chitin and hydroxyl groups that can sequester positively charged heavy metals via various mechanisms including ion exchange, precipitation, and complexation and can adsorb heavy metals (Gupta and Diwan 2017). EPS produced by various bacteria including *B. firmus* and *Arthrobacter* ps-5 has been reported which adsorbs various heavy metals (Zhang et al. 2017). Kazy et al. (2002) also studied EPS synthesized by copper-tolerant *P. aeruginosa* strains which showed increase in EPS production in response to higher Cu concentration. EPS can uptake metal ions by different strategies including homogenous consortial EPS, heterogeneous consortial EPS, dead biomass EPS, immobilized EPS, and modified EPS. Metal biosorption by homogenous consortial EPS mainly deals with the EPS from single bacterial culture which is reported by *Methylobacterium organophilum* that can efficiently remove Cu and Pb ions on producing EPS (Gupta and Diwan 2017). Similarly, *Herminiimonas arsenicoxydans* can remove As ions through EPS interaction (Gupta and Diwan 2017). Metal biosorption by heterogeneous consortial EPS means the use of EPS produced by bacterial consortia. For example, activated sludge-mixed cultures can reduce Zn, Cu, and Cr by approximately 85–95% (Gupta and Diwan 2017). Similarly, gram-negative bacterial consortia can reduce the concentration of Zn, Pb, Cr, Cu, Cd, and Co approximately by 75–85% (Gupta and Diwan 2017). Dead biomass EPS also can adsorb metal ions such that EPS of dead biomass of *Ochrobactrum anthropi* could remove cadmium ions. Immobilized EPS is produced by immobilizing bacterial cells to solid surfaces which stimulate EPS production. It was observed that immobilizing of *Paenibacillus polymyxa* in agar beads stimulates the EPS production which can adsorb lead ions. The activity of EPS can also be enhanced by chemical modification including acetylation, phosphorylation, carboxymethylation, methylation, and sulphonylation. This modified EPS is involved in higher sorption of metals in comparison to unmodified ones (Gupta and Diwan 2017).

9.5.3 Efflux of Toxic Ions from Bacterial Cells

Elimination of heavy metals from bacterial cell relies on an energy-dependent ion efflux mechanism. This efflux of toxic ions from the cytoplasm is mainly performed by bacterial cells with three different proteins, which are (i) resistance-nodulation-cell division (RND superfamily) proteins, (ii) cation diffusion facilitators (CDF family), and (iii) P-type ATPases (Fig. 9.2b). The RND protein family is a group of protein containing RND protein with membrane fusion protein family (MFP) and outer membrane factors (OMF) which form an efflux protein complex and involved in heavy metal resistance by transporting the heavy metals from the cytoplasm, cytoplasmic membrane, or periplasm across the outer membrane directly to the outer surface (Nies 2003). This efflux system is referred to as CBA efflux systems or CBA transporters. *Ralstonia metallidurans* is a gram-negative bacterium that offers resistance to many metal ions including Zn, Co, Cd, and Ni due to the presence of two plasmids, namely, pMOL28 and pMOL30. In pMOL30, the structural gene region contains genes for the OMF CzcC, the MFP CzcB, and the CzcA protein of the RND family which form an operon *czcCBA*. Due to the presence of these genes, *R. metallidurans* offer resistance to Ni and Co. This system can minimize the cytoplasmic and periplasmic concentrations of heavy metal cations; hence, the cations get removed before they enter into the cell. In addition to *czcCBA*, *cnrCBA* and *ncc* are present in *R. metallidurans* which offer resistance to Cd, Co, and Ni. *cnrCBA* encodes for OMF CnrC, the MFP CnrB, and the RND protein CnrA. *ncc* contains regulatory gene region *nccYXH* followed by the structural region *nccCBA*, again encoding a putative outer membrane protein (*NccC*), a MFP (*NccB*), and RND protein (*NccA*) (Nies 2003). Cation diffusion facilitators (CDF family), also a protein family of metal transporter, are involved in providing resistance to Zn and other metal cations. The CDF-encoding gene *czcD* along with TrkA dehydrogenase is present in *B. subtilis* which offers resistance to metals such as Zn^{2+}, Co^{2+}, and Cd^{2+}. Another CDF protein, ZitB, was reported from *E.coli* which makes the strain resistant to Zn on minimizing the gathering of Zn^{2+} by potassium gradient in addition to the proton motive force (Nies 2003). P-type ATPase is another protein family which is driven by ATP hydrolysis. Since this protein family takes part in both import and export of metals, it plays an important role in metal homeostasis and detoxification of heavy metals by effluxing metal ions (Nies 2003).

9.5.4 Biotransformation

The biotransformation of heavy metals is a detoxification process where metals, as a result of biological action, undergo changes in valence and/or conversion into organometallic compounds. It can be achieved either enzymatically or by synthesizing and producing metal-binding proteins such as metallothioneins (MTs) (Fig. 9.2a).

Bacteria can transform toxic effect of metals to harmless one by oxidation, reduction, methylation, and alkylation (Valls and de Lorenzo 2002). Various bacteria can oxidize toxic organic compounds to harmless compounds with oxidoreductase enzymes. They do it by cleaving chemical bonds and catalyzing the transfer of electrons from donor to acceptor (Karigar and Rao 2011). The arsenite-oxidizing bacteria such as *A. faecalis* can tolerate toxic arsenite [As(III)] and oxidized arsenite [As(III)] to arsenate [As(V)] and finally, As(V) forms insoluble sulfides upon exposure to H_2S (Valls and de Lorenzo 2002). Bacteria also reduce and precipitate some of the metals enzymatically, for example, iron-oxidizing bacteria can reduce Fe(III) to Fe(II) abiotically (Lloyd 2003). Similarly, *Serratia marinorubra* can transform arsenate to arsenite and methylarsonate (Bentley and Chasteen 2002). Bacteria also reduce mercury (Hg$^+$) into less toxic and volatile mercury (Hg^o) by mercury reductase and thus released into the atmosphere. Furthermore, highly toxic mercury derivatives such as organomercurials get transformed to mercury (Hg$^+$) and finally to volatile Hg^o by enzymatic process of bacteria (Valls and de Lorenzo 2002).

Microbial methylation also plays an important role in the transformation of toxic metals, where methyl groups are enzymatically transferred to metals and as a result the metals get transformed into different metalloids with varying toxicity, solubility, and volatility. A range of bacteria including *Clostridia*, methanogens, and sulfate-reducing bacteria can methylate various metals including Pb, Cd, As, tin (Sn), selenium (Se), tellurium (Te), and Hg under anaerobic conditions. For example, on methylation of selenium, it gets transformed into volatile dimethyl selenide. Similarly, arsenic gets transformed into gaseous arsines and lead into dimethyl lead. Some bacteria including *Bacillus* sp., *Escherichia* sp., *Clostridium* sp., and *Pseudomonas* sp. can methylate Hg ion (Hg^{2+}) into more toxic methylmercury [(CH$_3$)Hg$^+$] which in turn is methylated into volatile metallic mercury (Hg^0) (Igiri et al. 2018). Furthermore, some bacteria transform methylmercury to volatile dimethylmercury which can be enzymatically reduced to volatile metallic mercury. Another methylation process in bacteria is conversion of phenylmercury to diphenylmercury (Barkay and Wagner-Dobler 2005). Alkylation is another process of biotransformation of metals where an alkyl group other than methyl group is directly bonded to some metals through a carbon atom, for example, As(C$_2$H$_5$)(CH$_3$)$_2$, As(C$_2$H$_5$)$_2$(CH$_3$), As(C$_2$H$_5$)$_3$, and Sb(C$_2$H$_5$)$_3$ (Krupp et al. 1996).

Toxicity of metals can also be removed by bacteria on utilizing metallothioneins (MTs). Metallothioncins are small, cysteine-rich metal-binding proteins that can sequester heavy metals intracellularly as complexes (Romaniuk et al. 2018). Biosynthesis of MTs is induced by different factors such as hormones, cytotoxic agents, and metals including Cd, Zn, Hg, Cu, An, Ag, Co, Ni, and bismuth (Bi). Based on cysteine content and structure, MTs are further classified into Cys-Cys, Cys-X-Cys, and Cys-X-X-Cys motifs (in which X denotes any other amino acid). Metallothioneins act as "storehouse" for zinc and can protect the bacterial cells from cadmium toxicity (Das et al. 2016). They can also scavenge free radical and join with harmful molecules like superoxide and hydroxide ions. After cysteine gets oxidized to cystine, the bound metal ions are released into the environment (Das et al. 2016). For example, *Rhizobium leguminosarum* can also sequester cadmium ions by glutathione which makes the strain cadmium resistant (Lima et al. 2006). *P. putida* can tolerate Cd initially by immobilization of cadmium in polyphosphate granules followed by producing cysteine-rich, low-molecular-weight protein, viz., Pseudothioneins (Higham et al. 1986). Similarly, strain of *P. diminuta* having silver-binding proteins can intracellularly reduce the toxic effect of silver (Ibrahim et al. 2001). *P. aeruginosa* strain WI-1 having metallothionein (*BmtA*) can reduce the toxic effect of Pb by intracellular sequestration (Naik et al. 2012).

9.5.5 Precipitation – Intracellular and Extracellular

Bacteria can precipitate metal compounds intracellularly and/or extracellularly, as, for instance, *G. metallireducens* can convert lethal Mn(IV) to Mn(II), poisonous U(VI) to U(IV), and Cr(VI) to less toxic Cr(III) (Igiri et al. 2018). Metal precipitation mainly occurs as a result of dissimilatory metal reduction, sulfide precipitation, and phosphate precipitation (Valls and de Lorenzo 2002). Dissimilatory metal reduction is concerned with the extracellular precipitation of metals which is unrelated to its intake by the action of microbial catalyst, for example, precipitation of uranium, selenium, chromium, technetium, and gold by various bacteria (Valls and de Lorenzo 2002). Sulfide precipitation is another mechanism of metal precipitation where sulfur-reducing bacteria (SRB) produce sulfide to form metal sulfide as precipitate which is followed by entrapment of the sulfide by the exopolymer (Valls and de Lorenzo 2002). For instance, U(VI), Cr(VI), Tc(VI), Pd(II), and As(V) get precipitated by SRBs. *Klebsiella planticola* can precipitate cadmium (Cd) as insoluble sulfides on liberating hydrogen sulfide from thiosulfate (Igiri et al. 2018). Similarly, *Vibrio harveyi* precipitates soluble lead (Pb^{2+}) as complex lead phosphate salt (Igiri et al. 2018). Metal precipitation is also achieved by the release of inorganic phosphate from organic phosphate donor molecules (Valls and de Lorenzo 2002). Metal precipitation by phosphate production is also catalyzed by the liberation of crystallized inorganic phosphate via the action of periplasmic acid phosphatase (PhoN) and also encounters a role for phosphate groups present in the lipopolysaccharide of crystal nucleation (Macaskie et al. 2000; Valls and de Lorenzo 2002). For instance, *Citrobacter* sp. strain isolated from metal-contaminated soil precipitates high levels of uranium, nickel, and zirconium through the formation of highly insoluble metal phosphates. Similarly, *B. thuringiensis* DM55 also precipitates Cd by phosphate production (Valls and de Lorenzo 2002). Iron-reducing bacteria such as *Geobacter* spp. and SRB like *Desulfuromonas* spp. are involved in the precipitation of harmful metals to less or nontoxic metals by an extracellular sequestration.

In addition, bacteria secrete a variety of other metal-complexing metabolites (e.g., siderophores, carboxylic acids, amino acids, surface-active chemical species, and phenolic compounds) to precipitate the heavy metals. Siderophores are low-molecular-weight coordination molecules and highly specific Fe(III) ligands that can bind and transport or shuttle Fe and are secreted by a wide variety of bacteria to aid Fe assimilation. In spite of their preference for iron, they can likewise chelate various other metals with variable affinities (Fig. 9.2a) (Schalk et al. 2011). The production of siderophores in bacteria can also be stimulated by trivalent metals like Al, Ga, and Cr (Schalk et al. 2011). Koedam et al. (1994) reported the production of siderophores in *P. aeruginosa* in the presence of high iron concentrations. The formation of stable complexes between siderophores and other metal cations (other than iron) is prevalent in bacterial biology (Schalk et al. 2011). For example, the complex formation between Ga^{3+}, Al^{3+}, and In^{3+} with hydroxamate siderophore desferrioxamine B (a siderophore) is between 10^{20} and 10^{28} M^{-1}, whereas that with Fe^{3+} is 10^{30} M^{-1} (Schalk et al. 2011). Braud et al. (2009) reported that the pyoverdine and pyochelin (siderophores) produced by *P. aeruginosa* are able to chelate metals like Ag^+, Al^{3+}, Cd^{2+}, $Co2+$, Cr^{2+}, Cu^{2+}, Eu^{3+}, Ga^{3+}, Hg^{2+}, Mn^{2+}, Ni^{2+}, Pb^{2+}, Sn^{2+}, Tb^{3+}, Tl^+, and Zn^{2+}.

Some bacteria are able to produce surface-active chemical species (i.e., biosurfactants) which help in solubilization and precipitation of metals. For example, rhamnolipids are a class of glycolipid produced by *P. aeruginosa* and several other bacterial strains which act as a biosurfactant (Valls and de Lorenzo 2002). The anionic biosurfactants capture the heavy metal ions through electrostatic or complexation methods. These complexations lead to an expansion in the obvious solvency of metals. In this way, the bioavailability of metals is affected through their decrease by basic metabolic results, which prompts the development of less soluble metal salts including phosphate and sulfide precipitates (Valls and de Lorenzo 2002). There are several other surface-active chemical species produced in the form of polysaccharides, proteins, lipopolysaccharides, lipoproteins, or complex mixtures by a wide range of bacterial strain. For example, *Acinetobacter* spp. produce high-molecular-weight emulsifiers to precipitate metal ions (Mosa et al. 2016).

9.6 Advanced Omics Strategies to Uncover the Truth

Omics strategies provide comprehensive and profound understanding of the underlying mechanism and adaptation strategy in microbial cell in response to metal stress.

9.6.1 Genomics

Genomics refers to the mapping, sequencing, and analysis of the complete set of genes of an organism which provides answer to many unanswered questions (Sharma et al. 2008). It is well known that metallotolerant bacteria can survive in metal-rich environment on expressing different genes such as *cadB*, *chrA*, *copAB*, *pbrA*, *merA*, and *NiCoT* for cadmium, chromium, copper, lead, mercury, and nickel (Das et al. 2016). Genomics provides remarkable information about the various genes involved in resisting metal toxicity which makes us to understand their capacity to grow in metal-rich environment and their interaction with various physical and biological factors (Sharma et al. 2008). In that direction, genomes of many metallotolerant bacteria have already been sequenced (Sharma et al. 2008). During the last few years, the genome of metallotolerant bacteria such as *Sinorhizobium meliloti*, *Mesorhizobium amorphae*, and *Agrobacterium tumefaciens* has been sequenced (Xie et al. 2015). Genome sequencing of *Enterobacter cloacae* B2-DHA isolated from the Hazaribagh tannery areas in Bangladesh provides information on the presence of chromium and other heavy metal resistance genes including *chrR* and *chrA* which make the bacteria to survive in metal-rich environment (Aminur et al. 2017). The genome sequence of hydrocarbon degrading *Geobacillus thermodenitrificans* NG80-2, isolated from a deep oil reservoir in Northern China, provides information that the bacteria have a number of genes which make this organism capable in tolerating a number of contaminated environments including oil reservoirs (Feng et al. 2007). *Lysinibacillus sphaericus* OT4b.31, a native Colombian strain, is generally applied in bioremediation of heavy-metal-polluted environment. The genome sequencing of this bacterium was found to have sphaericolysin B354, the coleopteran toxin Sip1A, and heavy metal resistance clusters of *nik*, *ars*, *czc*, *cop*, *chr*, *czr*, and *cad* operons which support the bacteria to tolerate the metal toxicity (Pena-Montenegro and Dussan 2013). Furthermore, the genome sequencing and annotation of *Halomonas zincidurans* strain B6^T isolated from a deep-sea heavy metal-rich sediment from the South Atlantic Mid-Ocean Ridge provide information that the bacteria encode 31 different genes in relation to heavy metal resistance especially Zn which makes it a candidate for bioremediation of heavy metal-contaminated environments (Huo et al. 2014). The bacterial species *P. putida* is well known for biodegradation of organic compounds (Wu et al. 2011; Yang et al. 2019). Nevertheless, *P. putida* ATH-4 isolated from soil sediments at the "Prat" Chilean military base located in Greenwich Island, Antarctica, is found to be resistant to mercury/tellurite. Interestingly, it showed tellurite resistance only when it was allowed to grow in the presence of mercury, suggesting a cross-resistance mechanism (Rodriguez-Rojas et al. 2016). Further, it is also revealed that the bacterium can resist the toxicity of Cd^{2+}, Cu^{2+}, CrO_4^{2-}, and SeO_3^{2-}. The genome sequencing of *P. putida* ATH-4 provides information that the bacterium possesses more tRNA gene sequences than other known *P. putida* genomes which reflect the adaptation of the bacterium to extreme environmental conditions. On the other hand, in the ATH-43 genome sequence on using IS finder tool, 13 IS elements along with 21 transposases and 17 integrases were found which increase our understanding of its capability to horizontal gene transfer of metal resistance among others (Rodriguez-Rojas et al. 2016). All these information gathered from genomics provide holistic information about the interaction of metallotolerant bacteria with the environment (Sharma et al. 2008).

9.6.2   Transcriptomics

Though genomics provides information about the genes involved in resistance to metal stress, there are different stress response systems which get activated in response to metal stress which can be understood by transcriptome analysis (Peng et al. 2018). Transcriptome analysis of *E. coli* and *B. subtilis* conferred that there are three membrane stress-related regulons, i.e., *cpxRA*, *rpoE*, and *basRS* which get activated in response to metal stress (Hobman et al. 2007). Gene *cpxRA* can enhance the production of membrane chaperons and protease which mitigate periplasmic stress, whereas gene *basRS* controls the biogenesis of capsular- and lipopolysaccharides. Gene *rpoE* gets activated on the introduction of defect of the outer membrane protein assembly. It can restore the protein assembly of the outer membrane by activating the production of chaperon and by upregulating the expression of β-barrel assembly machinery (Peng et al. 2018). The transcriptome analysis of *P.*

putida KT2440 in response to different dose of Zn revealed that with increasing stress, genes responsible to metal homeostasis, cell envelope structure, antioxidant enzyme, and basic cellular metabolism get affected. At lowest dose, genes associated with transportation of metal and membrane homeostasis were influenced. And at an intermediate level, both the above mentioned genes are highly expressed along with the expression of genes associated with oxidative stress and genes for amino acid metabolism. At the higher level of zinc stress, zinc ions can induce the generation of reactive oxidative stress with induction of alkylhydroperoxide reductase and ferredoxin-NADPH reductase which become essential for the maintenance of optimum levels of NADPH. Moreover, at the highest dose, a gene responsible for Fe-S cluster biogenesis gets induced with induction of glyoxylate cycle (Peng et al. 2018). In the case of genus *Sphingobium*, on exposure to high Ni concentration, about 118 genes are differentially expressed. Out of them, 90 were upregulated genes, and a cluster including genes coding for nickel and other metal ion efflux systems (similar to either *cnrCBA*, *nccCBA*, or *cznABC*) and for a NreB-like permease is also found (Volpicella et al. 2017).

9.6.3 Proteomics

Proteomics is suitable to reveal useful physiological profiles of bacteria at protein level (Zhai et al. 2017). The two-dimensional gel electrophoresis (2-DE) and two-dimensional difference gel electrophoresis (2-D DIGE) are the commonly used techniques of proteomics profiling. However, due to the presence of certain limitations, isobaric tags for relative and absolute quantitation (iTRAQ) are more competent which can allow the reliable quantitative description of differentially regulated proteins in complex systems (Zhai et al. 2017). Comparative proteomics easily identify the changes in expression of protein in response to metal stress and also provide information related to molecular mechanisms of tolerance to metal ions (Zivkovic et al. 2018). *P. aeruginosa* is one of the promising candidates for bioremediation which can tolerate high level of Cd by extracellular biosorption, bioaccumulation, biofilm formation, controlled production of siderophore, enhanced respiration, and modified protein profile. The mechanism behind it can easily be understood by proteome profiling of these bacteria (Zivkovic et al. 2018). The resistance of Cd in *P. aeruginosa* is mainly attributed to upregulation of metalloproteins in particular interest to denitrification proteins which are mainly located in the periplasm. These denitrification proteins are overexpressed but not active in exposure to Cd toxicity which suggests their protective role. They also observed the downregulation of siderophore which is regulated by ferric uptake regulation protein (FUR) which showed the effect of Cd on the iron homeostasis (Singh et al. 2011; Zivkovic et al. 2018). It was observed that *Klebsiella pneumonia* isolated from contaminated water sample from river Mula, Pune, is resistant to cobalt (Co^{2+}) and lead (Pb^{2+}). Proteome profiling showed the overexpression of two important proteins, viz., DNA gyrase A and L-isoaspartate protein carboxymethyltransferase type II in response to metal stress (Bar et al. 2007). DNA gyrase A plays an important role in replication, transcription, recombination, and DNA repair. The overexpression of this protein help us to understand that the organism tried to survive in the presence of metal toxicity by expressing many genes on modifying the cellular processes including transcription and replication. Nonetheless, L-isoaspartate protein carboxymethyltransferase type II is mainly responsible to repair and/or degradation of damaged proteins. The overexpression of this protein revealed that bacterial cells have mechanism to allow repair of protein in response to cobalt stress (Bar et al. 2007). Proteomics analysis of *Acidithiobacillus ferrooxidans* also revealed up- and downregulation of four different proteins in response to high potassium concentration. The upregulation of proteins results in the thickening of glycocalyx layer which may be involved in survival of this bacterium in response to metal stress. However, downregulation of ATP synthase F1 delta subunit and ATP synthase F1 beta subunit proteins is associated with the decreased transportation of metal into the cell (Ouyang et al. 2013).

The iTRAQ analyses of *Lactobacillus plantarum* CCFM8610 which is strongly resistant to Cd and *L. plantarum* CCFM191 which is sensitive to Cd provide useful information about the underlying mechanism in response to Cd toxicity. Both the strains displayed physiological alterations in response to Cd. It was observed that 27 proteins were differently regulated in *L. plantarum* CCFM8610 on exposure to Cd, and 111 proteins were changed in *L. plantarum* CCFM191 in response to Cd stress (Zhai et al. 2017). The strong resistant of *L. plantarum* CCFM8610 to Cd is mainly attributed to specific energy-conservation survival mode, mild induction of its cellular defense and repair system, an enhanced biosynthesis of hydrophobic amino acids in response to Cd, inherent superior Cd binding ability and effective cell wall biosynthesis ability, a tight regulation on ion transport, and several key proteins, including prophage P2b protein 18, CadA, mntA, and lp_3327 (Zhai et al. 2017). Proteomics can also provide useful information of the impact of plant growth-promoting bacterial (PGPB) inoculation in plant for microbe-assisted phytoremediation. The total protein extract of maize plant, grown in normal and peripheral soils with or without PGPB, showed variation in protein profiling and protein responsible for various activities. In normal soil, 85 different maize proteins showed notable variation in response to inoculation of PGPB. The major proteins altered in maize with inoculation of PGPB were up-regulation of photosynthetic proteins which sustain the enhancement of chlorophyll a and total chlorophyll content of leaves, proteins in regulation-signal transduction, cellular metabolism, folding and degradation of protein. However, in peripheral soil, protein regulating the DNA repair, methionine biosynthesis, malate metabolic process, photosynthesis, and carbon fixation were upregulated in maize inoculated with PGPB and decreasing the activity of major antioxidant enzymes such as glutathione S-transferase (GST), catalase (CAT), peroxidase (POD), superoxide dismutase (SOD), and ascorbate peroxidase (APX) (Li et al. 2014).

9.6.4   Metabolomics

Metabolomics is a technology to determine and quantify metabolites involved in different life processes. Since bacteria can synthesize a wide range of metabolites to adopt different stress condition, identification and quantification of these metabolites provide a better understanding of stress biology in bacteria. The metabolic fingerprinting can be performed by different techniques such as nuclear magnetic resonance (NMR), MS, Fourier-transform ion cyclotron resonance mass spectrometry, or Fourier-transform infrared (FT-IR) spectroscopy. The identification and quantification of the metabolites can be done by NMR, GC-MS, liquid chromatography-mass spectrometry (LC-MS), capillary electrophoresis-mass spectrometry (CE-MS), gas chromatography-mass spectrometry (GC-MS), NMR, and FT-IR spectroscopy.

Metabolomics provides unprecedented access to the variations in cellular metabolic architecture in response to metal stress (Booth et al. 2015). It is well known that tellurium (Te) is one of the toxic metals which is harmful to both prokaryotes and eukaryotes. Tremaroli et al. (2009) isolated *P. pseudoalcaligenes* KF707 from soil which was found to be resistant to tellurite and also KF707 mutant (T5) with hyperresistance to tellurite. Metabolomics profiling showed a remarkable variation of metabolites of *P. pseudoalcaligenes* KF707 and T5 in response to tellurite. *P. pseudoalcaligenes* KF707 displayed variation in levels of several metabolites, i.e., increase in threonine, leucine, tyrosine, betaine, serine, lysine, isoleucine, alanine, arginine, valine, glutathione, and adenosine whereas decrease in glutamate, aspartate, glycine, histidine, tryptophan, and tyrosine in T5 with and without tellurite. This variation can be correlated with oxidative stress response, resistance to membrane perturbation, and extensive reconfiguration of cellular metabolism. Metabolomics can also provide useful information about the impact of plant growth-promoting bacterial (PGPB) inoculation in plant for microbe-assisted phytoremediation. The metabolite profiling of maize inoculated with PGPB revealed the upregulation of photosynthesis, hormone biosynthesis, and tricarboxylic acid cycle metabolites of maize which makes the plant to remediate metal-contaminated land as well as better growth and development of the plant in metal-contaminated land (Li et al. 2014). Similarly, in *E. coli* and *B. subtilis*, the synthesis of cysteine and histidine is upregulated on exposure to Zn (Hobman et al. 2007).

9.7   Evolution of Strategies

It is notable that most of the heavy metals are required in low level for smooth functioning of bacterial cell, but at high level, it becomes toxic to cell (Coombs and Barkay 2005). In this aspect, metal homeostasis genes play an important role to regulate the transportation of metals into and out of the cell. On increasing the exposure of heavy metals to bacterial cell, the metal homeostasis genes are gradually evolved to survive in metal-rich environment (Coombs and Barkay 2005). Bacteria also develop their genetic system and produce specific enzymes to sequester, remove, or transform the harmful effects of metal toxicity and adapt to that environment. Furthermore, it was observed that bacteria can adapt to changing environmental condition but it is poorly understood in relation to evolution (Hemme et al. 2016). The horizontal transfer of genes (HGT) for metal resistance present in plasmid and gene duplication provides a major contribution to understanding the evolution of bacterial genome in response to metal stress (Kandeler et al. 2000). However, identifying and quantifying such events remain elusive. To delineate the problem, the whole genomes sequencing of isolated bacteria from the environment can provide useful information on comparing to the reference. However, most of the metagenomes of environmental samples are found to be dominated by few reference genomes which also can provide information about the evolution of genome to adapt to contaminated environmental conditions.

The role of HGT has been well recognized for the transfer of antibiotic resistance genes in bacteria, but it is less explored in transfer of metal transporting genes. On performing cultivation-independent analyses of community genomic DNA and RNA from groundwater of contaminated sites, Hemme et al. (2016) observed the abundance of metal-resistant *Rhodanobacter*. The amplicon sequence analysis indicated that the genes coding for Fe^{2+}/Pb^{2+} permeases, most denitrification enzymes, and cytochrome c_{553} are not subjected to horizontal gene transfer (HGT). However, the numerous metal resistance genes, particularly $Co^{2+}/Zn^{2+}/Cd^{2+}$ efflux and mercuric resistance operon genes, are found to be mobile within *Rhodanobacter* populations which aid in understanding the dominance of *Rhodanobacter* populations in contaminated sites. Similarly, the conjugal transfer of *czc* genes of plasmid pDN705 found in *A.eutrophus* CH34 to *E. coli* confers resistance against cobalt, cadmium, and zinc in metal-contaminated sites (De Rore et al. 1994). Nongkhlaw et al. (2012) reported the evolution of ecologically important phenotype where HGT of P_{IB}-type ATPase genes among Firmicutes, Bacteroidetes and Proteobacteria in U rich soil from Domiasiat of India.

9.8   Ecophysiology and Application of Metallotolerant Bacteria

9.8.1   Bioremediation

Bioremediation of contaminated sites with bacteria or bacterial consortia has emerged as most probable alternative. The bioremediation potential of microorganisms mainly depends upon the response of microorganisms toward toxic heavy metals to remediate polluted environment (Bestawy et al. 2013). They can perform bioremediation both in aerobic and anaerobic conditions (Azad et al. 2014). The metallotolerant bacteria used various mechanisms for bioremediation, including biosorption, bioaccumulation, biomineralization, and biotransformation (Dixit et al. 2015).

In biosorption, bacteria can eliminate the metal ions on attracting them to cell membranes (Azad et al. 2014). Sorption of metals by bacterial cell is generally mediated by metallothioneins (a family of 0.5–14 kDa proteins) which are rich in cysteine residues as well as histidine residue also. These metallothioneins act as a scavenger of free radical and combine with harmful molecules like superoxide and hydroxide ions. After this, cysteine gets converted into cystine on oxidation, and thus the bound metals are released into the environment (Das et al. 2016). The process of biosorption is often coupled with enzymatic conversion of metals where on being adsorption of metals on bacterial cell, it gets acted upon by enzymes which can precipitate metals as salt (Das et al. 2016). Since the process of biosorption depends on the basic principle of adsorption, it also encounters difficulties in regard to change of pH and ionic strength (Diep et al. 2018). Again, this process of biosorption has limited life spans since it is not influenced by metabolic contribution and often uses dead biomass (Diep et al. 2018). Various bacterial strains such as *B. subtilis* and *Magnetospirillum gryphiswaldense* were tested as probable biosorbents. During the process of bioaccumulation, heavy metal ions pass across the cell membrane into the intracellular space by translocation pathway. On entering the intracellular space, the heavy metals are sequestered by proteins and peptide ligands. This type of metal uptake is referred to as active uptake (Al-Gheethi et al. 2015). In biotransformation process, the toxic heavy metals are converted to nontoxic form by redox conversions of inorganic forms and conversions from inorganic to organic form, and vice versa. Microbes can oxidize certain metals to obtain energy and/or can reduce some metals where they can utilize metals as a terminal electron acceptor (Niggemyer et al. 2001). In biomineralization, bacteria can remove toxic metals from solution by precipitating insoluble metal sulfide and phosphate.

Bioremediation potential of bacteria can also be upgraded basically by two methods. One of the techniques is to improve or redesign microorganisms to enhance the metal-accumulating properties of the cells, and another one is concerned with modifying the binding sites of the cell by developing commercial biosorbents using immobilization technologies or chemical modifications. According to many researchers, metallotolerant bacteria isolated from heavy metal-contaminated sites are more potent candidates of bioremediation due to their better adaptation and heavy metal resistance mechanism (Al-Gheethi et al. 2015). The bacteria *Lysinibacillus* sp., *Staphylococcus sciuri*, *B. fastidiosus*, *B. niacini*, *Clostridium* sp., and *Bacillus* sp. were reported to be tolerant to As, Cd, and Hg, which were also applied in bioremediation (Bhakta et al. 2018). Similarly, various gram-negative bacteria like *Enterobacter* sp., *Stenotrophomonas* sp., *Providencia* sp., *Chryseobacterium* sp., *Comamonas* sp., *Ochrobactrum* sp., and *Delftia* sp. were isolated from activated sludge, which appear to tolerate Cu, Cd, and Co, and can be efficiently utilized for bioaugmentation of activated sludge to treat the industrial effluents efficiently (Bestawy et al. 2013). Wastewater released from industries and domestic, commercial, or agricultural activities should be treated to detoxify the toxic effect of metals prior to release in water bodies or land surfaces. The heavy metal-resistant bacteria of the genus *Micrococcus* isolated from wastewater may be potentially used for bioremediation of sewage sludge, industrial wastes, and industrial effluent (Benmalek and Fardeau 2017). Marzan et al. (2017) isolated *Gemella* sp., *Micrococcus* sp., and *Hafnia* sp. from tannery effluent where *Gemella* sp. and *Micrococcus* sp. showed resistance to Pb, Cr, and Cd. These bacterial strains can be used as bioremediation agents in toxic tannery effluent treatment technology.

Phytoremediation is another type of bioremediation where hyperaccumulating plants are used to remediate contaminated sites. Plants can perform phytoremediation by different mechanisms, viz., phytoextraction, phytostabilization, phytovolatilization, and rhizofiltration (Rathore et al. 2017). However, the process of phytoremediation generally restricts the growth and biomass of the hyperaccumulating plant when the concentration of metal is very high in the contaminated soil. It can be easily overcome by introducing metallotolerant and plant growth-promoting microorganism to the plant (Tirry et al. 2018). These bacteria can alter the bioavailability of heavy metals to plant by acidification, by releasing chelating substances, and by changing the redox potentials (Whiting et al. 2001). The inoculation of endophytic bacteria in hyperaccumulating plants receives comparable importance due to their capacity to assist the plant growth and development in metal-contaminated sites by producing plant growth regulators, increasing the uptake of mineral nutrients and water, nitrogen fixation, and systemic resistance of plants against pathogens. Endophytic bacteria help the plants to survive in metal-rich sites by converting toxic metal ions to nontoxic forms. They can also degrade contaminants by producing various enzymes (Sharma et al. 2018). The endophytic bacteria belonging to the genera *Acinetobacter*, *Bacillus*, *Arthrobacter*, *Burkholderia*, *Clostridium*, *Enterobacter*, *Micrococcus*, *Paracoccus*, *Rhodococcus*, *Pseudomonas*, *Streptomyces*, *Staphylococcus*, etc. are reported as heavy metal-resistant bacteria which may be used to rehabilitate metal-contaminated sites (Sharma et al. 2018). Similarly, the rhizospheric bacteria in combination with plants play a significant role in phytoremediation of metal-contaminated soils by acidification, phosphate solubilization, releasing chelating agents, and redox changes. The heavy metal-resistant and plant growth-promoting bacterium *Cellulosimicrobium* sp., isolated from the rhizosphere of a contaminated region in the Plain of Sais, Fez (Morocco), enhances the growth of alfalfa plants in heavy metal-contaminated sites (Tirry et al. 2018). Similarly, *B. subtilis* SJ-101 was found to stimulate indole-3-acetic acid (IAA) production which enhances the growth of *Brassica juncea* in Ni-contaminated soil (Mishra et al. 2017). The As tolerant *B. licheniformis*, *Micrococcus luteus*, and *P. fluorescens* were reported to enhance the biomass of grapevines in contaminated sites (Mishra et al. 2017).

9.8.2   Bioleaching

Bioleaching is a process where the insoluble metal sulfides get converted to metal sulfate. The metals are released from metal sulfide mainly by two different mechanisms including direct and indirect bacterial leaching. During direct leaching, bacterial cell directly comes in contact with specific sites of crystal imperfection mineral sulfide surface, and sulfates get oxidized by different enzymatic pathways (Bosecker 1997). Microbes can solubilize metals by (i) acidolysis, where microbes produce acids by which it can leach metals; (ii) complexolysis, where microbes excrete biogenic agents which solubilize metal ions by ligand formation; and lastly (iii) redoxolysis, where microbes used oxidation and reduction reactions to solubilize metals (Monballiu et al. 2015). This method is widely used by bacteria for removal of heavy metals from metal-rich sites (Jeremic et al. 2016). The process of bioleaching is affected by various physicochemical and microbiological factors. Hence, it is important to maintain optimum growth of microorganisms so that they can leach the metals more efficiently (Monballiu et al. 2015). However, the growth and metabolic processes of microorganisms involved in bioleaching are adversely affected by the occurrence of heavy metals in bioleached materials. The metals can inhibit the enzyme activities, disrupt the membrane transport processes, and ultimately lead to inhibit the growth of microorganisms. Hence, it is better to utilize metallotolerant bacteria for the efficient bioleaching process (Monballiu et al. 2015). Metallotolerant bacteria can sequester metal ions extra- and/or intracellularly and carry metal resistance genes. They can adapt to metal-rich sites by active transportation of metal ions, interaction with extracellular polymeric substances (EPS), formation of cell surface complexes, and metal reduction to a less toxic state (Monballiu et al. 2015).

Most of the microorganisms used in bioleaching process are found to be belonging to the genus of *Thiobacillus* (*T. ferrooxidans, T. thiooxidans*, and *T. cuprinus*). They can oxidize sulfides, elemental sulfur, and thiosulfate to sulfate on utilizing sulfur as an energy source (Roy and Roy 2015). Hence, researchers are progressively focused on this group of bacteria isolated from metal-contaminated sites (mines and mine tailings) for their bioleaching potentials. Except these, many heterotrophic bacteria are also reported to be used for bioleaching. Pyrolusite is a mineral containing manganese dioxide and is important ore of manganese. This ore can be degraded by *Bacillus*, *Micrococcus*, *Pseudomonas*, *Achromobacter*, and *Enterobacter* by enzymatic reduction under both aerobic and micro-aerobic growth conditions (Roy and Roy 2015). Most of the metallotolerant bacteria belonging to the genus *Bacillus* are widely used for bioleaching. For example, *B. mucilaginosus* is one of the bacteria which is resistant to Cr, Ni, and As and can be used for its bioleaching (Monballiu et al. 2015). Furthermore, *T. ferrooxidans* and *P. Aeruginosa*, isolated from waste dump of magnesite and bauxite mines of Salem district in Tamil Nadu, are resistant to heavy metals Mn, Fe, Cu, Cr, and Hg and these two bacteria effective in bioleaching process (Mathiyazhagan and Natarajan 2011).

9.8.3   Biomining

Biomining is a process where microorganisms are used to oxidize iron and sulfur to recover metals from minerals containing copper, gold, and uranium (Valenzuela et al. 2006). In this process, metal tolerant acidophilic bacteria have a special advantage due to their metal resistance mechanisms (Jeremic et al. 2016). Mostly bacteria from   the   genus   of   *Acidithiobacillus*,   *Leptospirillum*,   *Acidimicrobium*,

Ferromicrobium, *Sulfobacillus*, and *Thiomonas* are reported as most potent for biomining (Valenzuela et al. 2006). They have the capacity to withstand extremely acidic environment and can tolerate high metal concentrations. For example, *A. ferrooxidans*, which is most studied biomining bacterium, can tolerate Cu, As, Zn, Cd, and Ni in the concentration of 800 mM, 84 mM, 1071 mM, 500 mM, and 1000 mM, respectively. Similarly, *A. caldus*, *Cupriavidus metallidurans*, *Thiomonas cuprina*, *Thiomonas arsenitoxydans*, *Metallosphaera sedula*, and *Sulfolobus solftaricus* can tolerate some of the heavy metals (Navarro et al. 2013). For that, they have genes related to metal tolerance. It was reported that *A. ferrooxidans* ATCC 23270 cells have at least ten genes relating to Cu homeostasis. These genes are upregulated on introducing this bacterium to a high level of Cu concentration (5–25 mM or higher) (Orell et al. 2009).

9.8.4   Metabolic Engineering

Metabolic engineering is a tool to develop or redesign bacteria to detoxify specific metal more specifically. It is performed by recombinant DNA/RNA technology and can be successfully used to decontaminant heavy metals from contaminated sites, e.g., GE *E. coli* strain JM109 containing *merA* gene removes mercury from contaminated site by expressing metallothioneins and polyphosphate kinase. By the same token, GE *E. coli* containing *arsR* gene removes arsenic from the contaminated site by bioaccumulation of As, and GE *Ralstonia metallidurans* and GE *Caulobacter* sp. strain JS4022/p723-6H removes Cr and Cd respectively from industrial wastewater, respectively (Azad et al. 2014). *Deinococcus geothermalis*, a radiation-resistant thermophilic bacterium, has been genetically engineered for bioremediation purposes on introducing Hg(II) resistant *mer* operon of *E. coli* (Dixit et al. 2015). The metal resistance capacity of metallotolerant bacteria can also be enhanced by inducing mutation with different mutagenic agents. It is observed that *Pseudomonas* sp., a bacterium isolated from soil and factory effluents, tolerate Cu and Zn more specifically when induced by mutagenic agents such as acriflavine, acridine orange, and ethidium bromide (Shakibaie et al. 2008). The Cd^{2+}tolerance is enhanced in *P. aeruginosa* on mutation of *cad* operon by acridine orange and acriflavine (Kermani et al. 2010).

9.9   Conclusion and Future Perspectives

Environmental niches have become more prone to heavy metal pollution due to their long persistent and nondegradable nature. It may be due to the growing industrialization and human activities related to mining, disposal or leakage of industrial wastes, the use of pesticide, and sewage sludge which become hazardous to both ecological and human health. In such condition, where growth of most of the life forms including microorganisms is difficult, metallotolerant bacteria have the capacity to adapt and survive in various metalliferous environments. They develop various cope-up mechanisms and metal-resistant genotypes to adapt to this unfavorable condition. Different omics approaches shed light on the detail molecular mechanism behind it. In addition, due to their metal tolerance capacity, these bacteria can be exploited for the remediation of heavy metal polluted sites, biomining, bioleaching, and various biotechnological approaches. Since a vast diversity of metallotolerant bacteria are prevalent in metalliferous environment, on exploring the diversity and genetic makeup of these bacteria, there is a possibility to discover novel metal-resistant genes and proteins which make them to survive and adapt more efficiently to these extreme environmental conditions. On understanding the mechanism of metal tolerance potentials of metallotolerant bacteria, it becomes also possible to transform bacteria and/or transfer the gene to other potent bacterial strains for enhanced bioremediation and other areas of biotechnological approaches.

Acknowledgments Financial support from the DBT-RA Program in Biotechnology and Life Sciences is gratefully acknowledged by DB. KB gratefully acknowledges the financial assistance from the Science and Engineering Research Board (SERB), Department of Science and Technology (DST), Government of India, New Delhi (Sanction No. PDF/2017/002174).

References

Abbas S, Ahmed I, Kudo T et al (2015) A heavy metal tolerant novel bacterium, *Bacillus malikii* sp. nov., isolated from tannery effluent wastewater. Antonie Van Leeuwenhoek 108(6):1319–1330

Al-Gheethi AAS, Lalung J, Noman EA et al (2015) Removal of heavy metals and antibiotics from treated sewage effluent by bacteria. Clean Techn Environ Policy 17:2101–2123

Aminur R, Björn O, Jana J et al (2017) Genome sequencing revealed chromium and other heavy metal resistance genes in *E. cloacae* B2-Dha. J Microb Biochem Technol 9:5

Ayangbenro AS, Babalola OO (2017) A new strategy for heavy metal polluted environments: a review of microbial biosorbents. Int J Environ Res Public Health 14(1)

Azad MAK, Amin L, Sidik NM (2014) Genetically engineered organisms for bioremediation of pollutants in contaminated sites. Chin Sci Bull 59(8):703–714

Bar C, Patil R, Doshi J et al (2007) Characterization of the proteins of bacterial strain isolated from contaminated site involved in heavy metal resistance-a proteomic approach. J Biotechnol 128(3):444–451

Barkay T, Wagner-Dobler I (2005) Microbial transformations of mercury: potentials, challenges, and achievements in controlling mercury toxicity in the environment. Adv Appl Microbiol 57:1–52

Benmalek Y, Fardeau ML (2017) Isolation and characterization of metal-resistant bacterial strain from wastewater and evaluation of its capacity in metal-ions removal using living and dry bacterial cells. Int J Environ Sci Technol 13:2153–2162

Bentley R, Chasteen TG (2002) Microbial methylation of metalloids: arsenic, antimony, and bismuth. Microbiol Mol Biol Rev 66(2):250–271

Bestawy EE, Helmy S, Hussien H et al (2013) Bioremediation of heavy metal-contaminated effluent using optimized activated sludge bacteria. Appl Water Sci 3:181–192

Bhakta JN, Lahiri S, Bhuiyna FA et al (2018) Profiling of heavy metal(loid)-resistant bacterial community structure by metagenomic-DNA fingerprinting using PCR–DGGE for monitoring and bioremediation of contaminated environment. Energ Ecol Environ 3:102

Bhaskar PV, Bhosle NB (2006) Bacterial extracellular polymeric substance (EPS): a carrier of heavy metals in the marine food-chain. Environ Int 32:191–198

Booth SC, Weljie AM, Turner RJ (2015) Metabolomics reveals differences of metal toxicity in cultures of *Pseudomonas pseudoalcaligenes* KF707 grown on different carbon sources. Front Microbiol 6:827

Borremans B, Hobman JL, Provoost A et al (2001) Cloning and functional analysis of the pbr lead resistance determinant of *Ralstonia metallidurans* CH34. J Bacteriol 183:5651–5658

Bosecker K (1997) Bioleaching: metal solubilization by microorganisms. FEMS Microbiol Rev 20:591–604

Braud A, Hannauer M, Milsin GLA et al (2009) The *Pseudomonas aeruginosa* pyochelin-iron uptake pathway and its metal specificity. J Bacteriol 191:5317–5325

Chojnacka K (2010) Biosorption and bioaccumulation–the prospects for practical applications. Environ Int 36:299–307

Coombs JM, Barkay T (2005) New findings on evolution of metal homeostasis genes: evidence from comparative genome analysis of Bacteria and Archaea. Appl Environ Microbiol 71(11):7083–7091

Costa PS, Reis MP, Ávila MP et al (2015) Metagenome of a microbial community inhabiting a metal-rich tropical stream sediment. PLoS One 10(3):e0119465

Crupper SS, Worrell V, Stewart GC et al (1999) Cloning and expression of *cad*D, a new cadmium resistance gene of *Staphylococcus aureus*. J Bacteriol 181:4071–4075

Cui X, Wang Y, Liu J et al (2015) *Bacillus dabaoshanensis* sp. nov., a Cr(VI)-tolerant bacterium isolated from heavy-metal-contaminated soil. Arch Microbiol 197(4):513–520

Das S, Dash HR, Chakraborty J (2016) Genetic basis and importance of metal resistant genes in bacteria for bioremediation of contaminated environments with toxic metal pollutants. Appl Microbiol Biotechnol 100(7):2967–2984

Dash HR, Mangwani N, Das S (2014) Characterization and potential application in mercury bioremediation of highly mercury-resistant marine bacterium *Bacillus thuringiensis* PW-05. Environ Sci Pollut Res 21(4):2642–2653

De Rore H, Top E, Houwen F et al (1994) Evolution of heavy metal resistant transconjugants in a soil environment with a concomitant selective pressure. FEMS Microbiol Ecol 14(3):263–273

Deng X, He J, He N (2013) Comparative study on Ni^{2+} -affinity transport of nickel/cobalt permeases (NiCoTs) and the potential of recombinant *Escherichia coli* for Ni^{2+} bioaccumulation. Bioresour Technol 130:69–74

Diep P, Mahadevan R, Yakunin AF (2018) Heavy metal removal by bioaccumulation using genetically engineered microorganisms. Front Bioeng Biotechnol 6:157

Dixit R, Wasiullah MD et al (2015) Bioremediation of heavy metals from soil and aquatic environment: an overview of principles and criteria of fundamental processes. Sustainability 7:2189–2212

Fashola M, Ngole-Jeme V, Babalola O (2016) Heavy metal pollution from gold mines: environmental effects and bacterial strategies for resistance. Int J Environ Res Public Health 13(11):E1047

Feng L, Wang W, Cheng J et al (2007) Genome and proteome of long-chain alkane degrading *Geobacillus thermodenitrifi* cans NG80-2 isolated from a deep-subsurface oil reservoir. Proc Natl Acad Sci 104(13):5602–5607

Giovanella P, Cabral L, Costa AP et al (2017) Metal resistance mechanisms in Gram-negative bacteria and their potential to remove Hg in the presence of other metals. Ecotoxicol Environ Saf 140:162–169

Gogada R, Singh SS, Lunavat SK et al (2015) Engineered *Deinococcus radiodurans* R1 with NiCoT genes for bioremoval of trace cobalt from spent decontamination solutions of nuclear power reactors. Appl Microbiol Biotechnol 99(21):9203–9213

González-Sánchez A, Cubillas CA, Miranda F et al (2018) The *ropAe* gene encodes a porin-like protein involved in copper transit in *Rhizobium etli* CFN42. Microbiol Open 7(3):e00573

Gupta P, Diwan B (2017) Bacterial exopolysaccharide mediated heavy metal removal: a review on biosynthesis, mechanism and remediation strategies. Biotechnol Rep 13:58–71

Gupta K, Chatterjee C, Gupta B (2012) Isolation and characterization of heavy metal tolerant gram-positive bacteria with bioremedial properties from municipal waste rich soil of Kestopur canal (Kolkata), West Bengal, India. Biologia 67(5):827–836

Gupta A, Joia J, Sood A et al (2016) Microbes as potential tool for remediation of heavy metals: a review. J Microb Biochem Technol 8(4):364–372

Handelsman J (2004) Metagenomics: application of genomics to uncultured microorganisms. Microbiol Mol Biol Rev 68(4):669–685

He LY, Zhang YF, Ma HY et al (2010) Characterization of copper-resistant bacteria and assessment of bacterial communities in rhizosphere soils of copper-tolerant plants. Appl Soil Ecol 44:49–55

Hemme CL, Green SJ, Rishishwar L et al. (2016) Lateral gene transfer in a heavy metal-contaminated-groundwater microbial community. mBio 7(2):e02234–15

Higham DP, Sadler PJ, Scawen MD (1986) Cadmium-binding proteins in *Pseudomonas putida*: pseudothioneins. Environ Health Perspect 65:5–11

Hobman JL, Yamamoto K, Oshima T (2007) Transcriptomic responses of bacterial cells to sublethal metal ion stress. In: Nies DH, Silver S (eds) Molecular microbiology of heavy metals. Springer, Heidelberg/Berlin, pp 73–115

Holt JG, Krieg NR, Sneath PHA et al (1994) Bergey's manual of determinative bacteriology, 9th edn. Lippincott Williams & Wilkins, Baltimore

Huo YY, Li ZY, Cheng H et al (2014) High quality draft genome sequence of the heavy metal resistant bacterium *Halomonas zincidurans* type strain B6T. Stand Genomic Sci 9:30

Ibrahim Z, Ahmad WA, Baba AB (2001) Bioaccumulation of silver and the isolation of metal-binding protein from *P. diminuta*. Braz Arch Biol Technol 44(3):223–225

Igiri BE, Okoduwa SIR, Idoko GO et al (2018) Toxicity and bioremediation of heavy metals contaminated ecosystem from tannery wastewater: a review. J Toxicol 2568038:1–16

Iyer A, Mody K, Iha B (2005) Biosorption of heavy metals by a marine bacterium. Mar Pollut Bull 50(3):340–343

Jeremic S, Beškoski VP, Djokic L et al (2016) Interactions of the metal tolerant heterotrophic microorganisms and iron oxidizing autotrophic bacteria from sulphidic mine environment during bioleaching experiments. J Environ Manag 172:151–161

Kandeler E, Tscherko D, Bruce KD et al (2000) Structure and function of the soil microbial community in microhabitats of a heavy metal polluted soil. Biol Fertil Soils 32:390–400

Karigar CS, Rao SS (2011) Role of microbial enzymes in the bioremediation of pollutants: a review. Enzyme Res 805187:1–11

Kazy SK, Sar P, Sen AK et al (2002) Extracellular polysaccharides of a copper-sensitive and a copper-resistant *Pseudomonas aeruginosa* strain: synthesis, chemical nature and copper binding. World J Microbiol Biotechnol 18(6):583–588

Kermani AJN, Ghasemi MF, Khosravan A et al (2010) Cadmium bioremediation by metal-resistant mutated bacteria isolated from active sludge of industrial effluent. Ira J Environ Health Sci Eng 7(4):279–286

Koedam N, Wittouck E, Gaballa A et al (1994) Detection and differentiation of microbial siderophores by isoelectric focusing and chrome azurol S overlay. Biometals 7(4):287–291

Krupp EM, Grümping R, Furchtbar URR et al (1996) Speciation of metals and metalloids in sediments with LTGC/ICP-MS. Fresenius J Anal Chem 354:546–549

Li K, Pidatala VR, Shaik R et al (2014) Integrated Metabolomic and proteomic approaches dissect the effect of metal-resistant Bacteria on maize biomass and copper uptake. Environ Sci Technol 48:1184–1193

Lima AIG, Corticeiro SC, Figueira EMAP (2006) Glutathione-mediated cadmium sequestration in *Rhizobium leguminosarum*. Enzyme Microb Technol 39(4):763–769

Lloyd JR (2003) Microbial reduction of metals and radionuclides. FEMS Microbiol Rev 27(2–3):411–425

Macaskie LE, Bonthrone KM, Yong P et al (2000) Enzymically mediated bioprecipitation of uranium by a *Citrobacter* sp.: a concerted role for exocellular lipopolysaccharide and associated phosphatase in biomineral formation. Microbiol 146:1855–1867

Maidak BL, Cole JR, Parker CT Jr et al (1999) A new version of the RDP (ribosomal database project). Nucleic Acids Res 27:171–173

Marzan LW, Hossain M, Mina SA et al (2017) Isolation and biochemical characterization of heavy-metal resistant bacteria from tannery effluent in Chittagong city, Bangladesh, Bioremediation viewpoint. Egypt J Aquat Res 43:65–74

Mathiyazhagan N, Natarajan D (2011) Bioremediation on effluents from magnesite and bauxite mines using *Thiobacillus* spp. and *Pseudomonas* spp. J Bioremed Biodegrad 2:1

Mishra J, Singh R, Arora NK (2017) Alleviation of heavy metal stress in plants and remediation of soil by rhizosphere microorganisms. Front Microbiol 8:1706

Monballiu A, Cardon N, Nguyen MT et al (2015) Tolerance of chemoorganotrophic bioleaching microorganisms to heavy metal and alkaline stresses. Bioinorg Chem Appl 861874:1–9

Mosa KA, Saadoun I, Kumar K et al (2016) Potential biotechnological strategies for the Cleanup of heavy metals and metalloids. Front Plant Sci 7:303

Mustapha MU, Halimoon N (2015) Screening and isolation of heavy metal tolerant bacteria in industrial effluent. Procedia Environ Sci 30:33–37

Naik MM, Pandey A, Dubey SK (2012) *Pseudomonas aeruginosa* strain WI-1 from Mandovi estuary possesses metallothionein to alleviate lead toxicity and promotes plant growth. Ecotoxicol Environ Safety 79:129–133

Navarro CA, von Bernath D, Jerez CA (2013) Heavy metal resistance strategies of acidophilic Bacteria and their acquisition: importance for biomining and bioremediation. Biol Res 46(4):363–371

Nies DH (2003) Efflux-mediated heavy metal resistance in prokaryotes. FEMS Microbiol Rev 27:313–339

Niggemyer A, Spring S, Stackebrandt E et al (2001) Isolation and characterization of a novel As(V)-reducing bacterium: implications for arsenic mobilization and the genus *Desulfitobacterium*. Appl Environ Microbiol 67:5568–5580

Nongkhlaw M, Kumar R, Acharya C et al (2012) Occurrence of horizontal gene transfer of P IB -type ATPase genes among bacteria isolated from uranium rich deposit of Domiasiat in north East India. PLoS One 7(10):e48199

Olaniran AO, Balgobind A, Pillay B (2013) Bioavailability of heavy metals in soil: impact on microbial biodegradation of organic compounds and possible improvement strategies. Int J Mol Sci 14:10197–10228

Oliveira A, Pampulha ME, Neto MM et al (2009) Enumeration and characterization of arsenic-tolerant Diazotrophic Bacteria in a Long-term heavy-metal-contaminated soil. Water Air Soil Pollut 200:237–243

Orell A, Navarro CA, Jerez CA (2009) Copper resistance mechanisms of biomining bacteria and archaea living under extremely high concentrations of metals. Adv Mater Res 71-73:279–282

Ouyang J, Guo W, Li B et al (2013) Proteomic analysis of differential protein expression in *Acidithiobacillus ferrooxidans* cultivated in high potassium concentration. Microbiol Res 168(7):455–460

Oyetibo GO, Ilori MO, Obayori OS et al (2015) Metal biouptake by actively growing cells of metal-tolerant bacterial strains. Environ Monit Assess 187:525

Pena-Montenegro TD, Dussan J (2013) Genome sequence and description of the heavy metal tolerant bacterium *Lysinibacillus sphaericus* strain OT4b.31. Stand Genomic Sci 9(1):42–56

Peng J, Miao L, Chen X et al (2018) Comparative transcriptome analysis of *Pseudomonas putida* KT2440 revealed its response mechanisms to elevated levels of zinc stress. Front Microbiol 9:1669

Perry RD, Silver S (1982) Cadmium and manganese transport in *Staphylococcus aureus* membrane vesicles. J Bacteriol 150:973–976

Prabhakaran P, Ashraf MA, Aqma WS (2016) Microbial stress response to heavy metals in the environment. RSC Adv 6:109862–109877

Rainey FA, Ward-Rainey N, Kroppenstedt RM et al (1996) The genus *Nocardiopsis* represents a phylogenetically coherent taxon and a distinct actinomycete lineage, proposal of *Nocardiopsaceae* fam. nov. Int J Syst Bacteriol 46(4):1088–1092

Rathore SS, Shekhawat K, Dass A et al (2017) Phytoremediation mechanism in Indian mustard (*Brassica juncea*) and its enhancement through agronomic interventions. Proc Natl Acad Sci India Sect B Biol Sci:1–9

Robinson NJ, Gupta A, Fordham-Skelton AP et al (1990) Prokaryotic metallothionein gene characterization and expression: chromosome crawling by ligation-mediated PCR. Proc R Soc London B 242:241–247

Rodriguez-Rojas F, Tapia P, Castro-Nallar E, Undabarrena A, Muñoz-Díaz P, Arenas-Salinas M, Díaz-Vásquez W, Valdés J, Vásquez C (2016) Draft genome sequence of a multi-metal resistant bacterium *Pseudomonas putida* ATH-43 isolated from Greenwich Island. Antarctica Front Microbiol 7:1777

Rodriguez-Sanchez V, Guzmán-Moreno J, Rodríguez-González V et al (2017) Biosorption of lead phosphates by lead-tolerant bacteria as a mechanism for lead immobilization. World J Microbiol Biotechnol 33:150

Romaniuk K, Ciok A, Decewicz P et al (2018) Insight into heavy metal resistome of soil psychrotolerant bacteria originating from King George Island (Antarctica). Polar Biol 41:1319–1333

Román-Ponce B, Ramos-Garza J, Vásquez-Murrieta MS et al (2016) Cultivable endophytic bacteria from heavy metal(loid)-tolerant plants. Arch Microbiol 198(10):941–956

Rossello-Mora R, Amann R (2001) The species concept for prokaryotes. FEMS Microbiol 25:39–67

Roy S, Roy M (2015) Bioleaching of heavy metals by sulfur oxidizing bacteria: a review. Int Res J Environment Sci 4(9):75–79

Roychowdhury R, Roy M, Rakshit A et al (2018) Arsenic bioremediation by indigenous heavy metal resistant bacteria of fly ash pond. Bull Environ Contam Toxicol 101(4):527–535

Ryan RP, Monchy S, Cardinale M et al (2009) The versatility and adaptation of bacteria from the genus *Stenotrophomonas*. Nat Rev Microbiol 7:514–525

Saier MH (2016) Transport protein evolution deduced from analysis of sequence, topology and structure. Curr Opin Struct Biol 38:9–17

Sarma B, Acharya C, Joshi SR (2016) Characterization of metal tolerant *Serratia* spp. isolates from sediments of uranium ore deposit of domiasiat in Northeast India. Proc Natl Acad Sci India Sect B Biol Sci 86(2):253–260

Schaefer JK, Rocks SS, Zheng W et al (2011) Active transport, substrate specificity, and methylation of Hg(II) in anaerobic bacteria. Proc Natl Acad Sci U S A 108:8714–8719

Schalk IJ, Hannauer M, Braud A (2011) New roles for bacterial siderophores in metal transport and tolerance. Environ Microbiol 13(11):2844–2854

Schauer K, Gouget B, Carrière M et al (2007) Novel nickel transport mechanism across the bacterial outer membrane energized by the TonB/ExbB/ExbD machinery. Mol Microbiol 63(4):1054–1068

Shakibaie MR, Khosravan A, Frahmand A et al (2008) Application of metal resistant bacteria by mutational enhancement technique for bioremediation of copper and zinc from industrial wastes. Iran J Environ Health Sci Eng 5(4):251–256

Sharma P, Kumari H, Kumar M et al (2008) From bacterial genomics to metagenomics: concept, tools and recent advances. Indian J Microbiol 48:173–194

Sharma A, Kumar V, Handa N et al (2018) Potential of endophytic bacteria in heavy metal and pesticide detoxification. In: Egamberdieva D, Ahmad P (eds) Plant microbiome: stress response, Microorganisms for sustainability, vol 5. Springer, Singapore, pp 307–336

Shi Y, Yang H, Zhang T et al (2014) Illumina-based analysis of endophytic bacterial diversity and space-time dynamics in sugar beet on the north slope of Tianshan mountain. Appl Microbiol Biotechnol 98(14):6375–6385

Shreedhar S, Devasya RP, Naregundi K et al (2014) Phosphate solubilizing uranium tolerant bacteria associated with monazite sand of a natural background radiation site in south-west coast of India. Ann Microbiol 64:1683–1689

Singh S, Mulchandani A, Chen W (2008) Highly selective and rapid arsenic removal by metabolically engineered *Escherichia coli* cells expressing *Fucus vesiculosus* metallothionein. Appl Environ Microbiol 74:2924–2927

Singh RP, Singh RN, Srivastava AK, Kumar S, Dubey RC, Arora DK (2011) Structural analysis and 3D-modeling of FUR protein from *Bradyrhizobium japonicum*. J Appl Sci Environ Sanit 6:357–366

Subhashini DV, Singh RP, Manchanda G (2017) OMICS approaches: tools to unravel microbial systems. Directorate of Knowledge Management in Agriculture, Indian Council of Agricultural Research. ISBN: 9788171641703. https://books.google.co.in/books?id=vSaLtAEACAAJ

Teitzel GM, Geddie A, Long SK et al (2006) Survival and growth in the presence of elevated copper: transcriptional profiling of copper-stressed *Pseudomonas aeruginosa*. J Bacteriol 188(20):7242–7256

Tirry N, Joutey NT, Sayel H et al (2018) Screening of plant growth promoting traits in heavy metals resistant bacteria: prospects in phytoremediation. J Genet Eng Biotechnol 16(2):613–619

Tremaroli V, Workentine ML, Weljie AM et al (2009) Metabolomic investigation of the bacterial response to a metal challenge. Appl Environ Microbiol 75(3):719–728

Tse C, Ma K (2016) Growth and metabolism of extremophilic microorganisms. In: Rampelotto RH (ed) Biotechnology of extremophiles: advances and challenges. Springer, Cham

Turgay OC, Görmez A, Bilen S (2012) Isolation and characterization of metal resistant-tolerant rhizosphere bacteria from the serpentine soils in Turkey. Environ Monit Assess 184:515–526

Tyson GW, Chapman J, Hugenholtz P et al (2004) Community structure and metabolism through reconstruction of microbial genomes from the environment. Nature 428:37–43

Valenzuela L, Chi A, Beard S et al (2006) Genomics, metagenomics and proteomics in biomining microorganisms. Biotechnol Adv 24(2):197–211

Valls M, de Lorenzo V (2002) Exploiting the genetic and biochemical capacities of bacteria for the remediation of heavy metal pollution. FEMS Microbiol Rev 26(4):327–338

Vartoukian SR, Palmer RM, Wade WG (2010) Strategies for culture of 'unculturable' bacteria. FEMS Microbiol Lett 309(1):1–7

Villadangos AF, Ordóñez E, Pedre B et al (2014) Engineered coryneform bacteria as a bio-tool for arsenic remediation. Appl Microbiol Biotechnol 98:10143–10152

Volpicella M, Leoni C, Manzari C et al (2017) Transcriptomic analysis of nickel exposure in *Sphingobium* sp. ba1 cells using RNA-seq. Sci Rep 7:8262

Whiting SN, de Souza MP, Terry N (2001) Rhizosphere bacteria mobilize Zn for hyperaccumulation by *Thlaspi caerulescens*. Environ Sci Technol 35(15):3144–3150

Wu X, Monchy S, Taghavi S et al (2011) Comparative genomics and functional analysis of niche-specific adaptation in *Pseudomonas putida*. FEMS Microbiol Rev 35(2):299–323

Xie P, Hao X, Herzberg M et al (2015) Genomic analyses of metal resistance genes in three plant growth promoting bacteria of legume plants in northwest mine tailings. China J Environ Sci (China) 27:179–187

Yang YJ, Singh RP, Lan X, Zhang CS, Sheng DH et al (2019) Synergistic effect of *Pseudomonas putida* II-2 and *Achromobacter* sp. QC36 for the effective biodegradation of the herbicide quinclorac. Ecotoxicol Environ Saf. https://doi.org/10.1016/j.ecoenv.2019.109826

Yu P, Yuan J, Deng X et al (2014) Subcellular targeting of bacterial CusF enhances cu accumulation and alters root to shoot cu translocation in Arabidopsis. Plant Cell Physiol 55(9):1568–1581

Zhai Q, Xiao Y, Zhao J et al (2017) Identification of key proteins and pathways in cadmium tolerance of *Lactobacillus plantarum* strains by proteomic analysis. Sci Rep 7:1182

Zhang Z, Cai R, Zhang W, et al (2017) A novel exopolysaccharide with metal adsorption capacity produced by a marine bacterium *Alteromonas* sp. JL2810. Mar Drug 15(6):175

Zivkovic LI, Rikalovic M, Cvijovic GG et al (2018) Cadmium specific proteomic responses of a highly resistant *Pseudomonas aeruginosa* san ai. RSC Adv 8:10549

Host-Plant Interaction and Pathogenesis

Endophytic Actinomycetes-Mediated Modulation of Defense and Systemic Resistance Confers Host Plant Fitness Under Biotic Stress Conditions

10

Waquar Akhter Ansari, Ram Krishna,
Mohammad Tarique Zeyad, Shailendra Singh,
and Akhilesh Yadav

Abstract

Plants are highly susceptible to biotic stress causing deleterious impacts on growth of plants, cellular development, intrinsic biological mechanisms, and productivity. Endophytic microorganisms have integral role on plant ecosystem to combat these biotic stress conditions. Apart from that, endophytic actinomycetes are the natural symbionts of several plants that modulate the defense strategies as well as systemic resistance of the plants to confer the resistance to host plants under adverse conditions. The interaction between the endophytic actinomycetes and plants leads to several biochemical, physiological, and molecular events that are very complex to understand. It has now become vital to interpret and perceive the relationship between plant and endophytic actinomycetes in terms of protection against biotic stress conditions. Simultaneously, an understanding of the regulation of stress-ameliorating mechanisms in the plant due to the interaction with endophytic actinomycetes is also very crucial. Many genomics-, transcriptomics-, proteomics-, and metabolomics-based studies on plant-endophytic actinomycetes interaction are helpful to illustrate the exact happenings during biotic stress conditions. This chapter summarizes the importance of endophytic actinomycetes in amelioration

---

The original version of this chapter was revised. A correction to this chapter can be found at https://doi.org/10.1007/978-981-15-3028-9_13.

W. A. Ansari · S. Singh
National Bureau of Agriculturally Important Microorganism, Mau, Uttar Pradesh, India

R. Krishna
Directorate of Onion and Garlic Research (DOGR), Pune, India

M. T. Zeyad
Department of Agricultural Microbiology, Aligarh Muslim University, Aligarh, India

A. Yadav (✉)
Department of Biotechnology, Indian Institute of Technology, Roorkee, India

of the adverse effect of biotic stress in host plants. Moreover, the profound knowledge about plant-microbe interaction at the root level supports the multifarious role of endophytic actinomycetes in the alleviation of abiotic stress in the plant.

Keywords
Plant endophytes · Plant growth promotion · Metabolites · Interaction · Antagonism

10.1 Introduction

Microbial populations, for example, bacteria and fungi, that enter the plant tissue and do not show any negative effect on host plant are termed as endophytes (Schulz and Boyle 2006). The endophyte term was given by De Bary (1866). Endophytes produce metabolites which promote growth, repellents for insect and pest, and antimicrobial agents against phytopathogen (Rai et al. 2014a, b) and additionally produce secondary metabolites, largely applied in agricultural, pharmaceutical, and many other purposes. Actinobacteria are Gram-positive bacteria which are mostly filamentous and perform an important role in the production of humus. Hence, actinobacteria decompose the complex dead plants and animal materials, algae and fungi additionally and thus help in nutrients recycling (Sharma 2014). These microbes are the largest producers of various antibiotics against many deadly diseases. Out of 23,000 bioactive metabolites, 10,000 (45%) alone are produced by these actinobacteria, and the compounds which are reported most are from a single genus which is *Streptomyces* (Berdy 2012). Large number of medicinal plants are an important source of numerous bioactive compounds in association with these endophytes. Endophytes associates with these plants metabolic pathways and increases their own metabolic activity to produce bioactive compound corresponds with host plant (Eyberger et al. 2006; Kumar et al. 2013; Rai et al. 2014a, b; Subhashini et al. 2017).

Life on the earth is only possible because of plants as they provide human beings about 90% calories and 80% protein. In addition to humans, also vertebrates gain their food from plants indirectly or directly. Humans utilize about 3000 species of plants for their purposes, though, presently, world primarily depends on approximately 20 species of crops only for the supply of their bulk energy out of which 50% contributed by 8 cereal species crops. As per estimation, the earth can only support 5 billion mixed diet or 15 billion strict vegetarian populations, but the world population will reach up to 10 billion by 2050 (Ingram 2011). In natural environment, plants face many biotic stresses, and sometimes the severe infection causes up to 100% crop loss (Prasanna et al. 2015). Synthetic chemical-mediated management of biotic stresses causes ecosystem erosion by killing nontarget organisms and biomagnifications of toxic chemicals (Yang et al. 2019). Many microbes have biocontrol properties due to the virtue of competition, parasitism, or antibiosis against pathogens for space and nutrition (Vurukonda et al. 2018). A biotic stress response caused by microorganisms (viruses, bacteria, and fungi) on all over the plant is illustrated in Fig. 10.1. Such microbes colonize inside the plant and interfere with infection processes and hence may also induce disease resistance in plants; the quality of disease resistance might be the consequences of plant-microbe or microbe-microbe interaction (Vurukonda et al. 2018). *Actinomycetes* or *Actinobacteria* are well-known for such types of activities; actinobacteria are the locus between the true bacteria and fungi, typically having more than 50% GC contents in their genome. *Actinomycetes* are well-known for producing bioactive compounds (Barka et al. 2016), including antibiotics, antioxidants, antiviral agents, plant growth hormones, enzymes, etc., which play important roles in agriculture (Castillo et al. 2002; Verma et al. 2018; Singh et al. 2019). *Actinobacteria* is an important phylum in Bacteria domain and consists of six classes, viz., (1) *Actinobacteria*, (2) *Acidimicrobiia*, (3) *Coriobacteriia*, (4) *Nitriliruptoria*, (5) *Rubrobacteria*, and (6) *Thermoleophilia* (Barka et al. 2016; Gao and Gupta 2012). *Actinobacteria* are aerobic and Gram-positive and have a diverse morphological feature that may be unicellular to filamentous. The *Streptomyces* is the most important genus belonging to class *Actinobacteria*. They produce a huge range of bioactive compounds and produce approximately 80% of all natural agriculturally important products like insecticides and herbicides produced by actinobacteria (Berdy 2005). *Streptomyces* are distributed widely, from terrestrial to aquatic ecosystems, together with marine environments. Most commonly they inhabit the soils where they form semidormant spore and spend most of the life cycles. These comprise about 20–30% microbial community of the rhizosphere (Bouizgarne and Aouamar 2014).

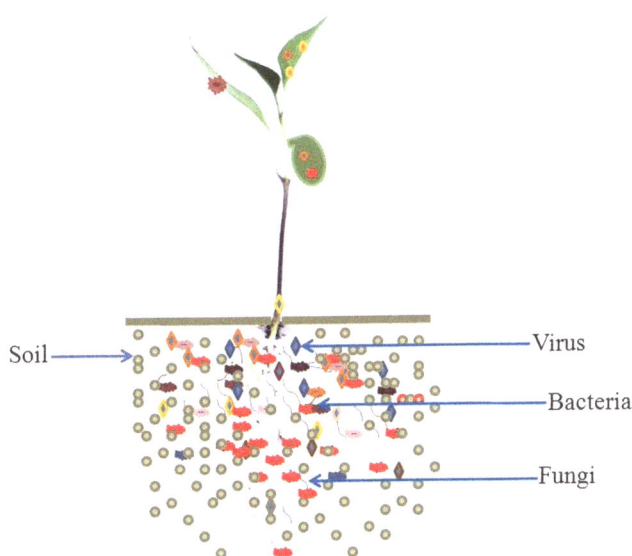

Fig. 10.1 Biotic stresses on the plant and key microorganisms viruses, bacteria, and fungi causing biotic stress

10.2 Plant Protection by Endophytic Actinomycetes

Many *Actinomycetes* are capable to establish close relationships with plants and colonize inside tissues without pathogenic properties and are called as endophytic *Actinomycetes* (Qin et al. 2009). Endophytic *Actinomycetes* perform a critical role in plant protection using bioactive compounds which may act as biocontrol agents, stress alleviators, or plant growth promoters, while, in turn, these endophytic *Actinomycetes* get protection and nutrition through host plant (Singh et al. 2017). Table 10.1 has displayed the representative reports on activity of endophytic actinomycoses in plant defense. Recently in the past two decades, antibiosis interest has increased, and mechanisms of biocontrol have been well understood (Whipps 2001). A number of metabolites having antibiotic properties produced by *Pseudomonas* like oomycin A, 2,4-diacetylphloroglucinol (DAPG), cyclic lipopeptide amphysin, pyrrolnitrin, aromatic polyketide pyoluteorin, and tropolone have been identified and characterized (de Souza et al. 2003). From other genera, like *Streptomyces, Bacillus,* and *Stenotrophomonas*, products such as kanosamine, macrolide oligomycin A, xanthobactin, and aminopolyol zwittermicin A have been identified (Compant et al. 2005; Prajakta et al. 2019). They also produce various enzymes that have the capability to degrade the cell walls of fungi (Watve et al. 2001). The endophytic actinobacteria inhibit plant pathogen growth by activities of antagonism between them, either directly (competition, antibiosis, and lysis) or indirectly (plant growth substances-mediated induction of plant defense system) (El-Tarabily and Sivasithamparam 2006; Hastuti et al. 2012). Additionally, endophytic *Actinobacteria* also inhibit pathogen growth by producing siderophores and chitinolytic enzymes. More than 90% of *Actinomycetes* are chitinolytic in nature. The property of producing chitinases by *Actinomycetes* and *Streptomycetes* makes them potential biocontrol agents. Hence this capability may be used for selecting and exploiting the chitinolytic biocontrolling microbial agents for fungal plant pathogens (De Boer et al. 2001). *Streptomycetes* produces another important compound known as allosamidin, an inhibitor of 18 families of chitinase (Suzuki et al. 2006). Consequently, allosamidin is considered as a potential antifungal secondary metabolite (Sakuda et al. 2013). In agricultural systems, *Streptomycetes* used as biocontrol agents against phytopathogens are becoming a more effective approach because of the high processing cost required for antimicrobial molecule purification. The siderophores produced by *Actinobacteria* have gained more importance because of their capability for plant growth promotion and antagonism enhancement against phytopathogens (Qin et al. 2011). The application of these *Actinobacteria* in consortia forms proved to provide better results than individual actinobacteria, as reported by El-Tarabily (2003). The inoculation of *Actinoplanes campanulatus* and *Micromonospora chalcea* to cucumber seedlings showed more effectiveness against damping-off and crown disease caused by *Pythium aphanidermatum*.

The consortia consisting of several *Streptomycetes* produce different siderophores which sequester rhizospheric iron, and consequently iron becomes unavailable to the rhizospheric microorganisms especially to the phytopathogens. Thus,

Table 10.1 Activity of endophytic actinomycetes in plant defense

Species/strain	Disease	Pathogens	Host plant	References
Streptomyces viridodiasticus	Basal drop disease	*Sclerotinia minor*	*Lactuca sativa*	El-Tarabily et al. (2000)
Streptomyces sp. KH-614, S. vinaceusdrappus	Blast	*Pyricularia oryzae*	*Oryza sativa*	Ningthoujam et al. (2009)
Streptomyces sp. AP77	Red rot	*Pythium porphyrae*	*Porphyra haitanensis*	Woo and Kamei (2003)
Streptomyces sp. S30	Damping-off	*Rhizoctonia solani*	*Solanum lycopersicum*	Cao et al. (2004)
S. halstedii	Blight	*Phytophthora capsica*	*Capsicum*	Liang et al. (2005)
S. violaceusniger XL-2	Wood rot	*Gloeophyllum trabeum*	Many	Shekhar et al. (2006)
S. ambofaciens S2	Anthracnose	*Colletotrichum gloeosporioides*	*Capsicum*	Heng (2015)
Streptomyces sp.	Damping-off	*Sclerotium rolfsii*	*Beta vulgaris*	Errakhi et al. (2007)
S. hygroscopicus	Leaf blight	*Colletotrichum gloeosporioides* and *Sclerotium rolfsii*	Many	Prapagdee et al. (2008)
Streptomyces sp.	Powdery mildew	*Oidium* sp.	*Lathyrus odoratus*	Sangmanee et al. (2009)
*Streptomyces spororaveus*RDS28	Collar or root rot, stalk rot, leaf spots, and botrytis blight	*Rhizoctonia solani, Fusarium solani, Fusarium verticillioides, Alternaria alternata, Botrytis cinerea*	Many	Al-Askar et al. (2011)
Streptomyces sp.	Wilt	*Fusarium oxysporum f.* sp. ciceri	*Cicer arietinum*	Gopalakrishnan et al. (2011)
Streptomyces sp.	Blight	*Xanthomonas campestris* pv. glycines	Soybean	Mingma et al. (2014)
Streptomyces sp. 5406	Soilborne diseases	Soilborne plant pathogens	*Gossypium*	Yin et al. (1965)
Streptomyces sp.	Seed infection	*Aspergillus* sp.	*Zea mays*	Bressan 2003
Streptomyces sp.	Silver scurf	*Helminthosporium solani*	*Solanum tuberosum*	Elson et al. (1997)
Streptomyces sp.	Brown spot	*Alternaria* spp.	*Nicotiana tabacum*	Gao et al. (2014)
Streptomyces sp. CBE	Stem rot	*Sclerotium rolfsii*	*Arachis hypogaea*	Adhilakshmi et al. (2013)

(continued)

Table 10.1 (continued)

Species/strain	Disease	Pathogens	Host plant	References
Streptomyces sp.	Root rot, blight, and fruit rot	*Alternaria brassicae, Colletotrichum gloeosporioides, Rhizoctonia solani, Phytophthora capsica*	*Capsicum frutescens*	Srividya et al. (2012)
Actinomycetes strains OUA3, OUA5, OUA18, and OUA40	Wilt	*Colletotrichum capsici and Fusarium oxysporum*	*Capsicum annuum*	Ashokvardhan et al. (2014)
Actinomycetes PACCH 277, PACCH129, PACCH225, PACCH24, and PACCH246, *Streptomyces hygroscopicus*	Stem rot	*Sclerotium rolfsii*	*Capsicum*	Pattanapipitpaisal and Kamlandharn (2012)
Streptomycetes indiaensis KJ872546	Wilt	*Fusarium oxysporum*	*Capsicum*	Jalaluldeen et al. (2014)
Micromonospora sp.	Leaf infection	*Botrytis cinerea*	*Medicago sativa*	
Saccharothrix algeriensis NRRL B-24137	Wilt	*Fusarium oxysporum f.* sp. *lycopersici* (FOLy)	*Solanum lycopersicum*	Merrouche et al. (2017)
S. cyaneus ZEA17I	Fungal infection	*Sclerotinia sclerotiorum* FW361	*Lactuca sativa*	Kunova et al. (2016)
Streptomyces humidus	Leaf spot	*Alternaria brassicicola*	*Brassica oleracea*	Hassan et al. (2017)

phytopathogens are unable to gain required iron quantities for their growth and development. Many *Streptomyces* spp. produce 6-prenylindole bioactive compound which significantly acts against phytopathogenic fungi like *Corynespora cassiicola, Colletotrichum orbiculare, Fusarium oxysporum*, and *Phytophthora capsici*. Other bioactive compounds like naphthomycins A and K and fistupyrone produced from different species of *Streptomyces* act against phytopathogenic fungi (Lu and Shen 2007). Many species of *Streptomyces* are capable of producing several secondary metabolites and cellulolytic enzymes that act on herbivorous insects directly and also have toxicity against insect pests as well as phytopathogens (Book et al. 2014). A number of anti-insect bioactive compounds like antimycin A, flavensomycin, piericidins, prasinons, and macro tetralines are produced by different species of *Streptomyces* which acts against many insect pests (Craveri and Giolitti 1957; Kido and Spyhalski 1950; Takahashi et al. 1968; Oishi et al. 1970, Box et al. 1973). *Streptomyces avermitilis* produces avermectins which are potential arthropods and nematodes controlling bioactive compounds (Turner and Schaeffer 1989). The avermectin mixtures based on commercial insecticides are commonly known as abamectin and are effective against phytophagous insects and directly act by contact or ingestion.

10.3 Biological Control Agents

Microbial biocontrol agents have the capability to react as antibiotic. Endophytic actinomycetes have attracted the attention as biocontrol agents due to their antifungal activity and colonization characteristics (Subhashini and Singh 2014; Vurukonda et al. 2018). They protect plants from several plant pathogens which are soilborne that include *Gaeumannomyces graminis* var. *tritici*, *Plectosporium tabacinum*, *Verticillium dahliae*, *Rhizoctonia solani*, *Fusarium oxysporum*, *Colletotrichum orbiculare*, and *Pythiuma phanidermatum* (Qin et al. 2011). The endophytic actinomycetes action mechanism mostly focuses on bioactive compound release, i.e., antibiotics and cell wall-degrading enzymes (El-Tarabily and Sivasithamparam 2006). Plant-induced systemic resistance (ISR)-inducing characteristics were also observed in endophytic actinobacteria. Seeds of *Arabidopsis thaliana* were inoculated with endophytic strain of *Micromonospora* sp. EN and *Streptomyces* sp. EN27 (Conn et al. 2008) which leads to enhanced tolerance in the leaves against plant pathogens *F. oxysporum* and *Erwinia carotovora* and induced defense gene expression linked with jasmonic acid/ethylene-dependent signaling pathways when the pathogens are absent. Furthermore, *Micromonospora* sp. strain EN43 culture filtrates activate the acquired resistance of the system and also the jasmonates/ethylene pathways. Hirsch and Valdes (2010) reported synthesis of two different secondary metabolites through endophytic actinobacteria, which provokes plant defense system. Antitrypanosomal alkaloids spoxazomicins released by the endophytic actinomycetes *Streptosporangium oxazolinicum* (Inahashi et al. 2011) have structural similarity with siderophores produced from *P. aeruginosa* and many nonribosomal peptide synthases (NRPS) and polyketide synthases (PKS) gene clusters (Miller et al. 2012; Brader et al. 2014). The endophytic actinobacterial association with plants produces bioactive compounds and a distinctive option to discover efficient drugs or fungicides for a potential appliance in plant defense. Although it is still uncertain about the actinomycetes contribute to disease control in plants. The presence of a high number of populations of actinomycetes in the soil could possibly control the fungal disease.

10.4 Active Principal Components of Actinomycetes Involved in Systemic Resistance for Plants

The obtained metabolites of secondary nature from actinomycetes have critical importance in the growth promotion of plants as well as growth suppression of plant pathogen. The advancement of organic influence against plant disease is acknowledged through a sturdy and eco-accommodating option as an agrochemical alternate. Advantageous microorganisms including fungi and bacteria that coexist in the rhizospheric zone of plants showed either direct or indirect or both mechanisms to act as a plant growth promoter. Among the various modes of actions, induced resistance is considered as one of the potential systems of plant defense through which the endophytes check the growth of pathogen inside the plant. Induced systemic resistance mediated by nonpathogenic, plant growth-promoting fungi (PGPF) and plant growth-promoting rhizobacteria (PGPR) was studied widely (Pieterse et al. 2014). Plant growth-promoting actinomycetes (PGPA) specifically *Streptomyces* are considered as one of the effective sources of many secondary metabolites which include antibiotics and cell wall-degrading enzymes that act on numerous bacterial and fungal pathogens (Nachtigall et al. 2011). Against *Sclerotinia minor*, El-Tarabily et al. (2000) reported about the chitinolytic action of *M. carbonacea* and *S. viridodiasticus* and observed a drastic reduction in the frequency of basal drop disease of lettuce.

10.5 Induced Systemic Resistance (ISR)

It is known from many decades that plants protect themselves against attacking pathogen by synthesizing numerous resistance compounds (Chester 1933). Furthermore, disease resistance even can be initiated or enhanced through bio-inoculation of biocontrol against the colonizing pathogen (da Rocha and Hammerschmidt 2005). However, in any condition, particular plants must possess particular gene of interest that gets expressed under the attack of disease pathogen. Correspondingly, the inducer or biocontrol agents additionally ought to be able to prompt protective compounds in plants. On the basis of nature of agent's acts as inducer and pathways of signaling concerned, resistance which gets activated after induction is categorized into two classes, i.e., ISR and SAR (van Loon et al. 1998). In their mode of action, SAR is distinct from ISR, as the first one is hypersensitive reaction or local necrotic inflammation in host plant induced due to pathogen which further results in growth retardation of attacking pathogen, while ISR is the improved degree of cautious reactions stimulated through plant growth-promoting rhizobacteria in light of pathogen assault (van Loon et al. 1998). In any case, biological regulators must release respective required inducers or elicitors alternately, which signal plant receptors for recognition mediated through complex metabolic pathway which finally actuates ISR. Metabolites, for example, chitin (Zhang et al. 2002), DAPG (Iavicoli et al. 2003), proteins and peptides (Harman et al. 2004), ergosterol (Kauss and Jeblick 1996), glucans (Mithofer et al. 1996), sphingolipids (Umemura et al. 2004), lipopolysaccharides (LPS) (Silipo et al. 2005), and many organic compounds of volatile nature (Ryu et al. 2004) formed, can perform as an elicitor for the period of ISR. Protein genes linked with pathogenesis and signaling of salicylic acid perform important functions in SAR (Hammerschmidt 1999). However, in the case of ISR, ethylene and jasmonic acid signaling takes part in main roles.

10.6   ISR Mechanisms Mediated Through PGPA

Plants have various insoluble protective basic hindrances which get induced against colonization of pathogen that include the development of cell wall positions, deposition of hydroxyproline-rich glycoproteins and callose, and collection of various phenolics, for example, lignin and suberin, minerals, silicon, and calcium (Ton et al. 2005). Under the development of hypersensitive responses, escalation of structural obstructions, and phytoalexin generation which induces many defense genes and encode peroxidase, catalase, superoxide dismutase, polyphenol oxidase, proteinase inhibitors, β-1,3-glucanase, chitinase, and lipoxygenase are perform a critical role in ISR (Pozo et al. 2005). ISR mechanisms mediated by strains of *P. fluorescens* was reported in the number of crop plants, which includes tomato, mango, groundnut, and rice. El-Tarabily et al. (2009) performed a study to check the biological effectiveness of endophytic actinomycetes, *Streptomyces spiralis*, *Micromonospora chalcea*, and *Actinoplanes campanulatus*, and observed these three endophytic actinomycetes formed elevated amounts of enzymes involved in cell wall degradation which include β-1,3, β-1,4, and β-1,6 glucanases, in that way considerably decreasing the damping-off occurrence in cucumber. Similarly, Cheng et al. (2014) reported that under greenhouse conditions, *S. felleus* YJ1 showed durable antagonistic action by synthesizing an increased amount of enzymes involved in plant defense against *S. sclerotiorum*. In a different experiment, Hasegawa et al. (2004) observed enhanced callose deposition results due to the inoculation of mountain laurel with *S. padanus*. These findings suggest that the detected cell wall appositions were the machinery following the increased resistance against diseases in plants that are pre-inoculated with *Streptomyces*. In Norway spruce needles, *Streptomyces* GB 4-2-mediated root inoculation induces systemic resistance against *B. cinerea*, in addition to the induction of the local defense (Lehr et al. 2008). Induction of cytosolic Ca^{2+} and biphasic oxidative burst by *Streptomyces* sp. were also reported (Baz et al. 2012). Conn et al. (2008) inoculated *Arabidopsis* seeds with endophytic actinobacteria (*Micromonospora* sp. AY291589, *Streptomyces* sp. AY148076, *Streptomyces* sp. AY148075, *Microbispora* sp. AY148073, and *Nocardioides albus* AY148081) to estimate the important pathway of SAR and JA/ET in defense genes induction against *Erwinia carotovora* and *F. oxysporum*. It showed that the genes involved in defense express in reactions to the treatment with streptomycetes, shared both pathways induced systematic resistance and systematic acquired resistance (Conn et al. 2008).

10.7   Conclusion

A large number of studies have been carried out by various groups of researchers presenting the capability of endophytic actinomycetes to elevate the plant development and synergistically to release the various compounds which impact plant growth and protection. Presented chapter discourses about actinomycetes as bacteria of helping nature, which are genuinely eminent for use as plant inoculants: they enhance plant-microorganism beneficial interaction in a manner that could prompt an expanded maintainable production of agricultural produce under differing conditions. This guarantee is primarily founded on utilization of environment-accommodating microorganisms which restrict pests and promote the growth and developments of plant. The utilization of biopesticides, biofertilizers, or consortia of plant advantageous microorganisms in the right combinations gives a potential answer for sustainable supportable agricultural practices. The informations presented in this chapter foundation the idea that upgraded formulation developments with interactive microbes may add to the enhanced plant growth and protection. On the other hand, these findings also emphasize the significance of proceeding research in this area, particularly targeting which till date were minimally utilized as inoculants to improve agrarian productivity and guarantee food security, in spite of the outstanding capability present in a large number of scientific publications so far.

References

Adhilakshmi M, Latha P, Paranidharan V, Balachandar D, Ganesamurthy K, Velazhahan R (2013) Biological control of stem rot of groundnut (L.) caused by Sacc. with actinomycetes. Arch Phytopathol Plant Protect 47(3):298–311

Al-Askar AA, Abdul K, Rashad YWM (2011) In vitro antifungal activity of Streptomyces spororaveus RDS28 against some phytopathogenic fungi. Afr J Agri Res 6(12):2835–2842

Ashokvardhan T, Rajithasri AB, Prathyusha P, Satyaprasad K (2014) Actinomycetes from *Capsicum annuum* L. rhizosphere soil have the biocontrol potential against pathogenic fungi. Int J Curr Microbiol Appl Sci 3:894–903

Barka EA, Vatsa P, Sanchez L, Gaveau-Vaillant N, Jacquard C et al (2016) Taxonomy, physiology, and natural products of Actinobacteria. Microbiol Mol Biol Rev 80(1):1–43

Baz M, Tran D, Kettani-Halabi M, Samri SE, Jamjari A et al (2012) Calcium- and ROS-mediated defence responses in BY2 tobacco cells by nonpathogenic sp. J Appl Microbiol 112:782–792

Berdy J (2005) Bioactive microbial metabolites. J Antibiot 58(1):1

Berdy J (2012) Thoughts and facts about antibiotics: where we are now and where we are heading. J Antibiot 65:385–395

Book AJ, Lewin GR, McDonald BR, Takasuka TE, Doering DT et al (2014) Cellulolytic Streptomyces strains associated with herbivorous insects share a phylogenetically linked capacity to degrade lignocellulose. Appl Environ Microbiol 80(15):4692–4701

Bouizgarne B, Aouamar AAB (2014) Diversity of plant-associated actinobacteria. In: Bacterial diversity in sustainable agriculture. Springer, Cham, pp 41–99

Box SJ, Cole M, Yeoman GH (1973) Prasinons A and B: potent insecticides from *Streptomyces prasinus*. Appl Environ Microbiol 26(5):699–704

Brader G, Compant S, Mitter B, Trognitz F, Sessitsch A (2014) Metabolic potential of endophytic bacteria. Curr Opin Biotechnol 1(27):30–37

Bressan W (2003) Biological control of maize seed pathogenic fungi by use of actinomycetes. BioControl 48(2):233–240

Cao L, Qiu Z, You J, Tan H, Zhou S (2004) Isolation and characterization of endophytic Streptomyces strains from surface-sterilized tomato (*Lycopersicon esculentum*) roots. Lett Appl Microbiol 39(5):425–430

Castillo UF, Strobel GA, Ford EJ, Hess WM, Porter H et al (2002) Munumbicins, wide-spectrum antibiotics produced by Streptomyces NRRL 30562, endophytic on *Kennedia nigricans*. Microbiol 148(9):2675–2685

Cheng G, Liu F, Huang Y, Yang H, Yao J et al (2014) Colonization of *Streptomyces felleus* YJ1 and its effects on disease resistant-related enzymes of oilseed rape. J Agric Sci 6:26–33

Chester KS (1933) The problem of acquired physiological immunity in plants. Q Rev Biol 8:275–324

Compant S, Duffy B, Nowak J, Clément C, Barka EA (2005) Use of plant growth-promoting bacteria for biocontrol of plant diseases: principles, mechanisms of action, and future prospects. Appl Environ Microbiol 71(9):4951–4959

Conn VM, Walker AR, Franco CM (2008) Endophytic actinobacteria induce defense pathways in *Arabidopsis thaliana*. Mol Plant-Microbe Interact 21:208–218

Craveri R, Giolitti G (1957) An antibiotic with fungicidal and insecticidal activity produced by Streptomyces. Natur 179(4573):1307

da Rocha AB, Hammerschmidt R (2005) History and perspectives on the use of disease resistance inducers in horticultural crops. Hort Technol 15:518–529

De Bary A (1866) Morphologie und Physiologie der Pilze, Flechten, und Myxomyceten. Hofmeister's in "handbook of physiological botany"

De Boer W, Gunnewiek PJK, Kowalchuk GA, Van VJA (2001) Growth of chitinolytic dune soil β-Subclass Proteobacteria in response to invading fungal hyphae. Appl Environ Microbiol 67(8):3358–3362

de Souza JT, de Boer M, de Waard P, van Beek TA, Raaijmakers JM (2003) Biochemical, genetic, and zoosporicidal properties of cyclic lipopeptide surfactants produced by *Pseudomonas fluorescens*. Appl Environ Microbiol 69(12):7161–7172

Elson MK, Schisler DA, Bothast RJ (1997) Selection of microorganisms for biological control of silver scurf (*Helminthosporium solani*) of potato tubers. Plant Dis 81(6):647–652

El-Tarabily KA (2003) An endophytic chitinase-producing isolate of Actinoplanes missouriensis, with potential for biological control of root rot of lupin caused by Plectosporium tabacinum. Aust J Bot 51(3):257–266

El-Tarabily KA, Sivasithamparam K (2006) Non-streptomycete actinomycetes as biocontrol agents of soil-borne fungal plant pathogens and as plant growth promoters. Soil Biol Biochem 38(7):1505–1520

El-Tarabily KA, Soliman MH, Nassar AH, Al-Hassani HA, Sivasithamparam K et al (2000) Biological control of *Sclerotinia minor* using a chitinolytic bacterium and actinomycetes. Plant Pathol 49:573–583

El-Tarabily KA, Nassar AH, Hardy GESJ, Sivasithamparam K (2009) Plant growth-promotion and biological control of *Pythium aphanidermatum* a pathogen of cucumber, by endophytic actinomycetes. J Appl Microbiol 106:13–26

Errakhi R, Bouteau F, Lebrihi A, Barakate M (2007) Evidences of biological control capacities of Streptomyces spp. against Sclerotium rolfsii responsible for damping-off disease in sugar beet (Beta vulgaris L.). World J Microbiol Biotechnol 23(11):1503–1509

Eyberger AL, Dondapati R, Porter JR (2006) Endophyte fungal isolates from *Podophyllum peltatum* produce podophyllo-toxin. J Nat Prod 69:1121–1124

Gao B, Gupta RS (2012) Phylogenetic framework and molecular signatures for the main clades of the phylum Actinobacteria. Microbiol Mol Biol Rev 76(1):66–112

Gao F, Wu Y, Wang M (2014) Identification and antifungal activity of an actinomycete strain against *Alternaria* spp. Span J Agri Res 12(4):1158–1165

Gopalakrishnan S, Pande S, Sharma M, Humayun P, Kiran BK et al (2011) Evaluation of actinomycete isolates obtained from herbal vermicompost for the biological control of Fusarium wilt of chickpea. Crop Protec 30(8):1070–1078

Hammerschmidt R (1999) Induced disease resistance: how do induced plants stop pathogens? Physiol Mol Plant Pathol 55:77–84

Harman GE, Howell CR, Viterbo A, Chet I, Lorito M (2004) Trichoderma species – opportunistic, avirulent plant symbionts. Nat Rev Microbiol 2:43–56

Hasegawa S, Meguro A, Nishimura T, Kunoh H (2004) Drought tolerance of tissue-cultured seedlings of mountain laurel (*Kalmia latifolia* L.) induced by an endophytic actinomycete. I. Enhancement of osmotic pressure in leaf cells. Actinomycetologica 18:43–47

Hassan N, Nakasuji S, Elsharkawy MM, Naznin HA, Kubota M et al (2017) Biocontrol potential of an endophytic Streptomyces sp. strain MBCN152-1 against *Alternaria brassicicola* on cabbage plug seedlings. Micro. Environment 32(2):133–141. https://doi.org/10.1264/jsme2

Hastuti RD, Lestari Y, Suwanto A, Saraswati R (2012) Endophytic Streptomyces spp. as biocontrol agents of rice bacterial leaf blight pathogen (*Xanthomonas oryzae* pv. *oryzae*). HAYATI J Biosci 19(4):155–162

Heng JLS (2015) Streptomyces ambofaciens S2-A potential biological control agent for Colletotrichum gleosporioides the causal agent for Anthracnose in red chilli fruits. J Plant Pathol Microbiol s1(suppl1):1–6

Hirsch AM, Valdés M (2010) Micromonospora: an important microbe for biomedicine and potentially for biocontrol and biofuels. Soil Biol Biochem 42(4):536–542

Iavicoli A, Boutet E, Buchala A, Me'traux JP (2003) Induced systemic resistance in Arabidopsis thaliana in response to root inoculation with *Pseudomonas fluorescens* CHA0. Mol Plant Microbe Interact 16:851–858

Inahashi Y, Iwatsuki M, Ishiyama A, Namatame M, Nishihara-Tsukashima A et al (2011) Spoxazomicins A–C, novel antitrypanosomal alkaloids produced by an endophytic actinomycete, *Streptosporangium oxazolinicum* K07-0460 T. J Antibiot 64(4):303

Ingram J (2011) A food systems approach to researching food security and its interactions with global environmental change. Food Security 3(4):417–431

Jalaluldeen SAM, Othman K, Ahmad RM, Abidin Z (2014) Isolation and characterization of actinomycetes with in-vitro antagonistic activity against *Fusarium oxysporum* from rhizosphere of chilli. Int J Enhanc Res Sci Technol Eng 3:54–61

Kauss H, Jeblick W (1996) Influence of salicylic acid on the induction of competence for H2O2 elicitation. Plant Physiol 111:755–763

Kido GS, Spyhalski E (1950) Antimycin A, an antibiotic with insecticidal and miticidal properties. Science (Washington) 112:172–173

Kumar A, Patil D, Rajamohanan PR, Ahmad A (2013) Isolation, purification and characterization of from endophytic fungus *Fusarium oxysporum* isolated from *Catharanthus roseus*. PLoS ONE 8(9):e71805

Kunova A, Bonaldi M, Saracchi M, Pizzatti C, Chen X (2016) Selection of Streptomyces against soil borne fungal pathogens by a standardized dual culture assay and evaluation of their effects on seed germination and plant growth. BMC Microbiol 16(1):272

Lehr NA, Schrey SD, Hampp R, Tarkka MT (2008) Root inoculation with a forest soil streptomycete leads to locally and systemically increased resistance against phytopathogens in Norway spruce. New Phytol 177:965–976

Liang J, Xue Q, Niu X, Li Z (2005) Root colonization and effects of seven strains of actinomycetes on leaf PAL and PPO activities of capsicum. Acta Botan Boreali-Occiden Sin 25(10):2118–2123

Lu C, Shen Y (2007) A novel ansamycin, naphthomycin K from *Streptomyces* sp. J Antibiot 60(10):649

Merrouche R, Yekkour A, Lamari L, Zitouni A, Mathieu F (2017) Efficiency of *Saccharothrix algeriensis* NRRL B-24137 and its produced antifungal dithiolopyrrolones compounds to suppress *Fusarium oxysporum*-induced wilt disease occurring in some cultivated crops. Arab J Sci Eng 42(6):2321–2327

Miller KI, Qing C, Sze DM, Neilan BA (2012) Investigation of the biosynthetic potential of endophytes in traditional Chinese anticancer herbs. PLoS One7(5):e35953

Mingma R, Pathom-aree W, Trakulnaleamsai S, Thamchaipenet A, Duangmal K (2014) Isolation of rhizospheric and roots endophytic actinomycetes from Leguminosae plant and their activities to inhibit soybean pathogen, *Xanthomonas campestris* pv. *glycine*. W J Microbiol Biotechnol 30(1):271–280

Mithofer A, Bhagwat AA, Feger M, Ebel J (1996) Suppression of fungal β-glucan-induced plant defense in soybean (*Glycine max* L.) by cyclic 1,3–1,6-β-glucans from the symbionts *Bradyrhizobium japonicum*. Planta 199:270–275

Nachtigall J, Kiluk A, Helaly S, Bull AT, Goodfellow M et al (2011) Atacamycins A-C, 22-membered antitumor macro-lactones produced by *Streptomyces* sp. C38. J Antibiot 64:775–780

Ningthoujam S, Sanasam S, Tamreihao K, Nimaich S (2009) Antagonistic activities of local actinomycete isolates against rice fungal pathogens. Afr J Microbiol Res 3(11):737–742

Oishi H, Takao S, Tsuneo O, Koji S, Toshiaki H et al (1970) Insecticidal activity of macrotetrolide antibiotics. J Antibiot 23:105–106

Pattanapipitpaisal P, Kamlandharn R (2012) Screening of chitinolytic actinomycetes for biological control of *Sclerotium rolfsii* stem rot disease of chilli. Songklanakarin J Sci Technol 34:387–389

Pieterse CMJ, Zamioudis C, Berendsen RL, Weller DM, Van Wees SCM, Bakker PAHM (2014) Induced systemic resistance by beneficial microbes. Annu Rev Phytopathol 52(1):347–375

Pozo MJ, van Loon LC, Pieterse CMJ (2005) Jasmonates signals in plant-microbe interactions. J Plant Growth Regul 23:211–222

Prajakta BM, Suvarna PP, Singh RP, Rai AR (2019) Potential biocontrol and superlative plant growth promoting activity of indigenous *Bacillus mojavensis* PB-35(R11) of soybean (*Glycine max*) rhizosphere. SN Appl Sci 1:1143. https://doi.org/10.1007/s42452-019-1149-1

Prapagdee B, Kuekulvong C, Mongkolsuk S (2008) Antifungal potential of extracellular metabolites produced by *Streptomyces hygroscopicus* against phytopathogenic fungi. Int J Biol Sci 4(5):330

Prasanna HC, Kashyap SP, Ram K, Sinha DP, Suresh R, Malathi VG (2015) Marker assisted selection of Ty-2 and Ty-3 carrying tomato lines and their implications in breeding tomato leaf curl disease resistant hybrids. Euphytica 204(2):407–418

Qin S, Li J, Chen HH, Zhao GZ, Zhu WY et al (2009) Isolation, diversity, and antimicrobial activity of rare actinobacteria from medicinal plants of tropical rain forests in Xishuangbanna. China Appl Environ Microbiol 75(19):6176–6186

Qin S, Xing K, Jiang JH, Xu LH, Li WJ (2011) Biodiversity, bioactive natural products and biotechnological potential of plant-associated endophytic actinobacteria. Appl Microbiol Biotechnol 89(3):457–473

Rai M, Agarkar G, Rathod D (2014a) Multiple applications of endophytic Colletotrichum species occurring in medicinal plants, in novel plant bioresources: applications in food, medicine and cosmetics. In: Gurib-Fakim A (ed) Novel plant bioresources. Wiley, Chichester, pp 227–236

Rai M, Rathod D, Agarkar G, Dar M, Brestic M, Marostica Junior MR (2014b) Fungal growth promotor endophytes: a pragmatic approach towards sustainable food and agriculture. Symbiosis 62:63–79

Ryu CM, Murphy JF, Mysore KS, Kloepper JW (2004) Plant growth-promoting rhizobacteria systemically protect *Arabidopsis thaliana* against cucumber mosaic virus by a salicylic acid and NPR1-independent and jasmonic acid-dependent signaling pathway. Plant J 39:381–392

Sakuda S, Inoue H, Nagasawa H (2013) Novel biological activities of allosamidins. Molecules 18(6):6952–6968

Sangmanee P, Bhromsiri A, Akarapisan A (2009) The potential of endophytic actinomycetes, (*Streptomyces* sp.) for the biocontrol of powdery mildew disease in sweet pea (*Pisum sativum*). As J Food Ag-Ind 93:e8

Schulz B, Boyle C (2006) What are endophytes? In: Schulz BJE, Boyle CJC, Sieber TN (eds) Microbial root endophytes. Springer-Verlag, Berlin, pp 1–13

Sharma M (2014) Actinomycetes: source, identification, and their applications. Int J Curr Microbiol Appl Sci 3(2):801–832

Shekhar N, Bhattacharya D, Kumar D, Gupta RK (2006) Biocontrol of wood-rotting fungi with *Streptomyces violaceusniger* XL-2. Can J Microbiol 52(9):805–808

Silipo A, Molinaro A, Sturiale L, Dow JM, Erbs G et al (2005) The elicitation of plant innate immunity by lipooligosaccharide of *Xanthomonas campestris*. J Biol Chem 280:33660–33668

Singh RP, Manchanda G, Li ZF, Rai AR (2017) Insight of proteomics and genomics in environmental bioremediation. In: Bhakta JN (eds) Handbook of research on inventive bioremediation techniques. IGI Global, Hershey. https://doi.org/10.4018/978-1-5225-2325-3

Singh RP, Manchanda G, Maurya I K, Maheshwari NK, Tiwari PK, Rai AR (2019) *Streptomyces* from rotten wheat straw endowed the high plant growth potential traits and agro-active compounds. Bio Agri Biotechnol 17:507–513. https://doi.org/10.1016/j.bcab.2019.01.014

Srividya S, Thapa A, Bhat DV, Golmei K, Dey N (2012) *Streptomyces* Sp. 9p as effective biocontrol against chilli soilborne fungal phytopathogens. Eur J Exp Biol 2(1):163–173

Subhashini DV, Singh RP (2014) Isolation of endophytic actinomycetes from roots and leaves of tobacco (*Nicotiana tabacum* L.). Annal Plant Pro Sci 22(2):458–459

Subhashini DV, Singh RP, Manchanda G (2017) OMICS approaches: tools to unravel microbial systems. Directorate of Knowledge Management in agriculture, Indian Council of Agricultural Research. ISBN:9788171641703. https://books.google.co.in/books?id=vSaLtAEACAAJ

Suzuki S, Nakanishi E, Ohira T, Kawachi R, Nagasawa H et al (2006) Chitinase inhibitor allosamidin is a signal molecule for chitinase production in its producing Streptomyces. J Antibiot 59(7):402

Takahashi N, Suzuki A, Kimura Y, Miyamoto S, Tamura S et al (1968) Isolation, structure and physiological activities of piericidin B, natural insecticide produced by a streptomyces. Agri Biol Chem 32(9):1115–1122

Ton J, Jakab G, Toquin V, Flors V, Iavicoli A et al (2005) Dissecting the β-aminobutyric acid-induced priming phenomenon in Arabidopsis. Plant Cell 17:987–999

Turner MJ, Schaeffer JM (1989) Mode of action of ivermectin. In: Campbell WC (ed) Ivermectin and abamectin. Springer, New York/Berlin

Umemura K, Tanino S, Nagatsuka T, Koga J, Iwata M et al (2004) Cerebroside elicitor confers resistance to Fusarium disease in various plant species. Phytopathology 94:813–819

van Loon LC, Bakker PAHM, Pieterse CMJ (1998) Systemic resistance induced by rhizosphere bacteria. Annu Rev Phytopathol 36:453–483

Verma JP, Jaiswal DK, Krishna R, Prakash S, Yadav J et al (2018) Characterization and screening of thermophilic Bacillus strains for developing plant growth promoting consortium from hot spring of Leh and Ladakh region of India. Front Microbiol 9:1293. https://doi.org/10.3389/fmicb.2018.01293

Vurukonda SS, Giovanardi D, Stefani E (2018) Plant growth promoting and biocontrol activity of *Streptomyces* spp. as endophytes. Int JMolSci 19(4):952

Watve MG, TickooR JMM, Bhole BD (2001) How many antibiotics are produced by the genus Streptomyces? Arc Microbiol 176(5):386–390

Whipps JM (2001) Microbial interactions and biocontrol in the rhizosphere. J Exp Bot 52(1):487–511

Woo JH, Kamei Y (2003) Antifungal mechanism of an anti-Pythium protein (SAP) from the marine bacterium *Streptomyces* sp. strain AP77 is specific for *Pythium porphyrae*, a causative agent of red rot disease in *Porphyra* spp. Appl Microbiol Biotechnol 62(4):407–413

Yang YJ, Singh RP, Lan X, Zhang CS, Sheng DH et al (2019) Synergistic effect of *Pseudomonas putida* II-2 and *Achromobacter* sp. QC36 for the effective biodegradation of the herbicide quinclorac. Ecotoxicol Environ Saf. https://doi.org/10.1016/j.ecoenv.2019.109826

Yin SY, Chang JK, Xun PC (1965) Studies in the mechanisms of antagonistic fertilizer "5406". IV. The distribution of the antagonist in soil and its influence on the rhizosphere. Acta Microbiol Sin 11:259–288

Zhang B, Ramonell K, Somerville S, Stacey G (2002) Characterization of early, chitin-induced gene expression in Arabidopsis. Mol Plant-Microbe Interact 15:963–997

Microbial Life in Stress of Oxygen Concentration: Physiochemical Properties and Applications

Atul K. Srivastava, Arvind Saroj, Ashish Nayak, Indrajeet Nishad, and Karmveer Gautam

Abstract

Anaerobic extremophiles are generally obligate anaerobes that strictly require anoxic environment; however, facultative anaerobe can tolerate both anaerobic and aerobic environmental conditions. Some of the anaerobic microbes are found in extreme habitat such as deep in the ocean or earth's crust; others are present in parts of varied landscape such as marshes, bogs, sewers, polar lakes, tundra, calderas, geothermal submarine vents, hot springs, deep-sea sediments, and deep subsurface rock. Further, mammalian, ruminant, and arthropod digestive tracts are also the houses of the anaerobic extremophiles. Besides facultative anaerobes and oxygen-tolerant anaerobes, obligate anaerobes require strictly anoxic environment for their existence. This chapter defines the different approaches required to understand the growth and isolation of the anaerobes and large-scale production of the most economically important biomolecules produced by obligate anaerobes. The physiological and biochemical factors accounting for the

Atul K. Srivastava and Arvind Saroj have been equally contributed to this chapter.

A. K. Srivastava
CCUBGA, Department of Microbiology, Indian Agricultural Research Institute, New Delhi, India

A. Saroj (✉) · I. Nishad
Department of Plant Pathology, CSIR-Central Institute of Medicinal and Aromatic Plants, Lucknow, India

A. Nayak
Microbial Genomics and Diagnostics Lab., Microbiology and Plant Pathology Division, Regional Plant Resource Centre, Bhubaneswar, Odisha, India

K. Gautam
Regional Plant Quarantine Station, Amritsar, Punjab, India

Directorate of Plant Protection Quarantine and Storage, Ministry of Agriculture and Farmer's Welfare, Govt. of India, Faridabad, Haryana, India

© Springer Nature Singapore Pte Ltd. 2020
R. P. Singh et al. (eds.), *Microbial Versatility in Varied Environments*,
https://doi.org/10.1007/978-981-15-3028-9_11

relative resistance of many strict anaerobes to oxygen and products of incomplete reduction are also described in this chapter.

Keywords

Anaerobes · Abiotic stress · Resistance · Genes · Commercial product

11.1   Introduction

One can understand the concept origin of life through the complex processes involved in "early microbial evolution." The first appearance of the life and diversification among primary microbial lineage during the primitive anaerobic condition of the earth lead to the complex eukaryotic organisms (Martin and Sousa 2016; Subhashini et al. 2017). Exergonic reactions without involving living organism lead to the release of geochemical methane near hydrothermal vents. This concept of methane production and biological production fits a methanogenic root for archaea and an autotrophic origin of microbial physiology (Martin and Sousa 2016). It is believed that photosynthetic organism's cyanobacteria were found approximately 3.5 billion years ago, which in turn is very complex in physiology to be the primary prokaryotes (Des Marais 2000). Cyanobacteria plays an important role to change the primitive earth atmosphere as anaerobic to aerobic due to its photosynthetic capabilities, further resulted in the evolution of other aerobic, anaerobic, and facultative organism.

According to Oparin-Haldane theory, life originated in the sea under reducing (free of oxygen) atmosphere. Hence, it is assumed that anaerobic bacteria are too close to the primitive forms of life. Many researchers have successfully demonstrated that the synthesis of organic matter is possible under such conditions in their laboratory experiments (Oro et al. 1990). The primary requirement for such synthesis included HCN, CO_2, and H_2S. The nature of these compounds is basically toxic for aerobic organisms and creates the environment for the escalation of anaerobic microbial species. Further, evolution of anaerobic organisms leads to the origination of anoxic phototrophs, and the studies of anaerobic bacteria regarding continued diversification revealed that it appeared before the development of aerobic photosynthesis. Later on, aerobic photosynthesis gave rise to an oxygen-rich atmosphere. However, the current book chapter highlights those anaerobes which live in CO_2-rich environment or anaerobic growth conditions and their physiological adaptation.

The extreme environmental condition leads to the diversity of anaerobic microorganisms. Places/regions with extreme temperature play an immense and major role in the diversity and ecology of anaerobes among other factors because such temperate places lead to the oxygen-deficient/oxygen-free environment. During 1979, the first thermo anaerobe was reported in thermal spring. Ever since, hyperthermophile anaerobes have been isolated from continental and submarine volcanic areas, such as solfatara fields, volcano vents, hydrothermal vents, and geothermal power plants.

11.2   Anaerobic Thermophiles

Most of the anaerobic microorganisms (anaerobic thermophiles, acidophiles, alkanophiles, psychrophiles, radiation-resistant) exist under geographically extreme conditions and do not require oxygen for their growth. These are known as anaerobic extremophiles (Fig. 11.1). The extremophiles are mainly prokaryotes (bacteria and archaea) but also some eukaryotes (Table 11.1). Based on the temperature in which anaerobic microorganisms exist, they may be classified in categories as optimum thermophiles (an optimal temperature 50–64 °C), extreme thermophiles (optimum temperature 65–80 °C), and Hyperthermophiles (optimum temperature +80 °C).

Anaerobic extreme thermophilic *Bacteria* and *Archaea* are widely distributed in all sorts of thermobiotic environments. Anaerobic extreme thermophiles living near hydrothermal vent tolerate the additional pressure generated by the water column pie and are known as piezophilic (Alain et al. 2002). The microorganisms exist near hydrothermal vent sites include *Thermodiscus*, *Archaeoglobus*, *Thermoproteus*, *Pyrococcus*, *Thermococcus*, and *Desulfurococcus*, which reduce sulfur or sulfate. The utilization of organic and inorganic compounds such as elemental sulfur, methane, and iron is used to classify anaerobic extreme thermophilic bacteria (Table 11.1).

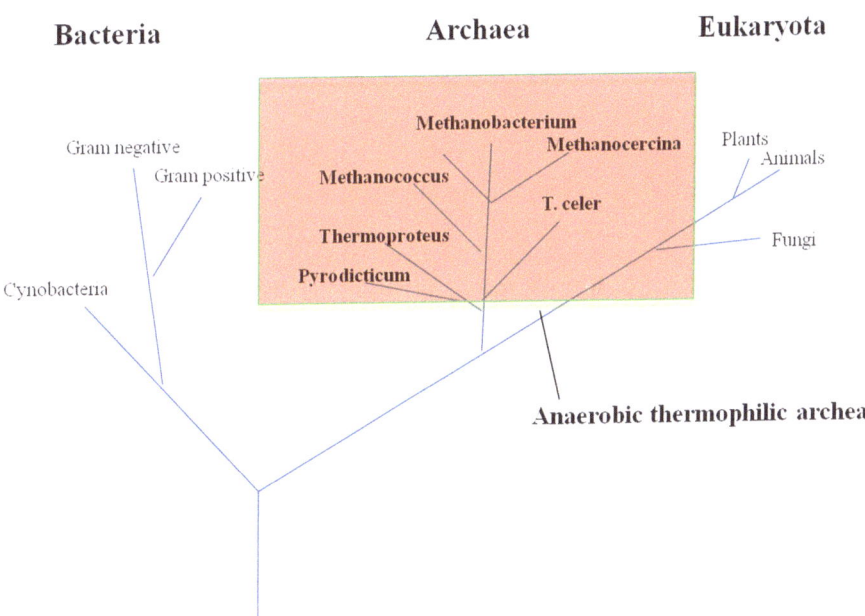

Fig. 11.1 Phylogenetic tree, highlighting possible evolutionary relatedness of anaerobic thermophilic *Archaea*

Table 11.1 Anaerobic species extremophiles

Species	Temperature range (°C)	Optimal temperature (°C)	References
Thermochromatium tepidum	34–57	48–50	Jun et al. (2017)
Moorella thermoacetica	45–65	55–60	Mock et al. (2014)
Thermoanaerobacter ethanolicus	37–78	69	Shao et al. (2016)
Carboxydothermus hydrogenoformans	40–78	70–72	Haddad et al. (2014)
Pelotomaculum thermopropionicum	45–65	55	Liu and Lu (2018)
Clostridium stercorarium	40–65	60	Broeker et al. (2018)
Methanocella conradii	37–60	55	Liu and Lu (2018)
Thermococcus barophilus	48–100	85	Thiel et al. (2014)
Pyrococcus furiosus	70–103	100	Keller et al. (2017)
Caldivirga maquilingensis	62–92	85	Lencina et al. (2012)

11.3 Anaerobic Psychrophiles

Anaerobic psychrophiles are those organisms which can survive and metabolize in low temperature ranging from −20 to +10 °C like glaciers, polar ice, permafrost, and sea ice waters (Clarke et al. 2013). The anaerobic psychrophiles control changes within macromolecules (protein) to retain the necessary fluidity or flexibility which is required to survive at low temperature. Major groups related to psychrophiles include alpha, beta, and gamma *Proteobacteria* in addition to relatives of the *Bacteroides*, *Thermus*, *Eubacterium*, and *Clostridium* (Segawa et al. 2003).

11.4 Anaerobic Ionizing Radiation-Resistant Microorganisms

Anaerobic ionizing radiation-resistant microorganisms can survive under the exposure of high gamma radiation. Archaeon isolated from a Northern Pacific hydrothermal vent, *Thermococcus gammatolerans* sp., showed resistance against ionizing radiation, during exposure of gamma rays at a dose of 30 kGy (Jolivet et al. 2003).

11.5 Anaerobic Thermoacidophiles

The Anaerobic thermoacidophiles have the ability to grow below 4.0 pH and high temperature 65–80 °C, such as genera *Thermoanaerobacter* and *Thermoanaerobacterium* found in the deep-sea vent (Prokofeva et al. 2005).

11.6 Anaerobic Alkalithermophiles

Alkalithermophiles represent those groups which are adapted to grow under high temperature and high alkaline conditions. Alkalithermophiles are representative of both archaea as well as bacteria. Growth of anaerobic *Clostridium thermoalcaliphilum* usually required pH ranging from 7.0 to 11.0, whereas the optimum pH was recorded as 9.6 to 10.1 (Engle et al. 1994).

11.7 Physiology of Anaerobic *Archaea* and *Bacteria*

The physiology of prokaryotic microorganisms differ from eukaryotic microbes and in part resulted in their survivability at extreme growth conditions like above 100 °C temperature (>100 °C), extreme salinity (saturated NaCl), pH (less than 2.0 and more than 10), and substrate stress. More importantly they do not utilize oxygen for respiration. This kind of extreme microbial growth conditions was found to be prevalent during primitive times but is limited in today's world. Some of the important and well-studied anaerobes are described given below.

11.7.1 Methanogen

The bacteria produce methane as a metabolic by-product in the anoxic condition known as methanogens. These methanogenic bacteria belong to a large and diverse group characterized by mainly three features: methane as the major by-product, strictly anaerobic, and belong to archaebacteria.

The prokaryotic methanogens are commonly found in wet wood in living trees, optimum pH for the growth ranges from 7.5–8.5 (Zeikus and Henning 1975). The sole source of carbon and energy for the growth and development of methanogenic bacteria depends upon production of methane. Methane can be utilized by few other aerobic bacteria as sole source of carbon, which ultimately breakdown the methane into carbon dioxide and water (Kunkel et al. 1998); hence, they are termed as methane-oxidizing bacteria.

11.7.1.1 Adaptation as Anaerobes and Biochemistry of Methane Production

The Methane-producing bacteria don't require oxygen for the respiration; in fact oxygen act as a growth limiting factor of methanogens. Therefore, carbon is the terminal electron acceptor instead of oxygen during methanogenesis. Carbon can occur in a small number of organic compounds, all with low molecular weights. The two well-defined pathways showed the use of inorganic carbon dioxide or acetic acid as terminal electron acceptors are as given below:

$$CO_2 + 4H_2 \rightarrow CH_4 + 2H_2O$$

$$CH_3COOH \rightarrow CH_4 + CO_2$$

Hydrogen acts as intermediates in the processes of methanogenic degradation of organic matter and serves as substrates for methanogenic archaea. However, it only contributes 33% during methanogenesis, while carbohydrates or associated forms of organic matter are degraded (Conrad 1999). However, during methanogenesis on the basis of temperature and pH, other small organic compounds have been also utilized as carbon source, like formate, methylamines, methanol, dimethyl sulfide, and methanethiol. The degradation of the methyl compounds is mediated by methyltransferases to give methyl-coenzyme M (Thauer 1998).

11.7.1.2 Methane-Oxidizing and Methane-Producing Bacteria
The methane is a primary component of oxygen-free water, marshes, soil, rumen of cattle, and gastrointestinal tract of mammals. Bacterial methanogenesis is a generalized process in most anaerobic environments. Thus, natural methane gas production can often be attributed to the growth of methanogens on specific energy sources that are formed as a result of microbial degradation of simple or complex organic matter or with the involvement of geochemical activities.

11.7.1.3 Oxidation and Production of Methane
Natural groupings of microorganisms, like those designated as parts in Bergey's Manual of Determinative Bacteriology (Buchanan and Gibbons 1974), are morphologically different as the methanogenic bacteria. However, all methanogenic species share few distinct unique or combine physiological properties. The world of methanogens no longer remained as a mysterious group due to the recent researches and advancements in the field of methanogen studies. The demand of energy in current scenario paves the way to better understand bacteria that produces natural gas; it could be the source of cheap energy as bioenergy at large scale (biogas).

   The bacteria which produce methane play the major role in regulating the breakdown (fermentation) of food under anoxic condition. The bacteria remove hydrogen gas through reduction of carbon dioxide to form methane. The low level of hydrogen favors the other microorganisms to grow faster due to the production of methane. Hence, collectively these consortia of microbes enhance the processes of fermentation more efficiently (Fig. 11.2).

11.8   Representative Life in Stress of Oxygen Concentration

11.8.1 *Clostridium*

The genus *Clostridium* belongs to the family *Clostridiaceae*, and it includes 203 species and 5 subspecies, out of which only few species are being reported as human pathogen. Among these species, 21 species have been placed to other genera, 5 have been reclassified within the genus, and 1 has been removed (Euzeby 2013).

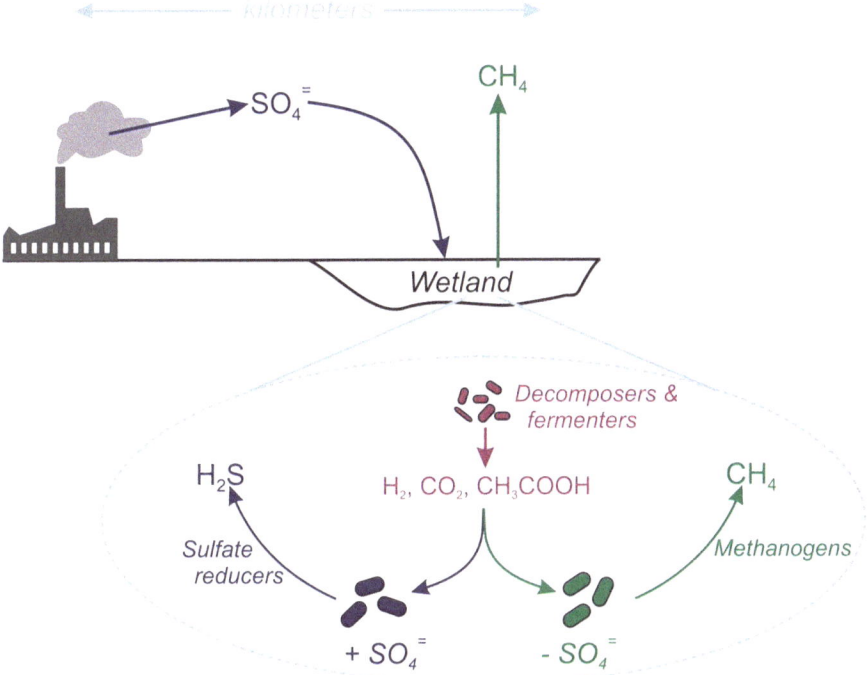

Fig. 11.2 Diagrammatic representation of methane production from different ways

The genus *Clostridium* have great genetic diversity such as strictly anaerobes, heterofermentative, spore-forming, and motile by peritrichous flagella. Generally *Clostridium* appear as Gram-positive but can be decolorized to show the appearance as Gram-negative or Gram-variable, with or without spore former, shape varies from either rods or cocci, and strictly anoxic bacteria. Generally, pathogenic strains are Gram-positive rods, but few are Gram variable; size varies from 0.3–2.0 × 1.5–20.0 μm which are more often arranged as pairs or short chains, with rounded or pointed ends. They are pleomorphic and differ significantly in their demand of oxygen (Finegold et al. 2002). The biotin and 4-aminobenzoate are crucially important as growth factors for the genus *Clostridium* at 37 °C temperature for optimum growth (Dolly et al. 2000).

Genus *Clostridium* can be isolated by following methods. These are (1) ability or inability to form endospores, (2) anoxic metabolism, (3) lack of ability to dissimilate sulfate reduction, and (4) presence of a Gram-positive cell wall (Andreesen et al. 1989).

Number of *Clostridia* species are saccharolytic in nature; hence, it can be easily isolated from acidic soils, because slightly acidic pH provides good opportunity for the growth of solvent-producing bacteria (Dolly et al. 2000).

The spores of *Clostridium* spp. play an immense important role in the field of medical and health sectors. For eample, two of the *Clostridium* spp. (*C. tetani* and

C. difficile) enhances the chance of hospital-acquired infections further resulted in severe helath complications (McFarland 1995).

However, few nonpathogenic species of *Clostridia* have great significance in gene therapy which provides the treatment of cancer by targeting the tumor cells. Bio-engineered strain of *Clostridia* proved to be good in targeting cancer cells and safe antitumor delivery system (Mellaert et al. 2008).

On the other hand, due to the anaerobic nature of *Clostridia*, it provides many economically important fermentation products, proteins, and enzymes. *C. acetobutylicum* is the most exploited microorganism for the production of acetone/butanol at large scale. Acetogenic clostridia (e.g., *C. thermoaceticum*, *C. thermoautotrophicum*, and *C. formicoaceticum*) have been studied as potential producers for calcium magnesium acetate as deicer. *C. thermohydrosulfuricum*, *C. thermocellum*, and *C. saccharolyticum* prove to be producer of alcohol. Some *Clostridium* species offer high-quality source of stable enzymes due to the tolerance of extreme conditions. Such as the main product, thermostable amylases and ethanol of starch fermentation by *C. thermosulfurogenes* and *C. thermohydrosulfuricum*. Fermentation of starch by *C. thermosaccharolyticum* resulted into pullalanase beside the production of thermostable amylases and ethanol beside this pullalanase also produce during this reaction. An enzyme Pullalanase is economically important industrial product generally used in sugar syrup production. *C. thermocellum* is the common cellulose-degrading species, accountable for the bioconversion of cellulose to useful economically important products like ethanol and biofuel (Minton and Clarke 1989).

11.8.2 *Propionibacterium*

Propionibacterium sp. are Gram-positive bacilli, non-motile, non-spore-forming, and catalase-positive, and size varies from 1 to 5 μm. *Propionibacterium* are considered as either anaerobic or facultative anaerobic bacteria. The studies of propionic acid bacteria (PAB) under anoxic conditions revealed that they are tiny and spherical (cocci) in shape. Moreover, in the presence of oxygen, it acts like pleomorphic bacteria (keep changing their shape). The optimum pH for the growth of PAB is reported around 7.0 (oscillate between 4.5 and 8.0) suitable for the production of propionic acid and vitamin B12. Generally, *Propionibacterium* sp. are mesophiles, though they can resist higher temperatures for some extent (few strains survive temperatures of up to 76 °C for 10 s). Nature of tolerance adopted by PAB to abovementioned stress may lead to the development of resistance to other factors (Benjelloun et al. 2007; Daly et al. 2010).

Genus *Propionibacterium* has extensive application in the food industry specially cheese industry like production of hard rennet cheese (Swiss Emmental cheese, The French Comte). Fermentation of lactate results in the production of acetic and propionic acid, which ultimately enhance the aroma of the final product and act as natural preservatives (Thierry and Maillard 2002; Thierry et al. 2005). Recent studies conducted by Cousin et al. (2016) reported that strain of *P. freudenreichii* subsp. *shermanii* can trigger apoptosis of the colon cancer cells. *P. freudenreichii* maintains microflora of the gastrointestine, by inspiring the growth of *Bifidobacterium*, and defends the organism/host by the secretion of bacteriocins, limiting the growth of certain pathogenic microorganisms. This aforementioned property recommends the use of PAB as probiotics for animal feeding.

11.8.3 *Porphyromonas*

Bacteria belonging to *Porphyromonas* sp. are strictly anaerobic, non-spore-forming, Gram-negative, and rod, culture on blood agar, and produce porphyrin pigment (dark brown/black). The *Porphyromonas* is common microflora of oral cavities of human body belongs to the Bacteroides genus. There are currently 16 valid *Porphyromonas* species that have been reported so far. The most common pathogenic strains of genus *Porphyromonas* are *P. gingivalis* and *P. endodontalis* that cause infection of periodontitis and endodontic in human, respectively (Kononen et al. 1996). Many studies revealed that fimbriae or fimbriae-like arrangement in bacteria has a significant role in the adhesion to the tooth surface.

11.8.4 Anaerobic Fungi

Prior to the finding of anaerobic fungi, it was believed that molds are not able to metabolize carbohydrates anaerobically (fermentation). In 1949, W. Foster completely refused the existence of anaerobic fungi, by stating that the difference between bacteria and fungi in the absence of anaerobic molds is either obligate or facultative. This belief held until the first recognition of anaerobic fungi in the mid-1970s (Orpin 1975). Orpin described that certain flagellated cells are zoospores of anaerobic fungi, earlier mistakenly identified as protozoan. Mostly the aquatic fungi posse's zoospore with single or bi flagella (Sparrow 1960) however, *N. frontalis* have multiple flagella up to 14 (Orpin 1975). Although, except multiple flagella stage rest of the life cycle of *N. frontalis* resembles to that of the chytrids. Since then, anaerobic fungi have been broadly identified and well established as gastrointestinal tract microflora of mammalian herbivores (Bauchop 1981; Joblin and Naylor 1989; Trinci et al. 1994). Nowadays, the significance and role of anaerobic fungus seek great attention to many research groups especially the complex mechanism involved in degrading the ingested cellulose and hemicellulose in mammalian herbivores. Such metabolism relies on symbiotic association of anaerobic fungi with another gut microorganism. However, anoxic fungi play the important role in the digestion of lignocellulosic plant material inside the rumen among other microbial consortia (Akin et al. 1988, 1990; Lee et al. 2000a, b).

Anaerobic fungi make some metabolic changes to adapt under the anoxygenic condition such as lack of mitochondria and other biochemicals/molecules which take part in oxidative phosphorylation (Yarlett et al. 1986; Youssef et al. 2013). Besides these hydrogenosomes, a specialized organelle present in anaerobic fungi helps in the glucose metabolism for the production of cellular energy under the anoxygenic condition (Brul and Stumm 1994; Muller et al. 2012).

Neocallimastigomycota represents anaerobic fungi that inhabit gastrointestinal tract and play a major role in plant material degradation. Anaerobic fungi degrade lignocellulosic plant material by producing a range of enzymes, mainly the powerful polysaccharide-degrading enzyme hydrogenase. However, recent research on *Orpinomyces* sp. toward carbohydrate-degrading enzymes revealed that there is a great diversity among them (Youssef et al. 2013). The study of *Orpinomyces* sp. genome sequence indicates that these genes come from rumen bacteria through horizontal gene transfer. Recent research pointed out that anaerobic fungi contain free enzymes and multiple enzyme system (Wilson and Wood 1992; Joblin et al. 2010). However, taxa of such phylum further need to be revised according to molecular phylogeny.

11.9 Techniques and Culture Media for Isolation of Microbes in Oxygen Stresses

Contribution of Hungate (1969) to study the rumen microbiology is immense especially designing of techniques and culture media for anaerobic rumen bacteria. The same culture methods with slight modifications (enzymes and antibiotics) are used to study anaerobic fungi (Bryant 1972; Miller and Wolin 1974). Instead of using oxygen-free N_2 as a reducing agent, media were prepared under oxygen-free CO_2. Lowe et al. (1985) described two kinds of media (with or without rumen fluid) for the isolation of rumen anaerobic fungi that had been widely accepted. The culture is maintained under an atmosphere of either 100% CO_2 or a combination of 85% N_2, 10% CO_2 and 5% H_2: 20–30% N_2 at 39 °C. Modified anaerobic glove box and Petri dish methods are significant even today to isolate and study anaerobic fungi (Leedle and Hespell 1980; Lowe et al. 1985). Batch culture explained by Theodorou et al. (1990) demonstrated the importance of antibiotics, range of pH 6.5–6.8, resazurin as a redox indicator, and incubation time from 2 to 10 days. Some other fermentors were also used for anaerobic fungi like continuous cultures (Hillaire and Jouany 1989) and fed-batch cultures (Tsai and Calza 1993). However, the most effective technique for isolation was illustrated by Theodorou et al. (2005) after modifying earlier described methods and techniques.

11.10 Application of Anaerobic Extremophiles

11.10.1 Biotechnology

Biotechnology is a technology that utilizes living organism or parts of this to develop different industrial product for betterment of human life. Along with other organism, extremophiles are important players of biotechnology, and therefore they have been used in biotechnology since long time for production of useful product in economic price and time (Fig. 11.3). New technologies and products are developed

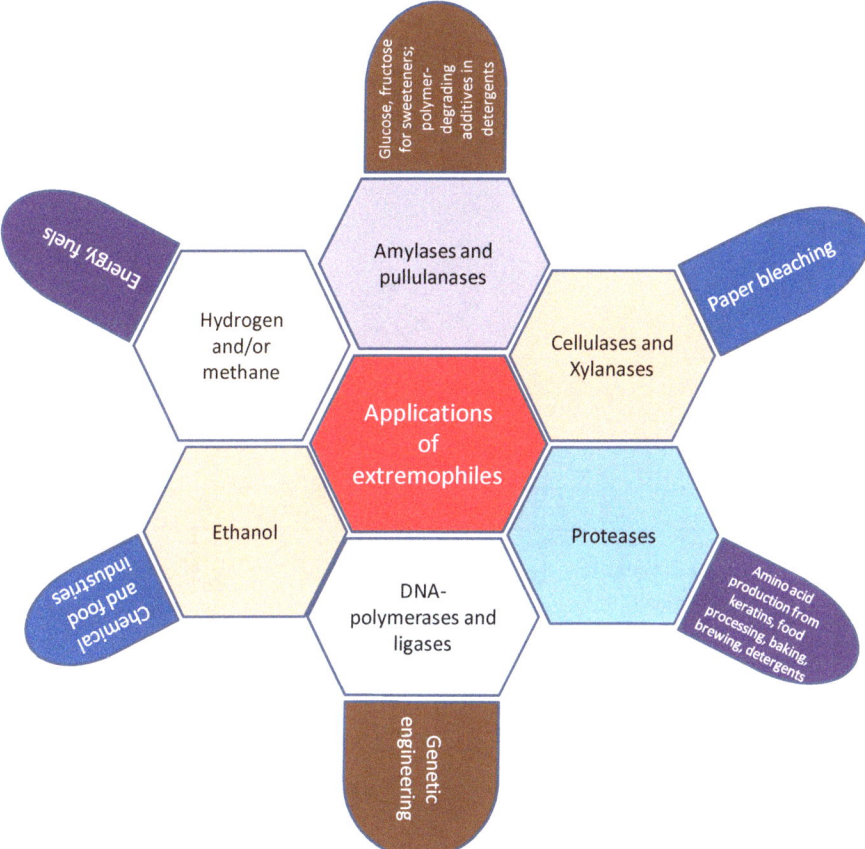

Fig. 11.3 Some important biotechnological application to explore extremophiles for human welfare

every year within the areas of medicine, agriculture, or industrial biotechnology (Coker 2016; Singh et al. 2017; Prajakta et al. 2019).

11.10.2 General Steps of Obtaining Product from Extremophiles at Large Scale

Extreme microorganism responsible for important enzymes/products is isolated and characterized, and particular gene of the respective product is cloned and cultured in laboratory labeled. The product is verified and optimized for the operational condition. The product is verified and optimized for the operational condition subsequently new organism could be use in the bioreactor for further large scale production (Fig. 11.4).

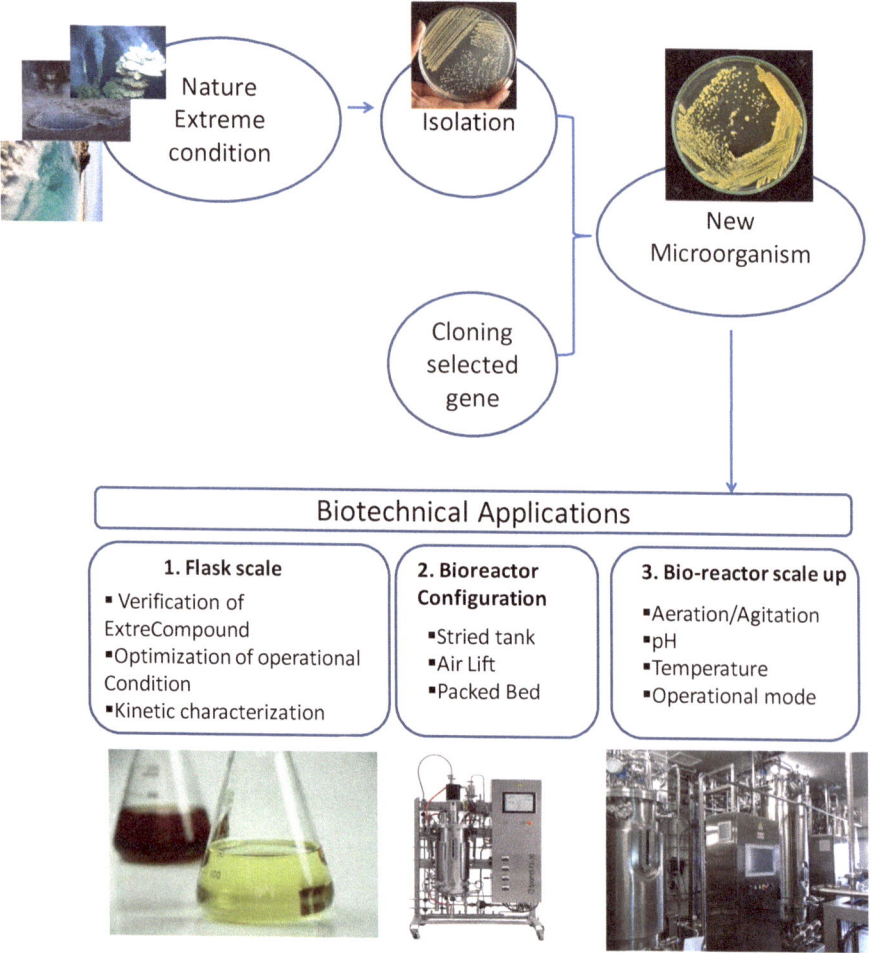

Fig. 11.4 Representational/symbolic flowchart; steps involved in the production of enzymes and other commercial product from extremophiles

Primary concern of using archaeal enzymes and metabolites at industrial level is the low production of the output of expensive fermentation processes, low biomass yields (Schiraldi et al. 2002; Zhang et al. 2019). Best way to address such problem is to develop extremophile conducive bioreactor with microfiltration. The microfiltration has been proven to be greatly resistant to the extreme working conditions suitable for the thermoacidophilic fermentation. Moreover, due to the high-level cell densities during the fermentation, frequent back flushing was proved to be enough to preserve 75% transmembrane flux of the maximum. The use of microfiltration is far better than earlier used cross-flow filtration, which fail to maintain even 20% flux after 30 h (Hayakawa et al. 1990). The assembled membrane bioreactor allowed a cell production that was 15- to 20-fold greater than the known best batch fermentation (Schiraldi et al. 1999).

11.10.3 Anaerobic Extremophiles and Its Commercial Exploitation

The major advantage for choosing enzymes from extremophiles are its worth of stability at extreme environment, little chance of contamination and easy to maintain other parameters of bioreactor (fermentation) (Dalmaso et al. 2015). The exploitation of extremophiles increases in current time, with the progress of research, development, and understanding of extremophiles' metabolic activity.

Biofuel Production
Elevation of temperatures and pH is frequent in several steps of biofuel production (Singh et al. 2014). Therefore, extremophiles are replacing the mesophilic organisms used in traditional methods. Recently certain thermophilic species such as *Thermotoga elfii* and *Caldicellulosiruptor saccharolyticus* are used for the larger-scale industrial-based system. The bacteria *Thermoanaerobacterium saccharolyticum* produces ethanol after the breakdown of complex carbohydrate, hemicellulose. Barnard et al. (2010) have revealed that extremophile fungi such as *Cyanidium caldarium* and *Galdieria sulphuraria* contain long-chain hydrocarbons like those found in petroleum.

DNA Polymerases
Polymerase chain reaction is the most important molecular technique comes into existence after the introduction of thermostable *Taq* polymerase, isolated from extremophile anaerobic *T. aquaticus*. Recent advancement and research regarding PCR leads to identification of other thermostable polymerase such as *Pyrococcus* and *Thermococcus* sp. (optimum temperature 90–100 °C) which have been introduced to the research industries (Antranikian 2009).

Carotenoids and Protease/Lipases
The Carotenoids are very stable molecules that habitually accompany the halophilic archaea and algae and have been isolated from the halophilic archeon such as *Halobacterium salinarum*. It has been reported for various application ranges from holography, production of artificial retinas, dyes, and more importantly in the renewable energy. Nowadays lipid-soluble canthaxanthin has been studied for use in food industries due to its antioxidant and safe property like additives and food dye. As with bacteriorhodopsin, halophilic archaea are the producers of choice with *Haloferax alexandrinus* being the preferred strain. Bacteriorhodopsin functions as a rudimentary form of photosynthesis, associated with membrane-bound retinal pigment and proton pump. The color of the β-carotene, i.e., red/orange pigment, seeks the attention of dye industries especially food industry.

Lactose-Free Milk
Nearly 3/4 population of the world suffers from lactose intolerance. Resultantly, a major population is avoiding ingesting lactose-free dairy products thoroughly that has been prepared by use of the β-galactosidase isolated from *Kluyveromyces lactis*.

The temperature of the dairy product is maintained within 5–25 °C for the stability and activity of the enzyme. However, such condition provides the suitable environment for the pathogens to grow and spoil the taste and flavor of the milk. This limitation can be overcome by the use of β-galactosidase isolated from a psychrophile. This enzyme would be active at low temperature and hydrolyze lactose throughout the entire process from production to shipment and storage which is significantly cost-effective and resulted into less chance of contamination.

Medical Applications

Anaerobic extremophiles produce antibiotics, antifungals, and antitumor molecules. *Halobacteriaceae* and *Sulfolobus* species generate antimicrobial peptides and diketopiperazines. The Diketopiperazines produced by *Haloterrigena hispanica* and *Natronococcus occultus* are used for treatment of pneumonia and cystic fibrosis.

Cyclodextrin Glycosyl Transferase (CGTase)

Cyclodextrin form the inclusion complexes along with different organic molecules which ultimately help in the ease of drug delivery system by improving the solubility of hydrophobic compounds. Thus, the application of cyclodextrin glycosyl transferase (CGTase) started at industrial level for the production of cyclodextrins.

11.11 Conclusions and Future Prospective

The knowledge of microbial diversity has been keep growing after the studies and thorough investigation of different extreme conditions such as high and low temeprature and anoxygenic environment. However, more extensive studies are required to investigate anaerobic diversity live in thermophilic, acidophilic, and halophilic environment, which can ultimately lead to the new species of bacteria and archaea.

Anaerobic bacteria and archaea have unique (unknown) physiological mechanism to counter the extreme environments by stabilizing their macromolecules (protein and lipid), in comparison to aerobic life. Extreme conditions are foremost requirement of the industrial biotechnology for the scale-up strategies of some products. More studies are required to understand molecular mechanism of anaerobes which will be later on exploited in industrial biotechnology for the human welfare. Especially, enzyme and lipid function of anaerobes at high temperature can ease the challenge of industrial biotechnology for large extent. However, cultural conditions required to scale up the production of the extremophiles are the major drawback at industrial level. Use of biotechnology and recombinant technology eases/address this limitation for the production of some extremozymes in large quantities up to some extent using other biological models such as *E. coli* and yeast. However, such limitations still persist for extremophiles. Therefore, utmost need is to develop new strategies to overcome such problems by involving anaerobic microorganism for the production of extremophile proteins at industrial level. This purpose can be only short out by collaboration between industries and research groups. Recent development/enhancement in the field of bioengineering and biotechnology paves the way to better understand the metabolism of anaerobic extremophilic microorganism for the purpose of its better exploitation at industrial level. Adaptation strategies used by these extremophiles involving unique enzyme system resulted into special metabolic products. In this context some enzymes like hydrolases are the best example with the ability to remain active in hyper-/ hypothermic conditions and enzymes active in high alkaline or acidic environment isolated from hydrothermal and volcano vents. Tolerance to such extreme conditions makes these enzymes valuable for the scale up of industrially important products. The gene pool of such anoxic extremophiles needs to be studied thoroughly using metagenomic and genetic engineering. Such enzyme system can be used to express in other microorganisms to facilitate large-scale production of economically important chemicals and proteins. The demand of biotech industry-oriented product in current scenario can only be achieved by the better understanding of anaerobic extremophilic microorganisms using different techniques like metagenomics, metabolomics, and recombinant technology.

References

Akin DE, Borneman WS, Windham WR (1988) Rumen fungi: morphological types from Georgia cattle and the attack on forage cell walls. BioSystem 21:385–339

Akin DE, Borneman WS, Lyon CE (1990) Degradation of leaf blades and stems by monocentric and polycentric isolates of ruminal fungi. Anim Feed Sci Technol 31:205–221

Alain K, Marteinsson VT, Miroshnichenko ML, Bonch-Osmolovskaya EA, Prieur JL (2002) Birrien, *Marinitoga piezophila* sp. nov., a rod-shaped, thermo-piezophilic bacterium isolated under high hydrostatic pressure from a deep-sea hydrothermal vent. Int J Syst Evol Microbiol 52:1331–1339

Andreesen JR, Bahl H, Gottschalk G (1989) Introduction to the physiology and biochemistry of the genus clostridium. Biotechnology handbooks. In: Nigel PM, Clarke DJ (eds) Clostridia. Plenum Press, New York/London

Antranikian G (2009) Extremophiles and biotechnology. In: Encyclopedia of Life Sciences (ELS). Wiley, Chichester. https://doi.org/10.1002/9780470015902.a0000391.pub2

Barnard D, Casanueva A, Tuffin M, Cowan D (2010) Extremophiles in biofuel synthesis. Environ Technol 31:8–9

Bauchop T (1981) The anaerobic fungi in rumen fibre digestion. Agri Environ 6:339–348

Benjelloun H, Rabe Ravelona M, Lebeault JM (2007) Characterization of growth and metabolism of commercial strains of propionic acid bacteria by pressure measurement. Eng Life Sci 7:143–148

Broeker J, Mechelke M, Baudrexl M, Mennerich D, Hornburg D et al (2018) The hemicellulose-degrading enzyme system of the thermophilic bacterium *Clostridium stercorarium*: comparative characterisation and addition of new hemicellulolytic glycoside hydrolases. Biotechnol Biofuels 11:229. https://doi.org/10.1186/s13068-018-1228-3

Brul S, Stumm CK (1994) Symbionts and organelles in anoxic protozoa and fungi. Trends Ecol Evol 9:319–324

Bryant MP (1972) Commentary on the Hungate technique for culture of anaerobic bacteria. American J Clinical Nutr 25:1324–1328

Buchanan RE, Gibbons NE (1974) Bergey's manual of determinative bacteriology. The Williams and Wilkins Co., Baltimore

Clarke A, Morris GJ, Fonseca F, Murray BJ, Acton E, Price HC (2013) A low temperature limit for life on earth. PLoS One 8:e66207

Coker JA (2016) Extremophiles and biotechnology: current uses and prospects [version 1; referees: 2 approved]. F1000Res 5:(F1000 Faculty Rev):396. https://doi.org/10.12688/f1000research.7432.1

Conrad R (1999) Contribution of hydrogen to methane production and control of hydrogen concentrations in methanogenic soils and sediments. FEMS Microbiol Ecol 28:193–202

Cousin FJ, Jouan-Lanhouet S, Théret N (2016) The probiotic *Propionibacterium freudenreichii* as a new adjuvant for TRAIL-based therapy in colorectal cancer. Oncotarget 7:7161–7178

Dalmaso GZL, Ferreira D, Vermelho AB (2015) Marine extremophiles: a source of hydrolases for biotechnological applications. Mar Drug 13:1925–1965

Daly DFM, McSweeney PLH, Sheehan JJ (2010) Split defect and secondary fermentation in Swiss-type cheeses: a review. Dairy Sci Technol 90:3–26

Des Marais DJ (2000) Evolution: when did photosynthesis emerge on Earth? Science 289:1703–1705

Dolly M, Spitia S, Silva E, Schwarz WH (2000) Isolation of mesophilic solvent producing strains from Colombian sources: physiological characterization, solvent production and polysaccharide hydrolysis. J Biotechnol 79:117–126

Engle M, Mandelco L, Wiegel J (1994) *Clostridium thermoalcaliphilum* sp. nov., an anaerobic and thermotolerant facultative alkaliphile. Int J Syst Bacteriol 44:111–118

Euzeby J (2013) List of new names and new combinations previously effectively, but not validly. Int J Syst Evol Microbiol 63:2365–2367

Finegold SM, Song Y, Liu C (2002) Taxonomy – general comments and update on taxonomy of clostridia and anaerobic cocci. AnaerobeAnaerobe 8:283–285

Foster JW (1949) Chemical activities of the fungi. Academic Press, New York

Haddad M, Vali H, Paquette J, Guiot SR (2014) The role of *Carboxydothermus hydrogenoformans* in the conversion of calcium phosphate from amorphous to crystalline state. PLoS One 9(2):e89480

Hayakawa K, Mizutani J, Wada K, Masai T, Yoshihara I, Mitsuoka T (1990) Effects of soybean oligosaccharides on human faecal flora. Microb Ecol Health Dis. http://www.microbecolhealthdis.net/index.php/mehd/article/view/7555/8897

Hillaire MC, Jouany JP (1989) Effects of rumen anaerobic fungi on the digestion of wheat straw and the end products of microbial metabolism studies in a semi continuous in vitro system. In: Nolan JV, Leng RA, Demeyer DI (eds) The roles of protozoa and fungi in ruminant digestion. Penambul Books, Armidale, pp 269–272

Hungate RE (1969) A roll tube method for the cultivation of strict anaerobes. Methods in Microbiol 3B:117–132

Joblin KN, Naylor GE (1989) Fermentation of woods by rumen anaerobic fungi. FEMS Microbiol Lett 65:119–122

Joblin K, Naylor G, Odongo N, Garcia M, Viljoen G (2010) Ruminal fungi for increasing forage intake and animal productivity. In: Odongo NE, Garcia M, Viljoen GJ (eds) Sustainable improvement of animal production and health. Food and Agriculture Organization of the United Nations, Rome, pp 129–136

Jolivet EL, Haridon S, Corre E, Forterre P, Prieur D (2003) *Thermococcus gammatolerans* sp. nov., a hyperthermophilic archaeon from a deep-sea hydrothermal vent that resists ionizing radiation. Int J Syst Evol Microbiol 53:847–851

Jun D, Huang V, Beatty JT (2017) Heterologous production of the photosynthetic reaction center and light harvesting 1 complexes of the thermophile *Thermochromatium tepidum* in the mesophile Rhodobacter sphaeroides and thermal stability of a hybrid Core complex. Appl Environ Microbiol 83(20):e01481–e01417

Keller MW, Lipscomb GL, Nguyen DM, Crowley AT, Schut GJ et al (2017) Ethanol production by the hyperthermophilic archaeon *Pyrococcus furiosus* by expression of bacterial bifunctional alcohol dehydrogenases. J Microbial Biotechnol 10(6):1535–1545

Kononen E, Valsanen ML, Finegold SM, Heine R, Jousimies-Somer H (1996) Cellular fatty acid analysis and enzyme profiles of *Porphyromonas catoniae* – a frequent colonizer of the oral cavity in children. Anaerobe 2(5):329–335

Kunkel A, Vorholt JA, Thauer RK, Hedderich R (1998) An Escherichia coli hydrogenase-3-type hydrogenase in methanogenic archaea. Eur J Biochem 252:467–476

Lee SS, Ha JK, Cheng K-J (2000a) Relative contributions of bacteria, protozoa, and fungi to in vitro degradation of orchard grass cell walls and their interactions. Appl Environ Microbiol 66:3807–3813

Lee SS, Ha JK, Cheng KJ (2000b) Influence of an anaerobic fungal culture administration on in vivo ruminal fermentation and nutrient digestion. Anim Feed Sci Technol 88:201–217

Leedle JAZ, Hespell RB (1980) Differential carbohydrate media and anaerobic replica plating techniques in delineating carbohydrate utilizing subgroups in rumen bacterial populations. Appl Environ Microbiol 39:709–719

Lencina AM, Ding Z, Schurig-Briccio LA, Gennis RB (2012) Characterization of the Type III sulfide:quinone oxidoreductase from Caldivirga maquilingensis and its membrane binding. Biochim Biophys Acta 1827(3):266–275

Liu P, Lu Y (2018) Concerted metabolic shifts give new insights into the syntrophic mechanism between propionate-fermenting *Pelotomaculum thermopropionicum* and hydrogenotrophic *Methanocella conradii*. Front Microbiol 9:1551. https://doi.org/10.3389/fmicb.2018.01551

Lowe SE, Theodorou MK, Trinci APJ, Hespell RB (1985) Growth of anaerobic rumen fungi on defined and semidefined media lacking rumen fluid. J Gen Microbiol 131:2225–2229

Martin WF, Sousa FL (2016) Early microbial evolution: the age of anaerobes. Cold Spring Harb Perspect Boil 8(2):a018127. https://doi.org/10.1101/cshperspect.a018127

McFarland LV (1995) Epidemiology of infectious and iatrogenic nosocomial diarrhea in a cohort of general medicine patients. Am J Infect Control 23:295–305

Mellaert LV, Barbe S, Anne J (2008) Clostridium spores as anti-tumour agents. Trends Microbiol 14:190–196

Miller TL, Wolin MJ (1974) A serum bottle modification of the Hungate technique for cultivating obligate anaerobes. Appl Microbiol 27:985–987

Minton NP, Clarke DJ (1989) In: Saha BC, Lamed R, Zeikus JG (eds) Clostridial enzymes. Plenum Press, New York

Mock J, Wang S, Huang H, Kahnt J, Thauer RK (2014) Evidence for a hexaheteromeric methylenetetrahydrofolate reductase in *Moorella thermoacetica*. J Bacterial 196(18):3303–3314

Muller M, Mentel M, van Hellemond JJ, Henze K, Woehle C et al (2012) Biochemistry and evolution of anaerobic energy metabolism in eukaryotes. Microbiol Mol Biol Rev 76:444–495

Oro J, Miller SL, Lazcano A (1990) The origin and early evolution of life on Earth. Ann Rev Earth Planet Sci 18(1):317–356

Orpin CG (1975) Studies on the rumen flagellate *Neocallimastix frontalis*. J Gen Microbiol 91:249–262

Prajakta BM, Suvarna PP, Singh RP, Rai AR (2019) Potential biocontrol and superlative plant growth promoting activity of indigenous *Bacillus mojavensis* PB-35(R11) of soybean (*Glycine max*) rhizosphere. SN Appl Sci 1:1143. https://doi.org/10.1007/s42452-019-1149-1

Prokofeva MI, Kublanov IV, Nercessian O, Tourova TP, Kolganova TV et al (2005) Cultivated anaerobic acidophilic/acidotolerant thermophiles from terrestrial and deep-sea hydrothermal habitats. Extremophiles 9:437–448

Schiraldi C, Marulli F, Martino IDLA, Rosa MD (1999) A microfiltration bioreactor to achieve high cell density in *Sulfolobus solfataricus* fermentation. Extremophiles 3:199–204

Schiraldi C, Giuliano M, De Rosa M (2002) Perspectives on biotechnological applications of archaea. Arch Microbiol 1(2):75–86

Segawa T, Miyamoto K, Ushida K, Agata K, Okada N et al (2003) Seasonal change in bacterial flora and biomass in mountain snow from the Tateyama mountains, Japan, analyzed by 16S rRNA gene sequencing and real-time PCR. Appl Environ Microbiol 71:123–130

Shao X, Zhou J, Olson DG, Lynd LR (2016) A markerless gene deletion and integration system for *Thermoanaerobacter ethanolicus*. Biotechnol Biofuels 9:100. https://doi.org/10.1186/s13068-016-0514-1

Singh RP, Singh RN, Manchanda G, Tiwari PK, Srivastava AK et al (2014) Biofuels: a way ahead. Ind J Appl Res 4(1):3–6

Singh RP, Manchanda G, Li ZF, Rai AR (2017) Insight of proteomics and genomics in environmental bioremediation. In: Bhakta JN (ed) Handbook of research on inventive bioremediation techniques. IGI Global, Hershey. https://doi.org/10.4018/978-1-5225-2325-3

Sparrow FKJR (1960) Aquatic phycomycetes, 2nd edn. University of Michigan Press, Ann Arbor

Subhashini DV, Singh RP, Manchanda G (2017) OMICS approaches: tools to unravel microbial systems. Directorate of Knowledge Management in Agriculture, Indian Council of Agricultural Research. ISBN:9788171641703. https://books.google.co.in/books?id=vSaLtAEACAAJ

Thauer RK (1998) Biochemistry of methanogenesis: a tribute to Marjory Stephenson. Microbiol-UK 144(9):2377–2406

Theodorou MK, Gill MK, King-Spooner C, Beever DE (1990) Enumeration of anaerobic chytridiomycetes as thallus forming units: a novel method for the quantification of fibrolytic fungal populations from the digestive tract ecosystem. Appl Environ Microbiol 56:1073–1078

Theodorou MK, Brookman J, Trinci APJ (2005) Anaerobic fungi. In: Makkar HPS, McSweeney CS (eds) Methods in gut microbial ecology for ruminants. Springer, Dordrecht, pp 55–66

Thiel A, Michoud G, Moalic Y, Flament D, Jebbar M (2014) Genetic manipulations of the hyperthermophilic piezophilic archaeon *Thermococcus barophilus*. Appl Environ Microbiol 80(7):2299–2306

Thierry A, Maillard MB (2002) Production of cheese flavour compounds derived from amino acid catabolism by *Propionibacterium freudenreichii*. Lait 82:17–32

Thierry A, Maillard MB, Richoux R, Kerjean JR, Lortal S (2005) *Propionibacterium freudenreichii* strains quantitatively affect production of volatile compounds in Swiss cheese. Lait 85:57–74

Trinci APJ, Davies DR, Gull K, Lawrence MI, Nielsen BB et al (1994) Anaerobic fungi in herbivorous animals. Mycol Res 98:129–152

Tsai KP, Calza RE (1993) Optimization of protein and cellulase secretion in *Neocallimastix frontalis* EB188. Appl Microbiol Biotechnol 39:477–482

Wilson CA, Wood TM (1992) The anaerobic fungus *Neocallimastix frontalis*: isolation and properties of a cellulosome-type enzyme fraction with the capacity to solubilize hydrogen-bond-ordered cellulose. Appl Microbiol Biotechnol 37:125–129

Yarlett N, Orpin CG, Munn EA, Yarlett NC, Greenwood CA (1986) Hydrogenosomes in the rumen fungus *Neocallimastix patriciarum*. Biochem J 236:729–739

Youssef NH, Couger MB, Struchtemeyer CG, Liggenstoffer AS, Prade RA et al (2013) The genome of the anaerobic fungus *Orpinomyces* sp. strain C1A reveals the unique evolutionary history of a remarkable plant biomass degrader. Appl Environ Microbiol 79:4620–4634

Zeikus JG, Henning DL (1975) *Methanobacterium arbophilicum* sp. *nov*. an obligate anaerobe isolated from wet wood of living trees. Anto van Leeu J Microbiol Serol 41:543–552

Zhang J, Wang E, Singh R, Guo C, Shang Y et al (2019) Grape berry surface bacterial microbiome: impact from the varieties and clones in the same vineyard from central China. J Appl Microbiol 126:204–214. https://doi.org/10.1111/jam.14124

Bacterial Metabolic Fitness During Pathogenesis

Saurabh Pandey, Nidhi Shukla, Shashi Shekhar Singh,
Deeksha Tripathi, Takshashila Tripathi, and Sashi Kant

Abstract

Pathogenic bacteria encounter hostile environments and experience diverse stresses from their initial moment of contact with the host. To survive within the host, bacteria improved metabolic fitness and altered virulence character. As in the case of respiratory pathogens, bacteria respond against an array of host-derived antimicrobial mediators, nitrosative stress, hyperosmolarity, and oxygen limitation. Further, in case of enteric pathogens, after ingestion, they must survive the acidic pH. Moreover, enteric pathogens living within the intestinal lumen encounter host-generated antimicrobial peptides, reactive oxygen radicals, bile salts, and free fatty acids and enhanced osmolality. Subsequently,

S. Pandey
Department of Biochemistry, School of Chemical and Life Sciences, Jamia Hamdard, New Delhi, India

N. Shukla
Department of Translational Hematology and Oncology, Cleveland Clinic Learner Research Institute, Cleveland, OH, USA

S. S. Singh
Department of Inflammation and Immunity, Cleveland Clinic Lerner Research Institute, Cleveland, OH, USA

D. Tripathi
Department of Microbiology, School of Life Sciences, Central University of Rajasthan, Ajmer, Rajasthan, India

T. Tripathi
Department of Neuroscience, Physiology and Pharmacology, University College London, London, UK

S. Kant (✉)
Department of Immunology and Microbiology, University of Colorado School of Medicine, Anschutz Campus, Aurora, CO, USA
e-mail: sashi.kant@cuanschutz.edu

© Springer Nature Singapore Pte Ltd. 2020
R. P. Singh et al. (eds.), *Microbial Versatility in Varied Environments*,
https://doi.org/10.1007/978-981-15-3028-9_12

inflammatory responses against pathogen recruit macrophages, neutrophils, and other phagocytic cells. These defense mechanisms increase oxidative and nitrosative stresses as well as sequestration of essential metals and nutrients. Bacteria respond to these by developing exquisite systems that sense these stresses and trigger appropriate mechanism for successful survival. Different environmental stimuli not only trigger adaptive responses but also modulate virulent genes at particular condition. An array of stress responses and their mechanism is consequently critical and necessary to understand bacterial pathogenesis.

In an effort of emphasizing the complexity of bacterial adaptation during the host infection, this chapter will focus on different stress-sensing and stress response mechanisms that finally improvise the bacterial metabolic fitness.

Keywords
Bacterial pathogens · Immunity · Autophagy · Defense strategies · Cell cycle

12.1 Introduction

Disease-causing pathogenic bacteria while thriving within host encounter diverse stresses. All successful pathogens are well equipped with its genetic, biochemical, and structural features that enable them to overcome the host-derived stresses and cause disease. These capabilities allow pathogens to not only withstand against precise stress situations but additionally to express their pathogenicity with the aid of modulating immune and survival signaling in a sophisticated way. The pathogens also have a series of adaptive mechanisms that safeguard from the host defense mechanisms. A comprehensive understanding of bacterial adaptation to host stress responses offers better understanding of host microenvironments, bacterial virulence, and stress resistance.

Pathogenic bacteria experience nutritional challenges, oxidative and nitrosative stress, envelope stress, oxygen limitation, hyperosmolarity, DNA damage, and change in temperature during infection and colonization within mammalian hosts. Bacteria not only counterattack these stresses and immune challenges but additionally activate their defense mechanisms that allow survival under these hostile environments. This chapter explains adaptation mechanisms of pathogenic bacteria during infection within the host.

12.2   Metabolic Adaptation Mechanisms in Intracellular Pathogenic Bacterium

The pathogens are protected against phagocytes and other extracellular immune defense. Though intracellular and extracellular pathogens often have no well-defined distinction, most pathogens live within infected host cells in some stages of their life cycle. For instance, the *Streptococcus pneumoniae* is an extracellular pathogen, but it can regularly be found intracellularly. The intracellular life does have its challenges. The host encountering intracellular pathogens fights back with innate and adaptive immunity and autophagy. After internalization, intracellular bacteria evade the immune system and replicate in specialized non-acidic vacuoles, for example, *Salmonella*, *Mycobacteria*, *Brucella*, *Legionella*, *Coxiella*, and *Chlamydia*. Some pathogens after escaping from their internalization vacuole reside in cytosol. This group of intracellular bacteria includes *Shigella*, *Listeria*, *Rickettsia*, and *Francisella* and escapes into the host cytosol and actively replicates (Ray et al. 2009).

After phagocytosis numerous pathogens use macrophages as their intracellular niche for survival and replication (Thakur et al. 2019). During evolution, pathogens have built up the capacity to instigate their own internalization even by non-phagocytic cells, along with epithelial and endothelial cells.

In general, the nutrient trafficking for intravacuolar bacteria is dependent on transporter proteins of host of pathogen origin, directly inserted into the membrane of vacuole or by fusion of endocytic vesicles. Majority of these vacuolar bacteria utilizes substrate-level phosphorylation to generate ATP by converting pyruvate to acetate. However, *Coxiella* and *Chlamydia* being obligate pathogens are unable to perform this ATP-generating reaction using pyruvate as a substrate in acidic pH range of 4.5–4.7. Alternatively, they harbor two ATP/ADP translocases that allow them to uptake ATPs from the host cells.

Additionally, several pathogenic bacteria successfully thrive in non-acidic vacuoles including *Salmonella*. *Salmonella*-containing vacuoles are remodeled by the several effectors secreted by the type 3 secretion system (T3SS) of the bacterium. These effector molecules allow the polymerization of an actin basket around the vacuoles by rearranging the actin cytoskeleton and thus help in regulation of bacterial virulence. Also these effectors are able to block the recruitment of NADPH oxidase to inhibit the production of a superoxide free radical. *Mycobacterium tuberculosis* has developed strategies to replicate successfully in macrophages because of having an exceptionally hydrophobic cell wall. As a consequence, it evades immune responses and persists within the host (Awuh and Flo 2017). It prevents fusion of phagosome and lysosome and gets nutrients from macrophages to persist intracellularly (BoseDasgupta and Pieters 2018).

12.2.1 Life in the Vacuole

Organisms survive from lysosomal degradation by resisting the endosome development pathway. Pathogens subvert phosphoinositide digestion to readjust character of endosomal layer, thus inhibiting endocytic pathway. *Legionella pneumophila* inhibits lysosomal execution by making ER-like compartment by obtaining ER-derived vesicles, while *Mycobacterium tuberculosis* and *Salmonella* resist endosomal maturation pathway (Xu and Luo 2013). Similarly, the intracellular pathogen *Chlamydia/Chlamydophila*, causing genital, visual, and respiratory diseases, redirect early vacuole from the endocytic pathway. They form metabolically idle rudimentary bodies for transmission, which subsequently change into metabolically dynamic reticulate bodies that, lastly, multiply in endosome-inferred vacuole or *Chlamydia*-containing vacuole (Wong et al. 2019) which diverge from the endocytic pathway (Nickel et al. 1999). *Chlamydia*-containing vacuole recruits Rab GTPases in some chlamydial species (Aeberhard et al. 2015).

Though escape from phago-lysosomal maturation is key mechanism for pathogen survival, some bacteria follow alternate mechanism as in the case of quasi-obligate intracellular pathogen *Coxiella burnetii*, causing Q fever. *C. burnetii* spreads among hosts through inhalation of contaminated aerosol (Gurtler et al. 2014). *Coxiella* infects macrophages, in addition to infecting epithelial cells, monocytes, and other cell types, thereby multiplying in an acidic parasitophorous vacuole (Stein et al. 2005). *Coxiella* has a way to escape the parasitophorous vacuole microbicidal conditions and its acidic pH 5.0 (Ghigo et al. 2012).

12.2.2 Escaping from Vacuole

To avoid the vacuolar acidification and subsequent destruction of inhabiting pathogen, immediate escape of pathogens from the phagosome to cytosol is necessary. Usually, pathogens have been detected in the cytosol within 30 min of internalization (de Souza Santos and Orth 2014). Escape of the pathogen is primarily driven by pH of the vacuole. Pathogens during escape use membrane-damaging toxins/enzymes to escape successfully (Radoshevich and Cossart 2018). For example, in *Shigella flexneri*, the plasmid-encoded T3SS effector proteins IpaB and IpaC are associated with escape from the phagosome, forming a IpaB-IpaC mediated by pore-forming complex (Mattock and Blocker 2017; Picking and Picking 2016).

12.2.3 Life in the Cytosol

Pathogens have ways to escape lysosomal degradation soon after engulfment. In the process, they ensure their intracellular survival in unfavorable microenvironment. As in the case of *Shigella*, pore-shaping effectors interfere with the phagosome

(Mattock and Blocker 2017). Free-living mycobacteria *Mycobacterium marinum* escape by exploiting Esx-1 secretion system and utilize the cytoplasm for replication (Bosserman et al. 2019). Thus, some pathogens have evolved the strategies to be exclusively cytosolic. Proficient cytosolic pathogens have evolved escape mechanisms exploiting actin-interceded motility and cell-to-cell communication junctions to spread (Wiesel and Oxenius 2012). By manipulating the host actin cytoskeleton, microbes impel themselves through the cell-cell junction, venture into contiguous cells, and spread without leaving the limits of the defensive intracellular compartment (Lamason and Welch 2017). By commandeering the host cell actin cytoskeletal apparatus, microscopic organisms can impel themselves via the cytosol, venture themselves into contiguous cells, and spread without leaving the limits of the defensive intracellular compartment (Colonne et al. 2016). In summary, in contrast with survival of cytosolic pathogens, the vacuolar survival seems to be safer choice for successful life cycle.

12.2.4 Evasion from Autophagy

Autophagy is a cellular destructive process which disposes cytoplasmic materials by lysosomal degradation (Hu et al. 2019). The autophagy pathway is modulated by pathogens, for example, the *Listeria* pore-forming toxin LLO triggers the autophagy. The PlcA and PlcB phospholipases support the *Listeria monocytogenes* to protect from decimation in autophagosomes (Cheng et al. 2018). Further, *L. monocytogenes* avoids autophagy in the cytosol; it might select the autophagic assembly to make an extra/intracellular compartment and cause disease progression. Also, during the early phases of infection, LLO of *L. monocytogenes* dephosphorylates histone H3 and deacetylates H4 that is independent of its pore-forming activity. These epigenetic modifications downregulate host inflammatory response (Hamon et al. 2007). By action of LLO, *Listeria* converts phagosome to *Listeria*-containing phagosome in which they can sustain (Mitchell et al. 2015; Vdovikova et al. 2017).

12.2.5 Cytoskeleton-Based Cell Motility

Cytosolic pathogens have been developed to activate host F-actin polymerization on their surface, delivering mechanical power that impels them into neighboring cells (Ireton 2013). For example, a surface protein in *Burkholderia*, *Shigella*, *Listeria*, and *Rickettsia* triggers actin-based motility by promoting actin polymerization. The F-actin fibers form cross-connecting structures as an unbending meshwork. This meshwork utilizes formation and engulfment of bacterial protrusions causing cell-to-cell spread of bacterium. The depolymerization of actin meshwork at the distal end appears as comet tail (Truong et al. 2014).

12.2.6 Modulation of Host Cell Autophagy

Modulation of host cell death pathway drives the pathogen survival within host. Intracellular pathogens evade extracellular immune surveillance by subverting host autophagy to save their replicative capability (Stewart and Cookson 2016). Organisms trigger cell survival pathways, for example, the NF-κB pathway by *Legionella* or the PI3K/Akt pathway by *Salmonella* (Creasey and Isberg 2012; Knodler et al. 2005), *Listeria* (Mansell et al. 2001), or *Rickettsia* (Joshi et al. 2004), by downregulation of apoptotic genes and the upregulation of key survival genes. This enables microbes to escape autophagy-mediated death and encourage their replication and disease spread.

12.2.7 Reprogramming of Host Cell Cycle

As another defense, intracellular pathogens regulate host cell cycle for their proliferation within host. Numerous pathogens including *Shigella* and *Salmonella* produce secretory toxins called cytolethal distending poisons, homologous to mammalian DNase I that can hinder the cell cycle at the G2/M phase (Lara-Tejero and Galan 2000). Also, *Chlamydia trachomatis* destabilizes the cell cycle proteins associated with the G2/M phase (Gargi et al. 2012) and controls the host cell cycle (Gargi et al. 2012). Genome-wide transcriptomic profiling demonstrated that intracellular pathogens make changes in host gene expression (Jenner and Young 2005). A large number of these changes impact infection-induced innate immune response as in the case of *Salmonella*. *M. tuberculosis* infection upregulates histone deacetylase which in turns causes the inhibition of IFN-γ-induced articulation of host macrophages (Chandran et al. 2015). Similarly, pathogens reprogram host cell cycle by epigenetic regulation. *Shigella* inhibits mitogen-activated protein kinase (MAPK) action by phosphatase action of its T3SS effector OspF (Schroeder and Hilbi 2008).

12.3 Metabolic Adaptation Mechanisms in Extracellular Pathogenic Bacterium

Bacterial pathogens can be characterized as either extracellular or intracellular on the basis of their life cycle in the host. Extracellular pathogens replicate outside of host cells. However, the condition seems more multifaceted with a dual way of life of bacterial pathogens.

12.3.1 Survival Measures for the Extracellular Pathogens: Defensive Strategies

To thrive inside the host, extracellular bacterial pathogens must defeat the host immune system. Pathogens devise a variety of mechanisms to get away with host defenses.

(a) *Molecular camouflage*

Bacterial pathogens hide their pathogenic properties from the host by molecular camouflage. This involves capsule generation, biofilm formation, and alteration of bacterial surface antigenic variation at bacterial surface. Capsule of extracellular polysaccharide coatings gives pathogens a physical barrier to protect them from host defense as in the case of both gram-positive and gram-negative pathogens. Certain bacterial pathogens contain capsules with host similar molecules and thus suppress immune response against them. Biofilms are microbial aggregation forms at the interface of solid-solid, solid-liquid, liquid-liquid, or liquid-gas and make infection more tolerant to antibacterial. *Pseudomonas aeruginosa* (causing cystic fibrosis (CF) in lungs) and *S. aureus* (forming biofilms on medical implants) are one of the most studied biofilm-forming microorganisms (Heidari et al. 2018). Extracellular bacteria can escape from recognition by hiding inside host (Nobbs et al. 2009). *S. aureus* exhibits a classic case of defense by generating a coagulase, where two of non-proteolytically prothrombins cause the polymerization of fibrin, thus resulting in clump arrangement (Abamecha et al. 2015; Liesenborghs et al. 2018). Inside this complex macromolecular structure, *S. aureus* can hide and escape phagocytic degradation. Similarly, bacteria such as *S. pneumonia* may change capsule, to skip from humoral or cell-mediated immune response. Antigenic variations of these proteins and the capsules enable *S. pneumonia* to keep away from host immune recognition. The facultative intracellular pathogens *N. gonorrhoeae* causing gonorrhea, similar to *N. meningitidis* causing meningitis, change surface proteins and display antigenic variabilities (Coureuil et al. 2013).

(b) *Pathogen modulates host immune system*

Extracellular bacterial pathogens guard themselves with the aid of altering host immune defenses for their survival. Pathogens summon pathways to meddle with the typical movement of supplement enactment, counteracting agent official, AMPs, and phagocytosis. The supplement framework is perplexing, and microorganisms have advanced various approaches to get away from its strong impacts; these incorporate tweak of supplement administrative proteins, direct association with supplement segments to forestall enactment, and enzymatic corruption of antibodies or supplement factors. At long last, some bacterial pathogens control the invulnerable framework by balancing the ordinary genius fiery reaction that is produced by the natural safe framework within the sight of a culpable creature, for example, the *Y. enterocolitica*-encoded protein, LcrV, modifies cytokine generation to guarantee *Y. enterocolitica* survival in vivo (Reithmeier-Rost et al. 2007).

12.3.2 Survival Measures for the Extracellular Pathogens: Offensive Strategies

Extracellular bacterial pathogens are loaded with hostile weapon to harm the host cells, tissues, and organs to guarantee their proliferation. The most noteworthy mechanisms of secretion of microbial toxins by extracellular pathogens are either discharge (by bacterial pathogens upon cell lysis) or release straightforwardly into the targeted host cell. Generation of toxins by extracellular bacterial pathogens increases the level of degradative proteins that impose hostile factors significant for the survival of bacterial pathogens in vivo.

(a) *Microbial toxins:* Microbial toxins are harmful to human host. They can function in two ways: first, during bacterial cell lysis, the toxins are re-released extracellularly (endotoxins), and second, pathogenic bacteria secrete toxins extracellularly into host cell (exotoxins). Numerous bacterial pathogens produce various toxins with these mechanism.

(b) *Degradative enzymes:* Degradative enzymes are not traditionally considered as toxins; they provide an important survival mechanism for extracellular pathogens. Large numbers of these degradative proteins can redirect host immune response by modulating immunoglobulins, extracellular matrix, cell membranes, and fibrin network, thus providing organism a chance to move away from site of disease with active immune challenges (Karlsson et al. 2018).

Extracellular pathogens escape from phagocytic events and thrive in the extracellular spaces, for example, *S. aureus, P. aeruginosa, or S. pyogenes.* However, the dual way of life in both intracellular and extracellular bacterial pathogens is quite complicated (Silva 2012), although many bacterial pathogens live extracellular life as second phase of their life cycle after establishing intracellular infection.

12.4 Impact of Metabolic Dynamics in Virulence and Pathogenesis of Bacteria

12.4.1 Metabolic Dynamics

Bacterial pathogens invade and thrive in the variable and hostile environment of their host and they rely on their proficient metabolic adaptation. The variable environment within the host ranges from oxidative stress due to host immune system, free radical oxygen, and nitrogen intermediates, various carbon sources to be used as a substrate for energy to the changes in the pH. These conditions thus make it necessary to eternally monitor and respond to the microenvironment in an appropriate way in order to establish the successful sustenance.

The important factor is to understand the functional and genomic condition of pathogen metabolism that exploits utilization of host's nutrient. Horizontal gene acquisition drives the metabolic functions to provide a selective advantage to the pathogen residing within the host. Some loss-of-function mutations that alter the metabolic abilities may also provide the selective advantage to the pathogen in the host.

12.4.2 Fight for Survival Among Invading Pathogen and Gut Microbiota

Within the host microenvironment, the occurrence of competition is very first and common step in order for the pathogen to thrive. One of the prerequisites to succeed within host and establish the pathogenesis is the ability of the pathogen to sense the available carbon sources. Carbon catabolite repression (CCR) is triggered in response to readily available energy source (glucose), and variation in amino acid sources is used to adopt to and regulate the virulence factor via stringent response through (p)ppGpp (Zhang et al. 2016). Hence, efficient processing of the available nutrients in the limiting conditions would help one species to outgrow compared to the other species (Rohmer et al. 2011). In order for the pathogen to adopt successfully in the host environment, they need to invade the niche that is already perfectly adopted by resident microbiota since the resident bacteria have their own mechanisms to protect their niche from the competitive bacterial species.

12.4.3 Challenging Environment Within the Host

One of the most important challenges pathogen faces is variation in the pH range within the host gut. Usually this variation ranges as low as pH 1.0 in the stomach to as high as pH 8.0 in the urine. Secondly, invading pathogen has to struggle with the already existing niches of the microbes which are perfectly adopted and more than ten times the number of human cells themselves in the body. The new estimates by Hans-Curt Flemming in their Nature Microbiology review in 2019 reveal that there is one to one ratio of host cells to microbiome are present in the human gut. A predominant aerobic bacterium *S. epidermidis* produces antimicrobial peptides which are toxic to invading pathogens *S. aureus* and *S. pyogenes* (Otto, M. 2010). Additionally, the resident microbes of the intestine form a heterogeneous complex ecosystem and ultimately affect the cytokine production by the spleen and bone marrow macrophages, promoting the defense against invading pathogens (Nicaise et al. 1999). Lactic acid-generating *Lactobacilli* species in the woman vagina is another example of this type which maintains the acidic pH of the vagina to inhibit the invasion of many pathogenic colonizers (Amabebe and Anumba 2018). Except for the mucosal surface within the host which is well oxygenated, the pathogens encounter oxygen-deficit environments in buccal cavity, the large intestine, female genital tract, abscesses, and damaged tissues. Bacteria need iron in sufficient amount to thrive within the host which is quite below the desired level within the host, subjecting bacteria to devise their own strategy to sustain in the iron scavenging conditions by upregulating the iron transporting genes (Moumene et al. 2017). Other environmental factors which are used in addition to nutrient sensing include temperature, ion concentration, pH, and oxygen, and these altogether determine the location of the pathogens within the host to adjust and successfully adjust the pathogenesis.

12.4.4 Metabolic Interaction Between Pathogen and Host: Dynamism

Metabolic modulation of the host is utilized by the pathogen to regulate the expression profile of their virulence determinants since the nutrient availability within the host is not always constant (Schaible and Kaufmann 2004). For example, iron availability within the host varies during the infection and after the production of host factors interacting during iron metabolism (Ibraim et al. 2019). Inflammation-induced sequestration of the iron occurs by host lactoferrin and lipochellin-2 (Raffatellu et al. 2009). These host proteins sequester the iron in response to infection to prevent iron acquisition by the pathogens (Skaar 2010). Bacterial pathogens thereby sense the iron depletion as a signal, and thus they subsequently modulate the production of virulence factors (Skaar 2010). For example, iron-activated global repressors Dtxr and Fur of the *Corynebacterium diphtheriae* and *Shigella* species, respectively, are well-studied examples of inhibition of diphtheria and *Shigella* toxin expression (Tao et al. 1994; Schaible and Kaufmann 2004). Invading pathogens sense the variation in nutrient availability within the host, and thus they accordingly modulate the expression of their virulence factors. Pathogens acquire the metabolic genes in the same way as the virulence genes in order to establish the successful infection within the host.

12.4.5 Colonizing New Territories: Contribution of Metabolic Genes

Pathogens need virulence factors to adapt to new niches (Schmidt and Hensel 2004), and they need new metabolic pathways to help them thrive in the new environment and to exploit the available host resources. Pathogenic genes helping to adapt to new metabolic environment are often located in the pathogenicity islands. These pathogenicity islands are absent in the non-pathogenic relatives but present in the pathogenic species, and they show the evidence of lateral transfer (Hacker and Kaper 2000).

Well-studied example of this is the tetrathionate respiration (as a terminal respiratory electron acceptor) in *S. enterica* serovar *typhimurium* infecting the intestinal lumen. This pathogen produces hydrogen sulfide (H_2S) in large amount, and the cecal mucosa protects itself by converting the highly toxic gas to thiosulfate ($S_2O_3^{2-}$). Intestinal inflammation is induced by *Salmonella* virulence factors which are located on pathogenicity islands, SPI1 and SPI2, causing the production of large quantity of reactive oxygen and nitrogen species. As a result, thiosulfate gets oxidized to tetrathionate ($S_2O_6^{2-}$), thereby selectively inhibiting the growth of coliforms. *Salmonella* now becomes able to use tetrathionate to utilize ethanolamine of 1, 2-propanediol as a source of carbon for the anaerobic growth in the intestinal lumen, thereby achieving a competitive edge over the gut microbes, allowing the pathogen to successfully establish itself within the host (Winter et al. 2010; Price-Carter et al. 2001).

12.4.6 Metabolic Adaptation Within the Host Microenvironment: A Characteristic to Pathogenicity

Different hosts have varying nutrient availability that makes it necessary for pathogens circulating among many hosts to adapt metabolically. For the pathogen, metabolic requirements are directly proportional to host infected and route of infection. The status of conservation of metabolic gene depends on its habitat. Nevertheless, metabolic potential of the pathogen is also dependent on effect of loss-of-function mutations offering survival benefits for a given niche (Rohmer et al. 2011).

12.4.7 Genome Reduction and Gene Loss: Common Adaptation Mechanisms

When a pathogen establishes itself a new niche that offers improved nutritional resources, detrimental metabolic pathways either get suppressed or can be even lost. It is also likely that virulence character can be altered because of loss of functional mutation in non-essential metabolic genes that reduces the metabolic demands of bacterium. Functional complementation of pseudogenes has proven that loss-of-function mutations could be enhancing the virulence, thereby increasing survival capacity. *Shigella* lacks lysine decarboxylase activity, but its closely related bacterium *E. coli* does have the functional lysine decarboxylase. When *Shigella* is complemented with lysine decarboxylase, its virulence capacity is attenuated due to reduction in the level of *Shigella* enterotoxin caused by lysine decarboxylase action (Maurelli et al. 1998).

Genome reduction is a known evolutionary mechanism in some bacteria offering them improved metabolic fitness and enhanced virulence. Pseudomonads infecting the airways of the cystic fibrosis patients have been shown to lose a variety of metabolic pathways. This is attributed to greater nutrients availability in the host airway microenvironment compared to the soil and water (Smith et al. 2006; Barth and Pitt 1996). A very interesting review was published by Ahmed et al. in 2008 in Nature Reviews Microbiology where he has shown that evolutionary success of *Mycobacterium tuberculosis* causing TB depends on the vertical genome reduction whereas *Helicobacter pylori* causing gastric cancer and ulcer depends on horizontal gene acquisition. These examples indicate how loss-of-function mutation, genome reduction, and horizontal gene acquisition by pathogen can impact its metabolic fitness and virulence.

12.5   Metabolic Adaptation and Bacterial Population Behavior

Why considering the population behavior is important than individual bacterial behavior in fitness dynamics? During infection when bacteria kill the host and when not depends on the metabolic fitness it offers. When killing the host ceases the chance of bacterial propagation, it will prefer not to kill the host. Its virulence will decrease and pathogen will not kill the host. Thus, sustenance of host increases the chance of bacterial propagation and evolutionary survival. The evolutionary success drives the direction of metabolic adaptation. This also emphasizes that population choose metabolic adaptation for their evolutionary success.

Further, cooperativity in the bacterial population is fortified by cell-cell communication. In bacterial population, cells collaborate with one another by exchange of signaling molecules, even in heterogeneous bacterial population in host like gut microbiome. In gut microbiome, multicellular behavior is regularly seen as constrained of resources, which may cause colony or biofilm formation which is an inclusive procedure for unicellular life forms to overcome environment challenges in collective and cooperative manner.

12.5.1 Synchronize Multicellular Behavior of Bacteria

E. coli colonies show impressive spatial association. Surface electron microscopy (SEM) uncovers zones inside colonies described by cells of particular sizes, shapes, and multicellular course of action (Shapiro 1987). Vertical areas through colony reveal stratification into layers of cells with various protein substances, a large number of which seem to be nonviable (Lyons and Kolter 2015; Shapiro 1992). Apoptosis of myxobacteria creates the zones of fruiting bodies leading to its rapid growth (Allocati et al. 2015). This shows an unforeseen limit with respect to cell division and making specific zones of differentiated cells in *E. coli* K12 colonies. *B. subtilis* can emerge with the potential to divide in different cell types when stimulated by extracellular signals from external environment. This capacity of *B. subtilis* to differentiate into different cell types offers survival advantages to the bacterium (Vlamakis et al. 2013).

Swarming is a group phenomenon far reaching among flagellated bacteria wherein the microbes relocate by and large over a strong surface and show expanded protection from antimicrobials. Swarming microbes have been grouped into two classes: strong swarmers that can explore over any agar surface, especially crosswise over hard agar (1.5% agar or more), and mild swarmers having capacity to swarm on the surface of relatively soft agar (0.5% to 0.8% agar). The swarming motility of any bacterial population is dependent on their aggregation. Single isolated bacterium cannot swarm over agar, but population of swarmers can do that. The capsule made up of polysaccharide in *P. mirabilis* facilitates the swarming, whereas mutants lacking capsule formation cannot swarm (Zablotni et al. 2018). *S. liquefaciens* swarmer cell differentiation is regulated by ectopic expression of the FlhDC (Soo et al. 2008). A second degree of swarming control in *Serratia* includes acyl-homoserine lactones (AHL) signaling. Mutant lacking functional swrI homologue of luxI loses swarming capabilities (Liu et al. 2011). *Serratia* swarm colonies produce cyclic peptide surfactants basic to motility (Daniels et al. 2004).

12.5.2 Adjustive Preferred Position from Multicellular Participation

As per developmental view point, bacteria are diffusely disseminated over numerous metabolic, regulatory as well as micro-evolutionary and genome modification includes just as favorable circumstances of development rate and population size. Ordinarily in culture media, microbes are provided with basic substrates. In nature, numerous bacteria catabolize large organic polymers, requiring the purposeful activity of numerous cells. For instance, myxobacteria use a "wolf pack" methodology to assault and lyse their prey life forms by freeing stomach-related extracellular proteins and engrossing the cell substance (Munoz-Dorado et al. 2016). *Myxococcus xanthus* shows autoaggregation caused due to environmental stress. When energy resources are plentiful, myxobacteria show single species swarm formation, while under resource-limiting condition, it shows the multicellular development (Vaksman and Kaplan 2015; Bretl and Kirby 2016).

Numerous agents can adequately cause secluded bacterial cells in suspension however are ineffectual against thick or dense population of similar bacteria. For instance, *S. aureus* is one of the most widely recognized hospital-acquired gram-positive bacteria (Tong et al. 2015). Biofilm development gives bacterium a survival benefit, for example, mechanism of quorum sensing or horizontal gene acquisition may increase tolerance to antimicrobials and cause immune evasion (Gebreyohannes et al. 2019). Multicellular defense has wide impact in survival of pathogens. Numerous microbes produce anti-toxins under the influence of intercellular communication and quorum sensing, in general.

12.5.3 Infection Dynamics and Impact on Host

Bacterial infection is great cause of worry for human health particularly with increasing antibiotic resistance (Ventola 2015). Clinical manifestation of diseases caused by pathogenic microorganisms shows interchange of factors between organisms and host. While the inborn and adaptive immune system effectively contains the disease, the acquisition of new factors by microbes makes significant impact on their fitness. Macrophages are primary phagocytes that are regularly targeted by pathogenic microorganisms for intracellular growth (Mitchell et al. 2015). A few species have advanced the mechanisms to modulate the host determinants for intrusion, replication, and proliferation within host (Bourdonnay and Henry 2016). The intracellular way of life protects bacteria from the adaptive immune response of the host (Thakur et al. 2019; Uribe-Querol and Rosales 2017). Intracellular bacteria additionally have better availability of micronutrients compared with other bacteria. Bacterial species living inside professional phagocytes like macrophages incorporate *Salmonella* (Di Russo and Samuel 2016; Hu et al. 2013), *L. monocytogenes* (Hanawa et al.1995), *M. tuberculosis*

(Mehra et al. 2013), *Shigella*, and some *E. coli* strains. Either acute or persistent infection can be established in the host without any clinical symptoms.

There are models based on fitness to explain infection outcomes:

(a) *Containment, growth inhibition, and bacterial clearance:* This occurs as a consequence of exceeding microbial death compared to their growth as well as replication inside macrophages. Here, growth is insufficient compared to clearance during the infection in the host.

(b) *Growth followed by persistent infection:* When bacterial growth and clearance within host are balanced in a way that maintains the state of infection, the bacteria harboring macrophages behave similar to Lotka-Volterra prey-predator dynamics, where oscillations of frequencies occur for both prey and predator.

(c) *Exhaustion of bacteria and its resource cells causing:* The third case of acute infection appears when bacterial growth is very high followed by sharp decline because of exhaustion of resource. In an extreme condition, both macrophage and, thus extinction of macrophage caused by bacterial overabundance diminish bacterial population.

The cross talk between host factors and pathogenic determinants makes it difficult to map the mechanism of pathogen infection. Microbial populations infect or colonize hosts and are subjected to selection pressure, and thus their metabolic adaptive dynamics evolve (Webb and Blaser 2002). Active research in bacterial metabolic adaptation has mostly focused on the genetics, biochemistry, and physiology of the interactions between microbe and host (Levin et al. 1999). The host metabolic adaptation dynamics are difficult to measure as there are no direct interactions.

12.6 Drug Resistance Menace, Reason, and Current Approaches

Emergence of antibiotic resistance and its spread in the nature is a huge threat to mankind and the wilderness. Use of antibiotics as a growth promoter in livestock in excess, increased international travel, poor hygiene, lack of best practices in use of antibiotics, and polluting water bodies by industrial effluent having antibiotics and household discharge containing remains of antibiotics accelerate the antibiotic resistance development over the globe. These factors collectively impart to the selection pressure which ultimately leads to the emergence of superbugs (multidrug and extreme drug resistance) infection (Hawkey et al. 2018). Moreover, certain pathogenic infections like pneumonia, tuberculosis, and salmonellosis are becoming hard to treat as the pathogenic strains are getting resistant and the antibiotics lose their effect. The sincere intervention efforts are required to address the issue of antibiotic resistance.

Most of the global pharmaceuticals assume that research for developing new antimicrobials are not financially advantageous as any new antibiotic will eventually lose its effect due to development of resistance. Long-term strategies are needed to devise the approaches for preventing drug resistance like improvising hygiene, running vaccination programs, devising alternative medicines, and making the public aware against misuse/overuse of antibiotics (Hughes and Karlen 2014).

Drug resistance is a sort of ecological calamity with enormous damage potential for humans and livestock. Various global organizations such as the Food and Agriculture Organization (FAO) and Centers for Disease Control and Prevention (CDC) are putting serious efforts at the global level to combat the calamity of drug resistance. Global Health Security Agenda (GHSA) and Antimicrobial Resistance Action Package are other such programs dealing with the global threat of developing drug resistance (Aslam et al. 2018; Brown et al. 2017).

Some countries are getting success in combating drug resistance by adopting certain approaches, viz., advancement in healthcare setup, restricted drug promotions, adopting health insurance policies, cautious use of antibiotics, and developing consistence disease control strategies (Sakeena et al. 2018; Van Boeckel et al. 2015).

12.7 Conclusion

In the last decade, impressive development has been made in our understanding of mechanisms of infection biology, though we only know the tip of the iceberg till yet. Every pathogen cooperates differently with its host. Pathogens utilize a mixed technique for intracellular survival and hijack the host defense mechanism in a sophisticated manner.

In a nutshell, bacterial evolutionary forces drive the direction of metabolic adaptation. This metabolic adaptation is quite fluidic and easily reshapes itself with appearance of new challenges. Also, the collective behavior of heterogeneous bacterial population and their interaction will contribute significantly in the metabolic adaptation. The better understanding of bacterial adaptation mechanisms can help us developing new treatments that instead of targeting single pathway which is a source of drug resistance, can provide a way to redirect the bacterial adaptation in a way decreasing their pathogenicity thereby weakening the infection born challenges for the host.

References

Abamecha A, Wondafrash B, Abdissa A (2015) Antimicrobial resistance profile of Enterococcus species isolated from intestinal tracts of hospitalized patients in Jimma, Ethiopia. BMC Res Notes 8:213. https://doi.org/10.1186/s13104-015-1200-2

Aeberhard L, Banhart S, Fischer M, Jehmlich N, Rose L et al (2015) The proteome of the isolated *Chlamydia trachomatis* containing vacuole reveals a complex trafficking platform enriched for retromer components. PLoS Pathog e1004883:11

Allocati N, Masulli M, Di Ilio C, De Laurenzi V (2015) Die for the community: an overview of programmed cell death in bacteria. Cell Death Dis 6:e1609. https://doi.org/10.1038/cddis.2014.570

Amabebe E, Anumba DOC (2018) The vaginal microenvironment: the physiologic role of lactobacilli. Front Med 5:181. https://doi.org/10.3389/fmed.2018.00181

Aslam B, Wang W, Arshad MI, Khurshid M, Muzammil S et al (2018) Antibiotic resistance: a rundown of a global crisis. Infect Drug Resist 11:1645–1658

Awuh JA, Flo TH (2017) Molecular basis of mycobacterial survival in macrophages. Cell Mol Life Sci 74:1625–1648

Barth AL, Pitt TL (1996) The high amino-acid content of sputum from cystic fibrosis patients promotes growth of auxotrophic *Pseudomonas aeruginosa*. J Med Microbiol 45:110–119

BoseDasgupta S, Pieters J (2018) Macrophage-microbe interaction: lessons learned from the pathogen Mycobacterium tuberculosis. Semin Immunopathol 40:577–591

Bosserman RE, Nicholson KR, Champion MM, Champion PA (2019) A new ESX-1 substrate in *Mycobacterium marinum* that is required for hemolysis but not host cell lysis. J Bacteriol 201(14):e00760–e00718. https://doi.org/10.1128/JB.00760-18

Bourdonnay E, Henry T (2016) Catch me if you can. Elife 26:5. https://doi.org/10.7554/eLife.14721

Bretl DJ, Kirby JR (2016) Molecular mechanisms of signaling in *Myxococcus xanthus* development. J Mol Biol 428:3805–3830

Brown L, Langelier C, Reid MJ, Rutishauser RL, Strnad L (2017) Antimicrobial resistance: a call to action! Clin Infect Dis 64(1):106–107

Chandran A, Antony C, Jose L, Mundayoor S, Natarajan K et al (2015) *Mycobacterium tuberculosis* infection induces HDAC1-mediated suppression of IL-12B gene expression in macrophages. Front Cell Infect Microbiol 5:90. https://doi.org/10.3389/fcimb.2015.00090

Cheng MI, Chen C, Engstrom P, Portnoy DA, Mitchell G (2018) Actin-based motility allows Listeria monocytogenes to avoid autophagy in the macrophage cytosol. Cell Microbiol 20:e12854

Colonne PM, Winchell CG, Voth DE (2016) Hijacking host cell highways: manipulation of the host actin cytoskeleton by obligate intracellular bacterial pathogens. Front Cell Infect Microbiol 6:107. https://doi.org/10.3389/fcimb.2016.00107

Coureuil M, Join-Lambert O, Lecuyer H, Bourdoulous S, Marullo S et al (2013) Pathogenesis of *meningococcemia*. Cold Spring Harb Perspect Med 3(6):a012393. https://doi.org/10.1101/cshperspect.a012393

Creasey EA, Isberg RR (2012) The protein SdhA maintains the integrity of the Legionella-containing vacuole. Proc Natl Acad Sci U S A 109:3481–3486

Daniels R, Vanderleyden J, Michiels J (2004) Quorum sensing and swarming migration in bacteria. FEMS Microbiol Rev 28:261–289

de Souza Santos M, Orth K (2014) Intracellular Vibrio parahaemolyticus escapes the vacuole and establishes a replicative niche in the cytosol of epithelial cells. MBio 5:e01506–e01514

Di Russo CE, Samuel JE (2016) Contrasting lifestyles within the host cell. Microbiol Spectr 4(1). https://doi.org/10.1128/microbiolspec.VMBF-0014-2015

Gargi A, Reno M, Blanke SR (2012) Bacterial toxin modulation of the eukaryotic cell cycle: are all cytolethal distending toxins created equally? Front Cell Infect Microbiol 2:124. https://doi.org/10.3389/fcimb.2012.00124

Gebreyohannes G, Nyerere A, Bii C, Sbhatu DB (2019) Challenges of intervention, treatment, and antibiotic resistance of biofilm-forming microorganisms. Heliyon 5:e02192. https://doi.org/10.1016/j.heliyon.2019.e02192

Ghigo E, Colombo MI, Heinzen RA (2012) The *Coxiella burnetii* parasitophorous vacuole. Adv Exp Med Biol 984:141–169

Gurtler L, Bauerfeind U, Blumel J, Burger R, Drosten C et al (2014) *Coxiella burnetii* – pathogenic agent of Q (query) fever. Transfus Med Hemother 41:60–72

Hacker J, Kaper JB (2000) Pathogenicity islands and the evolution of microbes. Annu Rev Microbiol 54:641–679

Hamon MA, Batsche E, Regnault B, Tham TN, Seveau S et al (2007) Histone modifications induced by a family of bacterial toxins. Proc Natl Acad Sci U S A 104:13467–13472

Hanawa T, Yamamoto T, Kamiya S (1995) Listeria monocytogenes can grow in macrophages without the aid of proteins induced by environmental stresses. Infect Immun 63:4595–4599

Hawkey PM, Warren RE, Livermore DM, McNulty CAM, Enoch DA et al (2018) Treatment of infections caused by multidrug-resistant Gram-negative bacteria: report of the British Society for Antimicrobial Chemotherapy/Healthcare Infection Society/British Infection Association Joint Working Party. J Antimicrob Chemother 73:iii2–iii78. https://doi.org/10.1093/jac/dky027

Heidari H, Hadadi M, Sedigh E-SH, Mirzaei A, Taji A et al (2018) Characterization of virulence factors, antimicrobial resistance patterns and biofilm formation of *Pseudomonas aeruginosa* and *Staphylococcus* spp. strains isolated from corneal infection. J Fr Ophtalmol 41:823–829

Hu M, Yang Y, Meng C, Pan Z, Jiao X (2013) Responses of macrophages against Salmonella infection compared with phagocytosis. In Vitro Cell Dev Biol Anim 49:778–784

Hu W, Chan H, Lu L, Wong KT, Wong SH et al (2019) Autophagy in intracellular bacterial infection. Semin Cell Dev Biol. https://doi.org/10.1016/j.semcdb.2019.07.014

Hughes D, Karlen A (2014) Discovery and preclinical development of new antibiotics. Ups J Med Sci 119:162–169

Ibraim IC, Parise MTD, Parise D, Sfeir MZT, de Paula CTL et al (2019) Transcriptome profile of *Corynebacterium pseudotuberculosis* in response to iron limitation. BMC Genomics 20:663. https://doi.org/10.1186/2Fs12864-019-6018-1

Ireton K (2013) Molecular mechanisms of cell-cell spread of intracellular bacterial pathogens. Open Biol 3:130079

Jenner RG, Young RA (2005) Insights into host responses against pathogens from transcriptional profiling. Nat Rev Microbiol 3:281–294

Joshi SG, Francis CW, Silverman DJ, Sahni SK (2004) NF-kappaB activation suppresses host cell apoptosis during Rickettsia rickettsii infection via regulatory effects on intracellular localization or levels of apoptogenic and anti-apoptotic proteins. FEMS Microbiol Lett 234:333–334

Karlsson CAQ, Jarnum S, Winstedt L, Kjellman C, Bjorck L et al (2018) *Streptococcus pyogenes* infection and the human proteome with a special focus on the immunoglobulin G-cleaving enzyme IdeS. Mol Cell Proteomics 17:1097–1111

Knodler LA, Finlay BB, Steele-Mortimer O (2005) The Salmonella effector protein SopB protects epithelial cells from apoptosis by sustained activation of Akt. J Biol Chem 280:9058–9064

Lamason RL, Welch MD (2017) Actin-based motility and cell-to-cell spread of bacterial pathogens. Cur Opinion Microbiol 35:48–57

Lara-Tejero M, Galan JE (2000) A bacterial toxin that controls cell cycle progression as a deoxyribonuclease I-like protein. Science 290:354–357

Levin BR, Lipsitch M, Bonhoeffer S (1999) Population biology, evolution, and infectious disease: convergence and synthesis. Science 283:806–809

Liesenborghs L, Verhamme P, Vanassche T (2018) *Staphylococcus aureus*, master manipulator of the human hemostatic system. J Thromb Haemost 16:441–454

Liu X, Jia J, Popat R, Ortori CA, Li J et al (2011) Characterisation of two quorum sensing systems in the endophytic *Serratia plymuthica* strain G3: differential control of motility and biofilm formation according to life-style. BMC Microbiol 11:26. https://doi.org/10.1186/1471-2180-11-26

Lyons NA, Kolter R (2015) On the evolution of bacterial multicellularity. Curr Opin Microbiol 24:21–28

Mansell A, Khelef N, Cossart P, O'Neill LA (2001) Internalin B activates nuclear factor-kappa B via Ras, phosphoinositide 3-kinase, and Akt. J Biol Chem 276:43597–43603

Mattock E, Blocker AJ (2017) How do the virulence factors of Shigella work together to cause disease? Front Cell Infect Microbiol 7:64

Maurelli AT, Fernandez RE, Bloch CA, Rode CK, Fasano A (1998) "Black holes" and bacterial pathogenicity: a large genomic deletion that enhances the virulence of *Shigella* spp. and enteroinvasive *Escherichia coli*. Proc Natl Acad Sci U S A 95:3943–3948

Mehra A, Zahra A, Thompson V, Sirisaengtaksin N, Wells A et al (2013) *Mycobacterium tuberculosis* type VII secreted effector EsxH targets host ESCRT to impair trafficking. PLoS Pathog 9:e1003734. https://doi.org/10.1371/journal.ppat.1003734

Mitchell G, Ge L, Huang Q, Chen C, Kianian S et al (2015) Avoidance of autophagy mediated by PlcA or ActA is required for *Listeria monocytogenes* growth in macrophages. Infect Immun 83:2175–2184

Moumene A, Gonzalez-Rizzo S, Lefrancois T, Vachiery N, Meyer DF (2017) Iron starvation conditions upregulate *Ehrlichia ruminantium* type IV secretion system, tr1 transcription factor and map1 genes family through the master regulatory protein ErxR. Front Cellula Inf Microbiol 7:535. https://doi.org/10.3389/fcimb.2017.00535

Munoz-Dorado J, Marcos-Torres FJ, Garcia-Bravo E, Moraleda-Munoz A, Perez J (2016) Myxobacteria: moving, killing, feeding, and surviving together. Front Microbiol 7:781. https://doi.org/10.3389/fmicb.2016.00781

Nicaise P, Gleizes A, Sandre C, Kergot R, Lebrec H et al (1999) The intestinal microflora regulates cytokine production positively in spleen-derived macrophages but negatively in bone marrow-derived macrophages. Eur Cytokine Netw 10:365–372

Nickel W, Weber T, McNew JA, Parlati F, Sollner TH, Rothman JE (1999) Content mixing and membrane integrity during membrane fusion driven by pairing of isolated v-SNAREs and t-SNAREs. Proc Natl Acad Sci U S A 96:12571–12576

Nobbs AH, Lamont RJ, Jenkinson HF (2009) *Streptococcus* adherence and colonization. Microbiol Mol Biol Rev 73:407–450

Otto M (2010) *Staphylococcus* colonization of the skin and antimicrobial peptides. Expert Rev Dermatol 5:183–195

Picking WL, Picking WD (2016) The many faces of IpaB. Front Cell Infect Microbiol 6:12

Price-Carter M, Tingey J, Bobik TA, Roth JR (2001) The alternative electron acceptor tetrathionate supports B12-dependent anaerobic growth of *Salmonella enterica* serovar *typhimurium* on ethanolamine or 1,2-propanediol. J Bacteriol 183:2463–2475

Radoshevich L, Cossart P (2018) Listeria monocytogenes: towards a complete picture of its physiology and pathogenesis. Nat Rev Microbiol 16:32–46

Raffatellu M, George MD, Akiyama Y, Hornsby MJ, Nuccio SP et al (2009) Lipocalin-2 resistance confers an advantage to *Salmonella enterica* serotype *Typhimurium* for growth and survival in the inflamed intestine. Cell Host Microbe 5:476–486

Ray K, Marteyn B, Sansonetti PJ, Tang CM (2009) Life on the inside: the intracellular lifestyle of cytosolic bacteria. Nat Rev Microbiol 7:333–340

Reithmeier-Rost D, Hill J, Elvin SJ, Williamson D, Dittmann S et al (2007) The weak interaction of LcrV and TLR2 does not contribute to the virulence of *Yersinia pestis*. Microbes Infect 9:997–1002

Rohmer L, Hocquet D, Miller SI (2011) Are pathogenic bacteria just looking for food? Metabolism and microbial pathogenesis. Trends Microbiol 19:341–348

Sakeena MHF, Bennett AA, McLachlan AJ (2018) Enhancing pharmacists' role in developing countries to overcome the challenge of antimicrobial resistance: a narrative review. Antimicrob Resist Infect Control 7:63. https://doi.org/10.1186/s13756-018-0351-z

Schaible UE, Kaufmann SH (2004) Iron and microbial infection. Nat Rev Microbiol 2:946–953

Schmidt H, Hensel M (2004) Pathogenicity islands in bacterial pathogenesis. Clin Microbiol Rev 17:14–56

Schroeder GN, Hilbi H (2008) Molecular pathogenesis of Shigella spp.: controlling host cell signaling, invasion, and death by type III secretion. Clin Microbiol Rev 21:134–156

Shapiro JA (1987) Organization of developing *Escherichia coli* colonies viewed by scanning electron microscopy. J Bacteriol 169:142–156

Shapiro JA (1992) Pattern and control in bacterial colony development. Sci Prog 76:399–424

Silva MT (2012) Classical labeling of bacterial pathogens according to their lifestyle in the host: inconsistencies and alternatives. Front Microbiol 3:71. https://doi.org/10.3389/fmicb.2012.00071

Skaar EP (2010) The battle for iron between bacterial pathogens and their vertebrate hosts. PLoS Pathog 6:e1000949

Smith EE, Buckley DG, Wu Z, Saenphimmachak C, Hoffman LR et al (2006) Genetic adaptation by *Pseudomonas aeruginosa* to the airways of cystic fibrosis patients. Proc Natl Acad Sci U S A 103:8487–8492

Soo PC, Horng YT, Wei JR, Shu JC, Lu CC et al (2008) Regulation of swarming motility and flhDC(Sm) expression by RssAB signaling in *Serratia marcescens*. J Bacteriol 190:2496–2504

Stein A, Louveau C, Lepidi H, Ricci F, Baylac P et al (2005) Q fever pneumonia: virulence of *Coxiella burnetii* pathovars in a murine model of aerosol infection. Infect Immun 73:2469–2477

Stewart MK, Cookson BT (2016) Evasion and interference: intracellular pathogens modulate caspase-dependent inflammatory responses. Nat Rev Microbiol 14:346–359

Tao X, Schiering N, Zeng HY, Ringe D, Murphy JR (1994) Iron, DtxR, and the regulation of diphtheria toxin expression. Mol Microbiol 14:191–197

Thakur A, Mikkelsen H, Jungersen G (2019) Intracellular pathogens: host immunity and microbial persistence strategies. J Immunol Res 2019:1356540

Tong SY, Davis JS, Eichenberger E, Holland TL, Fowler VG (2015) *Staphylococcus aureus* infections: epidemiology, pathophysiology, clinical manifestations, and management. Clin Microbiol Rev 28:603–661

Truong D, Copeland JW, Brumell JH (2014) Bacterial subversion of host cytoskeletal machinery: hijacking formins and the Arp2/3 complex. Bioessays 36:687–696

Uribe-Querol E, Rosales C (2017) Control of phagocytosis by microbial pathogens. Front Immunol 8:1368. https://doi.org/10.3389/fimmu.2017.01368

Vaksman Z, Kaplan HB (2015) *Myxococcus xanthus* growth, development, and isolation. Curr Protoc Microbiol 39:1–21

Van Boeckel TP, Brower C, Gilbert M, Grenfell BT, Levin SA et al (2015) Global trends in antimicrobial use in food animals. Proc Natl Acad Sci U S A 112:5649–5654

Vdovikova S, Luhr M, Szalai P, Nygard SL, Francis MK et al (2017) A novel role of *Listeria monocytogenes* membrane vesicles in inhibition of autophagy and cell death. Front Cell Infect Microbiol 7:154. https://doi.org/10.3389/fcimb.2017.00154

Ventola CL (2015) The antibiotic resistance crisis: part 1: causes and threats. Pharm Ther 40:277–283

Vlamakis H, Chai Y, Beauregard P, Losick R, Kolter R (2013) Sticking together: building a biofilm the *Bacillus subtilis* way. Nat Rev Microbiol 11:157–168

Webb GF, Blaser MJ (2002) Dynamics of bacterial phenotype selection in a colonized host. Proc Natl Acad Sci U S A 99:3135–3140

Wiesel M, Oxenius A (2012) From crucial to negligible: functional CD8(+) T-cell responses and their dependence on CD4(+) T-cell help. Eur J Immunol 42:1080–1088

Winter SE, Thiennimitr P, Winter MG, Butler BP, Huseby DL et al (2010) Gut inflammation provides a respiratory electron acceptor for *Salmonella*. Nature 467:426–429. https://doi.org/10.1038/nature09415

Wong WF, Chambers JP, Gupta R, Arulanandam BP (2019) Chlamydia and its many ways of escaping the host immune system. J Pathog 2019:8604958

Xu L, Luo ZQ (2013) Cell biology of infection by *Legionella pneumophila*. Microbes Infect 15:157–167

Zablotni A, Matusiak D, Arbatsky NP, Moryl M, Maciejewska A et al (2018) Changes in the lipopolysaccharide of *Proteus mirabilis* 9B-m (O11a) clinical strain in response to planktonic or biofilm type of growth. Med Microbiol Immunol 207:129–139

Zhang T, Zhu J, Wei S, Luo Q, Li L, Li S et al (2016) The roles of RelA/(p)ppGpp in glucose-starvation induced adaptive response in the zoonotic *Streptococcus suis*. Sci Rep 6:27169

Correction to: Endophytic Actinomycetes-Mediated Modulation of Defense and Systemic Resistance Confers Host Plant Fitness Under Biotic Stress Conditions

Waquar Akhter Ansari, Ram Krishna,
Mohammad Tarique Zeyad, Shailendra Singh,
and Akhilesh Yadav

Correction to:
Chapter 10 in: R. P. Singh et al. (eds.),
Microbial Versatility in Varied Environments,
https://doi.org/10.1007/978-981-15-3028-9_10

The book was inadvertently published with an incorrect spelling of the authors' names in Chapter 10 as Akhilesh Kumar Yadav and ShTarique Singh. The names have been corrected as Akhilesh Yadav and Shailendra Singh.

The updated online version of this chapter can be found at
https://doi.org/10.1007/978-981-15-3028-9_10